Picture-Perfect STEM Lessons, 3–5

Using Children's Books to Inspire STEM Learning

Picture-Perfect

STEM

Lessons, 3–5

Using Children's Books to Inspire STEM Learning

by Emily Morgan and Karen Ansberry

National Science Teachers Association

Arlington, Virginia

National Science Teachers Association

Claire Reinburg, Director
Rachel Ledbetter, Managing Editor
Amanda Van Beuren, Associate Editor
Donna Yudkin, Book Acquisitions Manager

ART AND DESIGN
Will Thomas Jr., Director
Linda Olliver, Cover, Interior Design, Illustrations

PRINTING AND PRODUCTION
Catherine Lorrain, Director

NATIONAL SCIENCE TEACHERS ASSOCIATION
David L. Evans, Executive Director
David Beacom, Publisher

1840 Wilson Blvd., Arlington, VA 22201
www.nsta.org/store
For customer service inquiries, please call 800-277-5300.

Copyright © 2017 by the National Science Teachers Association.
All rights reserved. Printed in the United States of America.
20 19 18 17   4 3 2 1

NSTA is committed to publishing material that promotes the best in inquiry-based science education. However, conditions of actual use may vary, and the safety procedures and practices described in this book are intended to serve only as a guide. Additional precautionary measures may be required. NSTA and the authors do not warrant or represent that the procedures and practices in this book meet any safety code or standard of federal, state, or local regulations. NSTA and the authors disclaim any liability for personal injury or damage to property arising out of or relating to the use of this book, including any of the recommendations, instructions, or materials contained therein.

PERMISSIONS
Book purchasers may photocopy, print, or e-mail up to five copies of an NSTA book chapter for personal use only; this does not include display or promotional use. Elementary, middle, and high school teachers may reproduce forms, sample documents, and single NSTA book chapters needed for classroom or noncommercial, professional-development use only. E-book buyers may download files to multiple personal devices but are prohibited from posting the files to third-party servers or websites, or from passing files to non-buyers. For additional permission to photocopy or use material electronically from this NSTA Press book, please contact the Copyright Clearance Center (CCC) (*www.copyright.com*; 978-750-8400). Please access *www.nsta.org/permissions* for further information about NSTA's rights and permissions policies.

Cataloging-in-Publication Data for this book and the e-book are available from the Library of Congress.
ISBN: 978-1-68140-331-1
e-ISBN: 978-1-68140-332-8

Contents

Preface ix
Acknowledgments xiii
About the Authors xv
Safety Practices for Science Activities..xvii

1 Why Use Picture Books to Teach STEM?......................................1

2 Reading Aloud 9

3 BSCS 5E Instructional Model 17

4 Connecting to the Standards 25

5 Science and Engineering Practices 47

6 The Inventor's Secret 61
Rosie Revere, Engineer and *The Inventor's Secret: What Thomas Edison Told Henry Ford*

7 Mesmerized 79
Mesmerized: How Ben Franklin Solved a Mystery That Baffled All of France and *Let's Think About the Power of Advertising*

8 Wind It Up 99
Clink and *Making Machines With Springs*

9 Light It Up! 115
How Things Work: Lightbulbs and *Orion and the Dark*

10 Burn 137
Burn: Michael Faraday's Candle and *National Geographic Kids: Wildfires*

11 From Edison to the iPod 161
Timeless Thomas: How Thomas Edison Changed Our Lives and *iPod and Electronics Visionary Tony Fadell*

12 Better Together 179
An Ambush of Tigers: A Wild Gathering of Collective Nouns and *Animals That Live in Groups*

13 Spider Science 203
Next Time You See a Spiderweb and *Nefertiti, the Spidernaut: The Jumping Spider Who Learned to Hunt in Space*

14 Bionic Animals 223
Winter's Tail: How One Little Dolphin Learned to Swim Again and *Biomedical Engineering and Human Body Systems*

15 From Seed to Tree 245
If You Hold a Seed and *Next Time You See a Maple Seed*

16 Hurricane! 263
Two Bobbies: A True Story of Hurricane Katrina, Friendship, and Survival and *Building Dikes and Levees*

17 Solving the Puzzle Under the Sea 289
Solving the Puzzle Under the Sea: Marie Tharp Maps the Ocean Floor and
How Mountains Are Made

18 Space Exploration 307
Boy, Were We Wrong About the Solar System! and *Space Exploration*

19 Star Stuff 333
Star Stuff: Carl Sagan and the Mysteries of the Cosmos and *Jump Into Science: Stars*

20 From Trash to Treasure 347
One Plastic Bag: Isatou Ceesay and the Recycling Women of The Gambia and
Recycling Crafts (Craft Attack!)

Appendix: Alignment With the *Next Generation Science Standards* 365

Index 381

Preface

Third-grade students listen as their teacher reads *Clink,* the humorous and heartwarming story of a lovable, worn-out, music-playing, toast-making house robot who sits discarded in the Robot Shoppe. Clink nearly gives up on finding a home—until the day he spies a boy who just might have the right one for him …

> *The shopkeeper handed Clink to the boy. "He's very old, and he's missing parts."*
>
> *The boy's eyes lit up. "He's perfect!"*
>
> *"I'm perfect?" thought Clink. It had been a very long time since anybody had thought he was perfect. Clink smiled. PLINK! POP!*
>
> *The boy ducked. "I'll take him!" he said.*

The third graders delight in this heartfelt tale of yearning to belong and to be accepted for who you are. It not only engages students on an emotional level but also provides a unique transition to a discussion of toys—how they work and how they are designed. After the read-aloud, the class explores wind-up toy robots. Students observe their robots' patterns of motion, graph the distance they travel, and predict their future motion. Then, they take the toys apart to see how they work. Through a nonfiction read-aloud they learn about the technology required to design and manufacture springs and how springs inside wind-up toys store energy that is released when the spring unwinds. Students discover that a surprising amount of science and engineering is packed inside a small wind-up toy! Next, students build their own wind-up spool cars and modify, test, and evaluate their designs. This activity addresses the engineering core idea that different solutions need to be tested to determine which solution best solves the problem, given the criteria and constraints. Students apply what they have learned by writing an instruction manual that explains how the wind-up car works, demonstrating their understanding of the physical science core idea of forces and motion. Finally, they reflect on what they have learned through a STEM at Home assignment. With a parent or adult helper, they watch a video called *Scientist Profile: Toyologist,* which is about an engineer who designs toys. Then, they draw a patent illustration for a wind-up toy of their own design. Through this engaging, hands-on lesson found in Chapter 8, "Wind It Up," students learn about the interdependence of science, technology, engineering, and mathematics in the toy-making industry—all within the context of a delightful, fictional story.

What Is Picture-Perfect STEM?

The Picture-Perfect Science program was developed to help K–5 teachers integrate science and reading in an engaging, kid-friendly way. Since the debut of the first book in the *Picture-Perfect Science Lessons* series in 2005, elementary teachers across the country have been using the lessons to integrate science and literacy. This new series of Picture-Perfect books, *Picture-Perfect STEM Lessons: Using Children's Books to Inspire STEM Learning,* follows the same philosophy and lesson format as the original books but adds an additional emphasis on the intersection of science, technology, engineering, and mathematics in the real world. *Picture-Perfect STEM Lessons, 3–5* contains 15 lessons for students in grade three through grade five, with embedded reading comprehension strategies to help them learn to read and read to learn while engaged in STEM activities. To help you set up a learning environment consistent with the principles of *A Framework for K–12 Science Education* (*Framework*; NRC 2012), the lessons are written in an easy-to-follow format of constructivist learning—the Biological Science Curriculum Study (BSCS) 5E Instructional Model (Bybee 1997, used with permission from BSCS; see Chapter 3 for more information). This learning cycle model allows students to construct their own understanding of scientific concepts as they cycle through the following phases: engage, explore, explain, elaborate, and evaluate. Although *Picture-Perfect STEM Lessons* is primarily a book for teaching STEM concepts, reading comprehension strategies and the *Common Core State Standards for English Language Arts* (NGAC and CCSSO 2010) are embedded in each lesson. These essential strategies can be modeled while keeping the focus of the lessons on STEM.

Use This Book Within Your Curriculum

We wrote *Picture-Perfect STEM Lessons* to supplement, not replace, your school's existing science or STEM program. Although each lesson stands alone as a carefully planned learning cycle based on clearly defined objectives, the lessons are intended to be integrated into a complete curriculum in which concepts can be more fully developed. The lessons are not designed to be taught sequentially. We want you to use *Picture-Perfect STEM Lessons* where appropriate within your school's current STEM program to support, enrich, and extend it. We also want you to adapt the lessons to fit your school's curriculum, your students' needs, and your own teaching style.

Special Features of This Book

Ready-to-Use Lessons With Assessments

Each lesson contains engagement activities, hands-on explorations, student pages, suggestions for student and teacher explanations, elaboration activities, assessment suggestions, opportunities for STEM education at home, and annotated bibliographies of more books to read on the topic. Assessments include poster sessions, writing assignments, design challenges, demonstrations, presentations, and multiple-choice and extended-response questions.

Background for Teachers

This section provides easy-to-understand background information for teachers to review before facilitating the lesson. Some information in the background section goes beyond the assessment boundary for students, but it is provided to give teachers a deeper understanding of the content presented in the lesson.

Time Needed

The information in this section helps you pace each lesson. We estimate a primary class period to be about 30–45 min.

Reading-Comprehension Strategies

Reading-comprehension strategies based on the book *Strategies That Work* (Harvey and Goudvis 2007) and specific activities to enhance comprehension are embedded throughout the lessons and clearly marked with an icon. Chapter 2 describes how to model these strategies while reading aloud to students.

Standards-Based Objectives

All lesson objectives are aligned to the *Framework* (NRC 2012) and are clearly identified at the beginning of each lesson. An alignment with the *Next Generation Science Standards* (NGSS Lead States 2013) is included in the appendix (p. 365). The lessons also incorporate the *Common Core State Standards for English Language Arts and Mathematics* (NGAC and CCSSO 2010). In a box titled "Connecting to the Common Core," you will find the Common Core subject the activity addresses as well as the grade level and standard number. You will see that writing assignments are specifically labeled with an icon.

STEM at Home

Each lesson also provides an extension activity that is intended to be done with a parent or other adult helper at home. Students write about what they learned about each topic and share their favorite part of the lesson. Then, together with their adult helper, they complete an activity to apply and extend the learning. If students are unable to complete the extension at home, the activities in this section also work well as in-class extensions.

Ideas for Further Exploration

A "For Further Exploration" box is provided at the end of each lesson to help you encourage your students to use the science and engineering practices in a more student-directed format. This box lists questions and challenges related to the lesson that students may select to research, investigate, or innovate. Students may also use the questions as examples to help them generate their own questions. After selecting one of the questions in the box or formulating their own questions, students can make predictions, design investigations to test their predictions, collect evidence, devise explanations, design solutions, examine related resources, and communicate their findings.

References

Bybee, R. W. 1997. *Achieving scientific literacy: From purposes to practices.* Portsmouth, NH: Heinemann.

Harvey, S., and A. Goudvis. 2007. *Strategies that work: Teaching comprehension for understanding and engagement.* 2nd ed. Portland, ME: Stenhouse Publishers.

National Governors Association Center for Best Practices and Council of Chief State School Officers (NGAC and CCSSO). 2010. *Common core state standards.* Washington, DC: NGAC and CCSSO.

National Research Council (NRC). 2012. *A framework for K–12 science education: Practices, crosscutting concepts, and core ideas.* Washington, DC: National Academies Press.

NGSS Lead States. 2013. *Next Generation Science Standards: For states, by states.* Washington, DC: National Academies Press. www.nextgenscience.org/next-generation-science-standards.

Children's Book Cited

DiPucchio, K. 2011. *Clink.* New York: Balzer + Bray.

Editor's Note

Picture-Perfect STEM Lessons, 3–5 builds on the texts of 30 children's picture books to teach STEM. Some of these books feature objects that have been anthropomorphized, such as a robot who longs for a friend. Although we recognize that many scientists and educators believe that personification, teleology, animism, and anthropomorphism promote misconceptions among young children, others believe that removing these elements would leave children's literature severely underpopulated. Furthermore, backers of these techniques not only see little harm in their use but also argue that they facilitate learning. Because *Picture-Perfect STEM Lessons, 3–5* specifically and carefully supports science and engineering practices, we, like our authors, feel the question remains open.

Acknowledgments

We would like to dedicate this book to the memory of Dr. Robert Yearout, who gave us the opportunity to present our first teacher workshop at the "Sharing What Works" Conference in Columbus, Ohio, in 2000. Dr. Yearout's leadership of the High Achievement in Math and Science Consortium, which we were both fortunate to be a part of for many years, provided us with opportunities and encouragement to grow as educators and advocates of science and math education. His selfless leadership style and utmost respect for the teaching profession continue to inspire us today.

We appreciate the care and attention to detail given to this project by Rachel Ledbetter, Wendy Rubin, and Claire Reinburg at NSTA Press.

And these thank-yous as well:

- To Linda Olliver for her "Picture-Perfect" illustrations
- To Tom Uhlman for his photography
- To Kim Stilwell for facilitating workshops to give us time to write and for sharing *Picture-Perfect Science Lessons* with teachers across the country
- To the staff and students of Blue Springs School District, Heritage Elementary, Indian Hill Elementary, and Mason City Schools for field-testing lessons and providing "photo ops"
- To Libby Beck and Nancy Smith for contributing photographs
- To Ken Roy for his thorough safety review
- To Ted Willard for answering all of our *Next Generation Science Standards* (*NGSS*) questions and creating his helpful *NGSS* guides
- To Bill Robertson, Debbie Rupp, and Rand Harrington for sharing their content knowledge
- To TeachEngineering for granting permission to use a modified version of its activity, "Protecting Our Cities With Levees" in Chapter 16, "Hurricane!"
- To Terri Collins, educational outreach coordinator and senior lecturer at Kennesaw State University, for contributing STEM activity ideas to this book and sharing Picture-Perfect Science with teachers
- To Dr. Joyce Poole, co-founder and co-director of ElephantVoices, for her research on elephant behavior
- To Derrick Campana, director of orthotics at Animal Ortho Care, LLC, for his help with Chapter 14, "Bionic Animals"
- To Kevin Rusnak, senior historian at the Air Force Research Laboratory, for reviewing Chapter 18, "Space Exploration"
- To Faye Harp, Lori Vanover, and the teachers at Heritage Elementary School for their help and inspiration for Chapter 20, "From Trash to Treasure"
- To Sienna Weinstein, a master's degree student in conservation biology in the Department of Environmental Studies at Antioch University New

Acknowledgments

England for reviewing the "Better Together" lesson

- To Jeff Morgan, a chemical engineer in the R&D product development department at Procter & Gamble, and Karl Vanderbeek, associate creative director in the industrial design department at Kaleidoscope Product Design and Innovation, for sharing their design process expertise

The contributions of the following reviewers are also gratefully acknowledged:

- Kevin Anderson
- Deborah Hanuscin
- Eileen LaTorre
- Brandy Whitney

About the Authors

Emily Morgan is a former elementary science lab teacher for Mason City Schools in Mason, Ohio, and seventh-grade science teacher at Northridge Local Schools in Dayton, Ohio. She served as a science consultant for the Hamilton County Educational Service Center and science leader for the High AIMS Consortium. She has a bachelor of science in elementary education from Wright State University and a master of science in education from the University of Dayton. She is also the author of the *Next Time You See* picture book series from NSTA Press. Emily lives in West Chester, Ohio, with her husband, son, and an assortment of animals.

Karen Ansberry is a former elementary science curriculum leader and fifth- and sixth-grade teacher at Mason City Schools in Mason, Ohio. She has a bachelor of science in biology from Xavier University and a master of arts in teaching from Miami University. Karen lives in historic Lebanon, Ohio, with her husband, two sons, two daughters, and two dogs.

Emily and Karen enjoy facilitating teacher workshops at elementary schools, universities, and professional conferences across the country. This is Emily and Karen's fourth book in the *Picture-Perfect Science Lessons* series. For more information on this series and teacher workshops, visit *www.pictureperfectscience.com*.

Safety Practices for Science Activities

With hands-on, process- and inquiry-based science activities, the teaching and learning of science today can be both effective and exciting. The challenge to securing this success needs to be met by addressing potential safety issues relative to engineering controls (ventilation, eye wash station, etc.), administrative procedures and safety operating procedures, and use of appropriate personal protective equipment (indirectly vented chemicals splash goggles meeting ANSI Z87.1 standard, chemical resistant aprons and gloves, etc.). Teachers can make it safer for students and themselves by adopting, implementing, and enforcing legal safety standards and better professional safety practices in the science classroom and laboratory. Throughout this book, safety notes are provided for science activities and need to be adopted and enforced in efforts to provide for a safer learning and teaching experience. Teachers should also review and follow local policies and protocols used in their school district and/or school (e.g., employer OSHA Hazard Communication Safety Plan and Board of Education safety policies).

Additional applicable standard operating procedures can be found in the National Science Teacher Association's "Safety in the Science Classroom, Laboratory, or Field Sites" (*www.nsta.org/docs/ SafetyInTheScienceClassroomLabAndField.pdf*). Students should be required to review the document or one similar to it for elementary-level students under the direction of the teacher. It is important to also include safety information about working at home for the "STEM at Home" activities. Both the student and the parent or guardian should then sign the document acknowledging procedures that must be followed for a safer working and learning experience in the classroom, laboratory, or field. The Council of State Science Supervisors also has a safety resource for elementary science activities titled "Science and Safety: It's Elementary!" Teachers can consult this document at *www.csss-science.org/ downloads/scisaf_cal.pdf*.

Please note that the safety precautions of each activity are based, in part, on use of the recommended materials and instructions, legal safety standards, and better professional practices. Selection of alternative materials or procedures for these activities may jeopardize the level of safety and therefore is at the user's own risk.

Why Use Picture Books to Teach STEM?

Think about a book you loved as a child. Maybe you remember the zany characters and rhyming text of Dr. Seuss classics such as *Green Eggs and Ham* or the delightful poems in Robert Louis Stevenson's *A Child's Garden of Verses*. Perhaps you enjoyed the page-turning suspense of Jon Stone's *The Monster at the End of This Book* or the powerful lessons in Shel Silverstein's *The Giving Tree*. Maybe your curiosity was piqued by the technical illustrations and fascinating explanations in *The Way Things Work* by David Macauley or the illustrated anthology *Childcraft: The How and Why Library*. Perhaps you dreamed of space travel after reading the classic adventure *You Will Go to the Moon* by Mae and Ira Freeman. You may have seen a little of yourself in *Madeline* by Ludwig Bemelmans, *Where the Wild Things Are* by Maurice Sendak, *Ramona the Pest* by Beverly Cleary, or *Curious George* by H. A. Rey. Perhaps your imagination was stirred by *Cloudy with a Chance of Meatballs* by Judi and Ronald Barrett or *A Wrinkle in Time* by Madeleine L'Engle. You most likely remember the warm, cozy feeling of having a treasured book such as Don Freeman's *Corduroy*, Margery Williams's *The Velveteen Rabbit*, or Robert Munsch's *Love You Forever* being read to you by a parent or grandparent. But chances are your favorite book as a child was not your fifth-grade science textbook. The format of picture books offers certain unique advantages over textbooks and chapter books for engaging students in a STEM lesson. More often than other books, fiction and nonfiction picture books stimulate students on both the emotional and intellectual levels. They

TEACHERS LOVE USING PICTURE BOOKS!

are appealing and memorable because children readily connect with the imaginative illustrations, vivid photographs, exciting experiences and adventures of characters, engaging storylines, fascinating information that supports them in their quest for knowledge, and warm emotions that surround the reading experience.

What characterizes a picture book? We like what *Beginning Reading and Writing* says: "Picture books are unique to children's literature as they are defined by format rather than content. That is, they are books in which the illustrations are of equal importance as or more important than the text in the creation of meaning" (Strickland and Morrow 2000, p. 137). Because picture books are more likely to hold children's attention, they lend themselves to reading-comprehension strategy instruction and to engaging students within an inquiry-based cycle of science instruction. "Picture books, both fiction and nonfiction, are more likely to hold our attention and engage us than reading dry, formulaic text. … Engagement leads to remembering what is read, acquiring knowledge and enhancing understanding" (Harvey and Goudvis 2000, p. 46). We wrote the *Picture-Perfect STEM Lessons* series so teachers can take advantage of the positive features of children's picture books by supplementing the traditional textbook or kit program with a wide variety of high-quality fiction and nonfiction STEM-related picture books.

Why STEM?

Turn on the television news, open a newspaper, or browse an internet news source, and you'll likely find a story about a new STEM initiative or program at a school, library, or museum—STEM is everywhere these days! Historically, these four disciplines (science, technology, engineering, and mathematics) have been taught independently (see the box on the next page for more details about each discipline), with engineering often overlooked in the elementary classroom. But over the past several years, STEM education has gained momentum as an interdisciplinary way of teaching that goes beyond what is being learned in these disciplines to include the *application* of what is being learned.

The U.S. Department of Education states in the STEM section of its website,

The current emphasis on STEM education is due to an ever-increasing demand for highly skilled workers, including women and minorities who are underrepresented in STEM-related fields. As our world becomes more technologically advanced and as problems become more complex and multidisciplinary, this demand will continue to grow. The STEM-capable workforce and STEM-literate society of America's future will meet challenges as varied as improving human health and well-being, harnessing clean energy, protecting national security, and succeeding in the global economy. In remarks on the "Educate to Innovate" initiative to improve STEM achievement, President Barack Obama said, "The key to meeting these challenges … will be reaffirming and strengthening America's role as the world's engine of scientific discovery and technological innovation. And that leadership tomorrow depends on how we educate our students today." (The White House, Office of the Secretary 2009)

As educators, we must be equipped to prepare our students for meeting the challenges of our rapidly changing world. Mariel Milano, a member of the *Next Generation Science Standards* writing team, explains it this way:

Students entering kindergarten this year will likely enter job fields upon graduation that have not yet been developed, using knowledge that has not been discovered and tools that have not yet been engineered. It will be the responsibility of elementary teachers to prepare their students for a changing world by arming them with the science and engineering background necessary to one day make informed choices and decisions. (Milano 2013, p. 10)

> ### The Four STEM Disciplines
>
> **Science** is the study of the natural world, including the laws of nature associated with physics, chemistry, and biology and the treatment or application of facts, principles, concepts, or conventions associated with these disciplines. Science is both a body of knowledge that has been accumulated over time and a process—scientific inquiry—that generates new knowledge. Knowledge from science informs the engineering design process.
>
> **Technology** comprises the entire system of people and organizations, knowledge, processes, and devices that go into creating and operating technological artifacts, as well as the artifacts themselves. Throughout history, humans have created technology to satisfy their wants and needs. Much of modern technology is a product of science and engineering, and technological tools are used in both fields.
>
> **Engineering** is both a body of knowledge—about the design and creation of human-made products—and a process for solving problems. This process is design under constraint. One constraint in engineering design is the laws of nature, or science. Other constraints include time, cost, available materials, ergonomics, environmental regulations, manufacturability, and repairability. Engineering uses concepts in science and mathematics as well as technological tools.
>
> **Mathematics** is the study of patterns and relationships among quantities, numbers, and shapes. Specific branches of mathematics include arithmetic, geometry, algebra, trigonometry, and calculus. Mathematics is used in science and in engineering.
>
> *Source:* National Academy of Engineering and National Research Council 2009.

So what exactly is meant by "STEM education"? A quick web search yields a long list of definitions, interpretations, and philosophies. But our approach to STEM education is simple: It involves making natural connections among the four STEM disciplines as students investigate and problem-solve within a meaningful context. In *Picture-Perfect STEM Lessons,* picture books provide this meaningful context. The books help engage and motivate students, introduce topics and establish themes, set up investigations and real-world problem-solving opportunities, spark creativity and innovation, and explain science and engineering concepts. Science and engineering standards provide the learning framework, while reading strategies, technology, and mathematics are used as tools within this framework to support and extend student learning. The lessons are written so that the connections among the four disciplines are natural, not forced. For example, mathematics is applied where it fits within the overall goal of the lesson (not simply to meet a mathematics objective). So you will not see all four STEM disciplines given equal emphasis in every lesson.

Our previous *Picture-Perfect Science Lessons* books are based on research that shows that integrating science with literacy makes science more meaningful to students and can lead to increases in achievement in both subjects. We believe that these benefits apply to STEM and literacy integration, because students are provided meaningful context in which to investigate, innovate, and communicate.

The Research
Context for Concepts
The wide array of high-quality STEM-related children's literature currently available can help you model reading-comprehension strategies while teaching STEM content in a meaningful context.

Chapter 1

Children's picture books, a branch of literature, have interesting storylines that can help students understand and remember concepts better than they would by using only textbooks, which tend to present science as lists of facts to be memorized (Butzow and Butzow 2000). In addition, the colorful pictures and graphics in picture books are superior to many texts for explaining abstract ideas (Kralina 1993). Many studies, including one by Van den Heuvel-Panhuizen, Elia, and Robitzsch (2014), show that reading picture books can have a positive influence on children's mathematical performance as well. As more and more content is packed into the school day and higher expectations are placed on student performance, teachers must make more efficient use of their time. Zemelman, Daniels, and Hyde (2012) suggest that connecting various content areas can lead to deep engagement as students read, write, talk, view, watch, explore, create, and interact around a topic. Although research is limited on the impact of picture books on integrated STEM learning, we feel that teaching STEM in conjunction with literacy may enhance learning in all areas.

More Depth of Coverage

Science textbooks can be overwhelming for many children, especially those who have reading problems. Textbooks often contain unfamiliar vocabulary and tend to cover a broad range of topics (Casteel and Isom 1994; Short and Armstrong 1993; Tyson and Woodward 1989). However, fiction and nonfiction picture books tend to focus on fewer topics and give more in-depth coverage of the concepts. It can be useful to pair an engaging fiction book with a nonfiction book to round out the science content being presented.

For example, the "Wind It Up" lesson in Chapter 8 features *Clink*, an endearing story about a toy robot that nobody wants to buy. It is paired with *Making Machines With Springs*, a nonfiction book that explains how springs can be used to store energy and make things move. The sweet story and engaging illustrations in *Clink* hook the reader, whereas the book *Making Machines With Springs* presents facts and background information. Together, they offer an engaging yet in-depth look at forces and motion and how engineering affects our everyday lives.

Improved Reading and Science Skills

Research by Morrow et al. (1997) on using children's literature and literacy instruction in the science program indicated gains in science as well as literacy. Romance and Vitale (1992) found significant improvement in the science and reading scores of fourth graders when the regular basal reading program was replaced with reading in science that correlated with the science curriculum. They also found an improvement in students' attitudes toward the study of science.

Opportunities to Correct Science Misconceptions

Students often have strongly held misconceptions about science that can interfere with their learning. "Misconceptions, in the field of science education, are preconceived ideas that differ from those currently accepted by the scientific community" (Colburn 2003, p. 59). Children's picture books, reinforced with hands-on inquiries, can help students correct their misconceptions. Repetition of the correct concept by reading several books, doing a number of experiments, and inviting scientists to the classroom can facilitate a conceptual change in children (Miller, Steiner, and Larson 1996).

But teachers must be aware that scientific misconceptions can be inherent in the picture books. Although many errors are explicit, some of the misinformation is more implicit or may be inferred from text and illustrations (Rice 2002). This problem is more likely to occur in fictionalized material. Mayer's (1995) study demonstrated that when both inaccuracies and science facts are presented in the same book, children do not necessarily remember the correct information.

Selection of Books

Each lesson in *Picture-Perfect STEM Lessons* focuses on *A Framework for K–12 Science Education* (*Framework*; NRC 2012). We selected fiction and nonfiction children's picture books that closely relate

to the *Framework*. An annotated "More Books to Read" section is provided at the end of each lesson. If you would like to select more children's literature to use in your science classroom, try the Outstanding Science Trade Books for Students K–12 listing, which is a cooperative project between the National Science Teachers Association (NSTA) and the Children's Book Council (CBC). The books are selected by a book-review panel appointed by NSTA and assembled in cooperation with CBC. Each year, a new list is featured in the March issue of NSTA's elementary school teacher journal *Science and Children*. See *www.nsta.org/ostbc* for archived lists.

When you select children's picture books for science instruction, you might consult with a knowledgeable colleague who can help you check them for errors or misinformation. Young and Moss (2006) describe five essential things to consider when selecting nonfiction trade books for science:

1. The authority of the author (i.e., the author's credibility and qualifications for writing the book)
2. The accuracy of the text, illustrations, and graphics
3. The appropriateness of the book for its intended audience (e.g., the book makes complex concepts understandable for young readers)
4. The literary artistry and quality of writing
5. The appearance or visual impact of the book

Using a rubric may also be valuable to help you make informed decisions about the science trade books you use in your classroom. One such tool that provides a systematic framework to simplify the trade book evaluation process is the Science Trade Book Evaluation Rubric, found in the article "Making Science Trade Book Choices for Elementary Classrooms" (Atkison, Matusevich, and Huber 2009).

Finding the Picture-Perfect Books

Each lesson includes a "Featured Picture Books" section with titles, author and illustrator names, publication details, and summaries of each book. The years and publisher names listed are for the most recent editions available—paperback whenever possible—as of the printing of *Picture-Perfect STEM Lessons, 3–5*.

All of the trade books featured in the lessons in this book are currently in print and can be found at your local bookstore or from an online retailer or library. NSTA Press has made available a tote bag full of all of the *Picture-Perfect STEM Lessons* books in one handy collection at a reduced cost, in addition to ClassPacks that contain the materials you need to do each lesson. You can purchase these items at the NSTA Science Store (*www.nsta.org/store*).

Considering Genre

Considering genre when you determine how to use a particular picture book within a STEM lesson is important. Donovan and Smolkin (2002) identify four different genres frequently recommended for teachers to use in their science instruction: story, non-narrative information, narrative information, and dual purpose. This book identifies the genre of each featured book at the beginning of each lesson. Summaries of the four genres, a representative picture book for each genre, and suggestions for using each genre within the Biological Sciences Curriculum Study (BSCS) 5E learning cycle follow. (Chapter 3 describes in detail the science learning cycle, known as the BSCS 5E Instructional Model, which we follow.)

Storybooks

Storybooks center on specific characters who work to resolve a conflict or problem. The major purpose of stories is to entertain, not to present factual information. The vocabulary is typically commonsense, everyday language. An engaging storybook can spark interest in a science topic and move students toward informational texts to answer questions inspired by the story. For example, "The Inventor's Secret" lesson in Chapter 6 begins with a read-aloud of *Rosie Revere, Engineer,* a story about a girl who secretly invents a variety of fantastic machines. The

charming story hooks the learners and engages them in an exploration of the design process.

Non-Narrative Information Books

Non-narrative information books are nonfiction texts that introduce a topic, describe the attributes of the topic, or describe typical events that occur. The focus of these texts is on the subject matter, not specific characters. The vocabulary is typically technical. Readers can enter the text at any point in the book. Many non-narrative information books contain features found in nonfiction such as a table of contents, bold-print vocabulary words, a glossary, and an index. Some research suggests that these types of books are "the best resources for fostering children's scientific concepts as well as their appropriation of science discourse" (Pappas 2006). Young children tend to be less familiar with this genre and need many opportunities to experience this type of text. Using non-narrative information books will help students become familiar with the structure of textbooks, as well as "real-world" reading, which is primarily nonfiction. Teachers may want to read only those sections that provide the concepts and facts needed to meet particular science objectives.

One example of non-narrative information writing is the book *TIME for Kids: Space Exploration*, which contains nonfiction text features such as a table of contents, bold-print words, insets, a glossary, and an index. This book is featured in Chapter 18, "Space Exploration." The appropriate placement of non-narrative information text in a science learning cycle is typically after students have had the opportunity to explore concepts through hands-on activities. At that point, students are engaged in the topic and are motivated to read the non-narrative informational text to learn more.

Narrative Information Books

Narrative information books, another subset of nonfiction books, are sometimes called *hybrid* books. They provide an engaging format for factual information. They communicate a sequence of factual events over time and sometimes recount the events of a specific case to generalize to all cases. When using these books within science instruction, establish a purpose for reading so that students focus on the science content rather than the storyline. In some cases, teachers may want to read the book one time through for the aesthetic components of the book and a second time to look for specific science content or engineering practices. *One Plastic Bag: Isatou Ceesay and the Recycling Women of The Gambia,* an example of a narrative information text, is used in Chapter 20, "From Trash to Treasure." This narrative tells the true story of a woman in The Gambia who, with the help of her friends, started a recycling movement that changed her community. The narrative information genre can be used at any point within a science learning cycle. This genre can be both engaging and informative.

Dual-Purpose Books

Dual-purpose books are intended to serve two purposes: present a story and provide facts. They employ a format that allows readers to use the book like a storybook or to use it like a non-narrative information book. Sometimes, information can be found in the running text, but more frequently information appears in insets and diagrams. Readers can enter on any page to access specific facts, or they can read the book through as a story. You can use the story component of a dual-purpose book to engage the reader at the beginning of the science learning cycle.

Dual-purpose books typically have little science content within the story. If the insets and diagrams are read, discussed, explained, and related to the story, these books can be very useful in helping students refine concepts and acquire scientific vocabulary after they have had opportunities for hands-on exploration.

Using Fiction and Nonfiction Texts

It can be useful to pair fiction and nonfiction books in read-alouds to round out the science or engineering content being presented. Because fiction books tend to be very engaging for students, they can be used to hook students at the beginning of a science lesson. But most of the reading people do in everyday life is nonfiction. We are immersed in informational text every day, and we must be able to comprehend it to be successful in school, at work, and in society. Nonfiction books and other informational text such as articles should be used frequently in the elementary classroom. They often include text structures that differ from stories, and the opportunity to experience these structures in read-alouds can strengthen students' abilities to read and understand informational text. Duke (2004) recommends four strategies to help teachers improve students' comprehension of informational text:

1. Increase students' access to informational text.
2. Increase the time students spend working with informational text.
3. Teach comprehension strategies through direct instruction.
4. Create opportunities for students to use informational text for authentic purposes.

Picture-Perfect STEM Lessons addresses these recommendations in several ways. The lessons expose students to a variety of nonfiction picture books, articles, and websites on science topics, thereby increasing access to informational text. Various tools (e.g., card sorts, anticipation guides, "Stop and Try It"; see Chapter 2 for a complete list of these tools) help enhance students' comprehension of the informational text by increasing the time they spend working with it. Each lesson includes instructions for explicitly teaching comprehension strategies within the learning cycle. The inquiry-based lessons provide an authentic purpose for reading informational text, as students are motivated to read or listen to find the answers to questions generated in the inquiry activities.

References

Atkison, T., M. N. Matusevich, and L. Huber. 2009. Making science trade book choices for elementary classrooms. *The Reading Teacher* 62 (6): 484–497.

Butzow, J., and C. Butzow. 2000. *Science through children's literature: An integrated approach.* Portsmouth, NH: Teacher Ideas Press.

Casteel, C. P., and B. A. Isom. 1994. Reciprocal processes in science and literacy learning. *The Reading Teacher* 47 (7): 538–544.

Colburn, A. 2003. *The lingo of learning: 88 education terms every science teacher should know.* Arlington, VA: NSTA Press.

Donovan, C., and L. Smolkin. 2002. Considering genre, content, and visual features in the selection of trade books for science instruction. *The Reading Teacher* 55 (6): 502–520.

Duke, N. K. 2004. The case for informational text. *Educational Leadership* 61 (6): 40–44.

Harvey, S., and A. Goudvis. 2000. *Strategies that work: Teaching comprehension to enhance understanding.* York, ME: Stenhouse Publishers.

Kralina, L. 1993. Tricks of the trades: Supplementing your science texts. *The Science Teacher* 60 (9): 33–37.

Martin, D. J. 1997. *Elementary science methods: A constructivist approach.* Albany, NY: Delmar.

Mayer, D. A. 1995. How can we best use children's literature in teaching science concepts? *Science and Children* 32 (6): 16–19, 43.

Milano, M. 2013. The *Next Generation Science Standards* and engineering for young learners: Beyond bridges and egg drops. *Science and Children* 51 (2): 10–16.

Miller, K. W., S. F. Steiner, and C. D. Larson. 1996. Strategies for science learning. *Science and Children* 33 (6): 24–27.

Morrow, L. M., M. Pressley, J. K. Smith, and M. Smith. 1997. The effect of a literature-based program integrated into literacy and science instruction with children from diverse backgrounds. *Reading Research Quarterly* 32 (1): 54–76.

National Academy of Engineering and National Research Council. 2009. *Engineering in K–12 education: Understanding the status and improving the prospects.* Washington, DC: National Academies Press.

National Research Council (NRC). 2012. *A framework for K–12 science education: Practices, crosscutting concepts, and core ideas.* Washington, DC: National Academies Press.

Pappas, C. 2006. The information book genre: Its role in integrated science literacy research and practice. *Reading Research Quarterly* 41 (2): 226–250.

Rice, D. C. 2002. Using trade books in teaching elementary science: Facts and fallacies. *The Reading Teacher* 55 (6): 552–565.

Romance, N. R., and M. R. Vitale. 1992. A curriculum strategy that expands time for in-depth elementary science instruction by using science-based reading strategies: Effects of a year-long study in grade four. *Journal of Research in Science Teaching* 29 (6): 545–554.

Short, K. G., and J. Armstrong. 1993. Moving toward inquiry: Integrating literature into the science curriculum. *New Advocate* 6 (3): 183–200.

Strickland, D. S., and L. M. Morrow, eds. 2000. *Beginning reading and writing.* New York: Teachers College Press.

Tyson, H., and A. Woodward. 1989. Why aren't students learning very much from textbooks? *Educational Leadership* 47 (3): 14–17.

U.S. Department of Education. 2009. Science, technology, engineering and math: Education for global leadership. www.ed.gov/stem.

Van den Heuvel-Panhuizen, M., I. Elia, and A. Robitzsch. 2014. Effects of reading picture books on kindergartners' mathematics performance. *Educational Psychology* 36 (2): 323–346.

Young, T. A., and B. Moss. 2006. Nonfiction in the classroom library: A literary necessity. *Childhood Education* 82 (4): 207–212.

Zemelman, S., H. Daniels, and A. Hyde. 2012. *Best practice: Bringing standards to life in America's classrooms.* 4th ed. Portsmouth, NH: Heinemann.

Children's Books Cited

Barrett, J., and R. Barrett. *Cloudy with a chance of meatballs.* New York: Atheneum Books for Young Readers.

Beaty, A. 2013. *Rosie Revere, engineer.* New York: Abrams Books for Young Readers.

Bemelmans, L. 1958. *Madeline.* New York: Penguin Young Readers Group.

Childcraft Editors. 1973. *Childcraft: The how and why library.* New York: World Book.

Cleary, B. 1968. *Ramona the pest.* New York: HarperCollins.

DiPucchio, K. 2011. *Clink.* New York: Balzer + Bray.

Dugan, C. 2013. *TIME for kids: Space exploration.* Huntington Beach, CA: Teacher Created Materials.

Freeman, D. 1968. *Corduroy.* New York: Viking Press.

Freeman, M., and I. Freeman. 1959. *You will go to the moon.* New York: Random House Children's Books.

L'Engle, M. 1963. *A wrinkle in time.* New York: Farrar, Straus & Giroux.

Macauley, D. 1988. *The way things work.* New York: Houghton Mifflin/Walter Lorraine Books.

Munsch, R. 1986. *Love you forever.* Scarborough, Ontario: Firefly Books.

Oxlade, C. 2015. *Making machines with springs.* Chicago: Heinemann Raintree.

Paul, M. 2015. *One plastic bag: Isatou Ceesay and the recycling women of The Gambia.* Minneapolis, MN: Millbrook Press.

Rey, H. A. 1973. *Curious George.* Boston: Houghton Mifflin.

Sendak, M. 1988. *Where the wild things are.* New York: HarperCollins.

Seuss, Dr. 1960. *Green eggs and ham.* New York: Random House Books for Young Readers

Silverstein, S. 1964. *The giving tree.* New York: HarperCollins.

Stevenson, R. L. 1957. *A child's garden of verses.* New York: Grosset & Dunlap.

Stone, J. 2003. *The monster at the end of this book.* New York: Golden Books.

Williams, M. 1922. *The velveteen rabbit.* New York: Doubleday & Company

Reading Aloud

This chapter addresses some of the research supporting the importance of reading aloud, tips to make your read-aloud time more valuable, descriptions of Harvey and Goudvis's six key reading strategies (2007), and tools you can use to enhance students' comprehension during read-aloud time.

Why Read Aloud?

Being read to is the most influential element in building the knowledge required for eventual success in reading (Anderson et al. 1985). It improves reading skills, increases interest in reading and literature, and can even improve overall academic achievement. A good reader demonstrates fluent, expressive reading and models the thinking strategies of proficient readers, helping build background knowledge and fine-tune students' listening skills. When a teacher does the reading, children's minds are free to anticipate, infer, connect, question, and comprehend (Calkins 2000). In addition, being read to is risk free. In *Yellow Brick Roads: Shared and Guided Paths to Independent Reading, 4–12,* Allen (2000) says, "For students who struggle with word-by-word reading, experiencing the whole story can finally give them a sense of the wonder and magic of a book" (p. 45).

Reading aloud is appropriate in all grade levels and for all subjects. Appendix A of the *Common Core State Standards for English Language Arts* (*CCSS ELA;* NGAC and CCSSO 2010) states that "children in the early grades—particularly kindergarten through grade 3—benefit from participating in rich, structured conversations with an adult in response to written texts that are read aloud, orally comparing and contrasting as well as analyzing and synthesizing" (p. 27). Reading aloud is important not only when children can't read well on their own but also after they have become proficient readers (Anderson et al. 1985). Allen (2000) supports this

READ-ALOUD TIME

view: "Given the body of research supporting the importance of read-aloud for modeling fluency, building background knowledge, and developing language acquisition, we should remind ourselves that those same benefits occur when we extend read-aloud beyond the early years" (p. 44). Likewise, the *CCSS ELA* advocate the use of read-alouds in upper elementary:

> *Because children's listening comprehension likely outpaces reading comprehension until the middle school years, it is particularly important that students in the earliest grades build knowledge through being read to as well as through reading, with the balance gradually shifting to reading independently. By reading a story or nonfiction selection aloud, teachers allow children to experience written language without the burden of decoding, granting them access to content that they may not be able to read and understand by themselves. Children are then free to focus their mental energy on the words and ideas presented in the text, and they will*

eventually be better prepared to tackle rich written content on their own. (NGAC and CCSSO 2010, Appendix A, p. 27)

Ten Tips for Reading Aloud

We have provided a list of tips to help you get the most from your read-aloud time. Using these suggestions can help set the stage for learning, improve comprehension of science material, and make the read-aloud experience richer and more meaningful for both you and your students.

1. Preview the Book

Select a book that meets your science objectives and lends itself to reading aloud. Preview it carefully before sharing it with your students. Are there any errors in scientific concepts or misinformation that could be inferred from the text or illustrations? If the book is not in story form, is there any nonessential information you could omit to make the read-aloud experience better? If you are not going to read the whole book, choose appropriate starting and stopping points before reading. Consider generating questions and inferences about the book in advance and placing them on sticky notes inside the book to help you model your thought processes as you read aloud.

2. Set the Stage

Because reading aloud is a performance, you should pay attention to the atmosphere and physical setting of the session. Gather the students in a special reading area, such as on a carpet or in a semicircle of chairs. Seat yourself slightly above them. Do not sit in front of a bright window where the glare will keep students from seeing you well or in an area where students can be easily distracted. You may want to turn off the overhead lights and read by the light of a lamp or use soft music as a way to draw students into the mood of the text. Establish expectations for appropriate behavior during read-aloud time, and, before reading, give the students an opportunity to settle down and focus their attention on the book.

3. Celebrate the Author and Illustrator

Tell students the names of the author and the illustrator before reading. Build connections by asking students if they have read other books by the author or illustrator. Increase interest by sharing facts about the author or illustrator from the book's cover or from library or internet research.

4. Read With Expression

Practice reading aloud to improve your performance. Can you read with more expression to more fully engage your audience? Try louder or softer speech, funny voices, facial expressions, or gestures. Make eye contact with your students every now and then as you read. This strengthens the bond between reader and listener, helps you gauge your audience's response, and cuts down on off-task behaviors. Read slowly enough that your students have time to build mental images of what you are reading, but not so slowly that they lose interest. When reading a nonfiction book aloud, you may want to pause after reading about a key concept to let it sink in and then reread that part. At suspenseful parts in a storybook, use dramatic pauses or slow down and read softly. This can move the audience to the edges of their seats!

5. Share the Pictures

Don't forget the power of visual images to help students connect with and comprehend what you are reading. Make sure that you hold the book in such a way that students can see the pictures on each page. Read captions if appropriate. In some cases, you may want to hide certain pictures so students can visualize what is happening in the text before you reveal the illustrator's interpretation.

6. Encourage Interaction

Keep chart paper and markers nearby in case you want to record questions or new information. Try providing students with "think pads" in the form of sticky notes to write on as you read aloud. Not only does this help extremely active children keep their hands busy while listening, but it also encourages students to interact with the text as they jot down questions or comments. After the read-aloud, have students share their questions and comments. You may want students to place their sticky notes on a class chart whose subject is the topic being studied. Another way to encourage interaction without taking the time for each student to ask questions or comment is to do an occasional "turn and talk" during the read-aloud. Stop reading, ask a question, allow thinking time, and then have each student share ideas with a partner.

7. Keep the Flow
Although you want to encourage interaction during a read-aloud, avoid excessive interruptions that may disrupt fluent, expressive reading. Aim for a balance between allowing students to hear the language of the book uninterrupted and providing them with opportunities to make comments, ask questions, and share connections to the reading. You may want to read the book all the way through one time so students can enjoy the aesthetic components of the story, and then go back and read the book for the purpose of meeting the science objectives.

8. Model Reading Strategies
As you read aloud, it is important that you help children access what they already know and build bridges to new understandings. Think out loud, model your questions for the author, and make connections to yourself, other books, and the world. Show students how to determine the important parts of the text or story, and demonstrate how you synthesize meaning from the text. Modeling these reading-comprehension strategies when appropriate before, during, or after reading helps students internalize the strategies and begin to use them in their own reading. Six key strategies are described in detail in the next section.

9. Don't Put It Away
Keep the read-aloud book accessible to students after you read it. They will want to get a close-up look at the pictures and will enjoy reading the book independently. Don't be afraid of reading the same book more than once—children benefit from the repetition.

10. Have Fun
Let your passion for books show. It is contagious! Read nonfiction books with interest and wonder. Share your thoughts, question the author's intent, synthesize meaning out loud, and voice your own connections to the text. When reading a story, let your emotions show—laugh at the funny parts and cry at the sad parts. Seeing an authentic response from the reader is important for students. If you read with enthusiasm, read-aloud time will become special and enjoyable for everyone involved.

We hope these tips will help you and your students reap the many benefits of read-alouds. As Miller (2002) writes in *Reading With Meaning: Teaching Comprehension in the Primary Grades,* "Learning to read should be a joyful experience. Give children the luxury of listening to well-written stories with interesting plots, singing songs and playing with their words, and exploring a wide range of fiction, nonfiction, poetry and rhymes. ... Be genuine. Laugh. Love. Be patient. You're creating a community of readers and thinkers" (p. 26).

Reading-Comprehension Strategies

Children's author Madeleine L'Engle (1995) says, "Readers usually grossly underestimate their own importance. If a reader cannot create a book along with the writer, the book will never come to life. The author and the reader ... meet on the bridge of words" (p. 34). It is our responsibility as teachers, no matter what subjects we are assigned to teach, to help children realize the importance of their own thoughts and ideas as they read. Modeling our own thinking as we read aloud is the first step. Becoming a proficient reader is an ongoing, complex process, and children need to be explicitly taught the strategies that good readers use. In *Strategies That Work,* Harvey and Goudvis (2007) identify six key reading strategies essential to achieving full understanding when we read. These strategies are used where appropriate in each lesson and are seamlessly embedded into the 5E Model. The strategies should be modeled as you read aloud to students from both fiction and nonfiction texts.

Research shows that explicit teaching of reading-comprehension strategies can foster comprehension development (Duke and Pearson 2002). Explicit teaching of the strategies is the initial step in the gradual-release-of-responsibility approach to delivering reading instruction (Fielding and Pearson 1994). During this first phase of the gradual-release method, the teacher *explains* the strategy, demonstrates *how* and *when* to use the strategy, explains *why* it is worth using, and *thinks aloud* to model the mental processes used by good readers. Duke (2004, p. 42) describes the process in this way:

> *I often discuss the strategies in terms of good readers, as in "Good readers think about what might be coming next." I also model the uses of comprehension strategies by thinking aloud as I read. For example, to model the importance of monitoring understanding, I make comments such as, "That doesn't make sense to me because ..." or "I didn't understand that last part—I'd better go back."*

Using the teacher-modeling phase within a science learning cycle will reinforce what students do during reading instruction, when the gradual-release-of-responsibility model can be continued. When students have truly mastered a strategy, they are able to apply it to a variety of texts and curricular areas and can explain how the strategy helps them construct meaning.

Descriptions of the six key reading-comprehension strategies featured in *Strategies That Work* (Harvey and Goudvis 2007) follow. The following icon highlights these strategies here and within the lessons: .

Making Connections

Making meaningful connections during reading can improve learners' comprehension and engagement by helping them better relate to what they read. Comprehension breakdown that occurs when reading or listening to expository text can come from a lack of prior information. These three techniques can help readers build background knowledge where little exists:

- *Text-to-self connections* occur when readers and listeners link the text to their past experiences or background knowledge.
- *Text-to-text connections* occur when readers and listeners recognize connections from one book to another.
- *Text-to-world connections* occur when readers and listeners connect the text to events or issues in the real world.

Questioning

Proficient readers ask themselves questions before, during, and after reading. Questioning allows readers to construct meaning, find answers, solve problems, and eliminate confusion as they read. Harvey and Goudvis (2007) write, "Questioning is the strategy that propels readers forward. When readers have questions, they are less likely to abandon the text" (p. 18). Asking questions not only is a critical reading skill but also is at the heart of scientific inquiry and can lead students into meaningful investigations. Questioning as a scientific practice is clearly articulated in *A Framework for K–12 Science Education*, which suggests that students ask questions based on observations to find more information or to design an investigation (NRC 2012).

Visualizing

Visualizing is the creation of mental images while reading or listening to text. Mental images are created from the learner's emotions and senses, making the text more concrete and memorable. Imagining the sensory qualities of things described in a text can help engage learners and stimulate their interest in the reading. When readers form pictures in their minds, they are also more likely to stick with a challenging text. During a reading, you can stop and ask students to visualize the scene. What sights, sounds, smells, and colors are they imagining?

Inferring

Reading between the lines, or inferring, involves a learner's merging clues from the reading with prior knowledge to draw conclusions and interpret the text. Good readers make inferences before, during, and after reading. Inferential thinking is also an important science skill and can be reinforced during reading instruction.

Determining Importance

Reading to learn requires readers to identify essential information by distinguishing it from nonessential details. Deciding what is important in the text depends on the purpose for reading. In *Picture-Perfect STEM Lessons,* each lesson's objectives determine importance. Learners read or listen to the text to find answers to specific questions, to gain understanding of concepts, and to identify misconceptions.

Synthesizing

In synthesizing, readers combine information gained through reading with prior knowledge and experience to form new ideas. To synthesize, readers must stop, think about what they have read, and contemplate its meaning before continuing on through the text. The highest level of synthesis involves those "aha!" moments when readers achieve new insight and, as a result, change their thinking.

Tools to Enhance Comprehension

We have identified several activities and organizers that can enhance students' science understanding and reading-comprehension in the lessons. These tools, which support the reading comprehension strategies from *Strategies That Work* listed in the previous section, are briefly described on the following pages and in more detail within the lessons.

Anticipation Guides

Anticipation guides (Herber 1978) are sets of questions that serve as a pre- and postreading activity for a text. They can be used to activate and assess prior knowledge, determine misconceptions, focus thinking on the reading, and motivate reluctant readers by stimulating interest in the topic. An anticipation guide should revolve around four to six key concepts from the reading that learners respond to before reading. They will be motivated to read or listen carefully to find the evidence that supports their predictions. After reading, learners revisit their anticipation guide to check their responses. In a revised extended anticipation guide (Duffelmeyer and Baum 1992), learners are required to justify their responses and explain why their choices were correct or incorrect.

CARD SEQUENCING

Card Sorts and Sequencing

Card sorts help learners understand the relationships among key concepts and help teach classification. They can also reveal misconceptions and increase motivation to read when used as a prereading activity. Learners are asked to sort words or phrases written on cards into different categories or sequence the events described on the cards. In an "open sort," learners sort the cards into categories of their own making or sequence events any way they wish. They can re-sort and re-sequence to help refine their understanding of concepts or events. In a "closed sort," the teacher gives them the categories for sorting or provides more information for correctly sequencing their cards.

Chunking

Chunking is dividing the text into manageable sections and reading only a section at any one time. This gives learners time to digest the information in a section before moving on. Chunking is also a useful technique for weeding out nonessential information when reading nonfiction books. Reading only those parts of the text that meet your learning objectives focuses the learning on what is important. Remember: Nowhere is it written that you must read nonfiction books cover to cover when doing a read-aloud. Feel free to omit parts that are inaccurate, out of date, or don't contribute in a meaningful way to the lesson.

New Vocabulary List

A new vocabulary list, sometimes called a *personal vocabulary list* (Beers and Howell 2004) is a "guess and check" type of visual representation. Students develop vocabulary as they draw and write predictions about the meanings of new words, read the words in context, and draw and write their definitions of the words.

Cloze Strategy

Cloze refers to an activity that helps readers infer the meanings of unfamiliar words. In the cloze strategy, key words are deleted in a passage. Students then fill in the blanks with words that make sense and sound right. Words can be printed on cards for the students to place in the blanks before reading a passage in order for students to predict where they go. Then, after reading the passage, students can move them if necessary.

Rereading

Nonfiction text is often full of unfamiliar ideas and difficult vocabulary. *Rereading* content for clarification is an essential skill of proficient readers, and you should model this frequently. Rereading content for a different purpose can aid comprehension. For example, you might read aloud a text for enjoyment and then revisit the text to focus on specific science content.

STOP AND TRY IT

Picture Walk

A picture walk consists of showing students the cover of a book and browsing through the pages in order, without reading the text. The purpose of this tool is to establish interest in the story and expectations about what is to come. It also reinforces the importance of using visual cues while reading. Students look at the pictures and talk about what they see, what may be happening in each illustration, and how the pictures come together to make a story. Some useful questions to ask during a picture walk are as follows:

- From looking at the cover, what do you think this book is about?
- What do you see?
- What do you think is happening?
- What do you think will happen next?
- What are you curious to know more about in the book?

Stop and Try It

Stop and Try It is a read-aloud format in which the teacher stops reading the text periodically to allow students to observe a demonstration or take part in a hands-on activity to better understand the content being presented. For example, in Chapter 8, "Wind It Up," we recommend that the teacher stop reading the book *Making Machines With Springs* at key points to allow students to perform some of the activities described in the book. This way, students have an experience that connects to the information they are learning from the book.

Turn and Talk

Learners pair up with a partner to share their ideas, explain concepts in their own words, or talk about a connection they have to the book. This method allows each child to respond so that everyone in the group is involved as either a talker or a listener. Saying, "Take a few minutes to share your thoughts with someone" gives students an opportunity to satisfy their needs to express their own thoughts about the reading. "Walk and talk" and "stretch and share" are variations of this strategy that incorporate movement.

Using Features of Nonfiction

Many nonfiction books include a table of contents, index, glossary, bold-print words, picture captions, diagrams, and charts that provide valuable information. Because children are generally more used to narrative text, they often skip over these text structures. It is important to model how to interpret the information these features provide the reader. To begin, show the cover of a nonfiction book and read the title and table of contents. Ask students to predict what they'll find in the book. Show students how to use the index in the back of the book to find specific information. Point out other nonfiction text structures as you read, and note that these features are unique to nonfiction. Model how nonfiction books can be entered at any point in the text, because they generally don't follow a storyline.

USING THE TABLE OF CONTENTS

How Do Picture Books Enhance Comprehension?

Students should be encouraged to read a wide range of print materials, but picture books offer many advantages when teaching reading-comprehension strategies. Harvey and Goudvis (2007) not only believe that interest is essential to comprehension but also maintain that, because picture books are extremely effective for building background knowledge and teaching content, instruction in reading-comprehension strategies during picture book read-alouds allows students to better access that content. In summary, picture books are invaluable for teaching reading-comprehension strategies because they are extraordinarily effective at keeping readers engaged and thinking.

References

Allen, J. 2000. *Yellow brick roads: Shared and guided paths to independent reading, 4–12.* Portland, ME: Stenhouse Publishers.

Anderson, R. C., E. H. Heibert, J. Scott, and I. A. G. Wilkinson. 1985. *Becoming a nation of readers: The report of the Commission on Reading.* Washington, DC: National Institute of Education, U.S. Department of Education.

Beers, S., and L. Howell. 2004. *Reading strategies for the content areas: An action toolkit, volume 2.* Alexandria, VA: Association for Supervision and Curriculum Development.

Calkins, L. M. 2000. *The art of teaching reading.* Boston: Pearson Allyn & Bacon.

Duffelmeyer, F. A., and D. D. Baum. 1992. The extended anticipation guide revisited. *Journal of Reading* 35 (8): 654–656.

Duke, N. K. 2004. The case for informational text. *Educational Leadership* 61 (6): 40–44.

Duke, N. K., and P. D. Pearson. 2002. Effective practices for developing reading comprehension. In *What research has to say about reading instruction,* ed. A. E. Farstrup and S. J. Samuels, 205–242. Newark, DE: International Reading Association.

Fielding, L., and P. D. Pearson. 1994. Reading comprehension: What works? *Educational Leadership* 51 (5): 62–67.

Harvey, S., and A. Goudvis. 2007. *Strategies that work: Teaching comprehension for understanding and engagement.* 2nd ed. Portland, ME: Stenhouse Publishers.

Herber, H. 1978. *Teaching reading in the content areas.* Englewood Cliffs, NJ: Prentice Hall.

L'Engle, M. 1995. *Walking on water: Reflections on faith and art.* New York: North Point Press.

Miller, D. 2002. *Reading with meaning: Teaching comprehension in the primary grades.* Portland, ME: Stenhouse Publishers.

National Governors Association Center for Best Practices and Council of Chief State School Officers (NGAC and CCSSO). 2010. *Common core state standards.* Washington, DC: NGAC and CCSSO.

National Research Council (NRC). 2012. *A framework for K–12 science education: Practices, crosscutting concepts, and core ideas.* Washington, DC: National Academies Press.

Children's Book Cited

Oxlade, C. 2015. *Making machines with springs.* Chicago: Heinemann Raintree.

BSCS 5E Instructional Model

The guided inquiries in this book are designed using the BSCS 5E Instructional Model, commonly referred to as the 5E Model (or the 5Es). Developed by the Biological Sciences Curriculum Study (BSCS), the 5E Model is a learning cycle based on a constructivist view of learning. Constructivism embraces the idea that learners bring with them preconceived ideas about how the world works. According to the constructivist view, "learners test new ideas against that which they already believe to be true. If the new ideas seem to fit in with their pictures of the world, they have little difficulty learning the ideas … if the new ideas don't seem to fit the learners' picture of reality then they won't seem to make sense. Learners may dismiss them … or eventually accommodate the new ideas and change the way they understand the world" (Colburn 2003, p. 59). The objective of a constructivist model, therefore, is to provide students with experiences that make them reconsider their conceptions. Then, students "redefine, reorganize, elaborate, and change their initial concepts through self-reflection and interaction with their peers and their environment" (Bybee 1997, p. 176). The 5E Model (Figure 3.1) provides a planned sequence of instruction that places students at the center of their learning experiences, encouraging them to explore, construct their own understanding of science and engineering concepts, and relate those understandings to other concepts. The phases of the 5E Model—engage, explore, explain, elaborate, and evaluate—are described here.

Figure 3.1. The BSCS 5Es as a Cycle of Learning

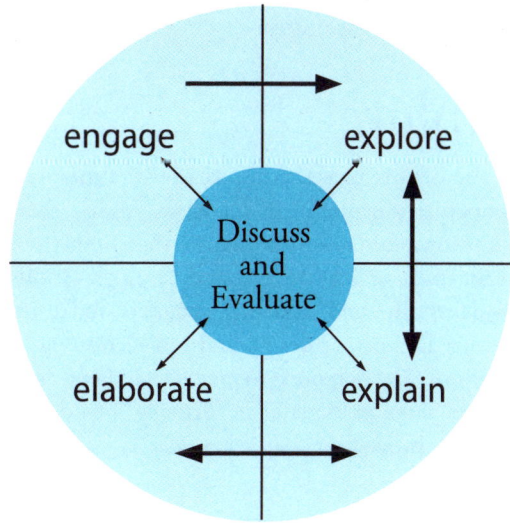

Source: Adapted from Barman, C. R. 1997. *The learning cycle revised: A modification of an effective teaching model.* Arlington, VA: Council for Elementary Science International.

Phases of the 5E Model

engage

The purpose of this introductory phase, *engage,* is to capture students' interest and draw them into a STEM lesson. Here, you can uncover what students know and think about a topic as well as determine their misconceptions. The engage phase can also serve to establish a theme for the lesson or set up opportunities for investigating or innovating in the explore phase. Engagement activities might include a reading, a demonstration, or other activity that piques students' curiosity.

Chapter 3

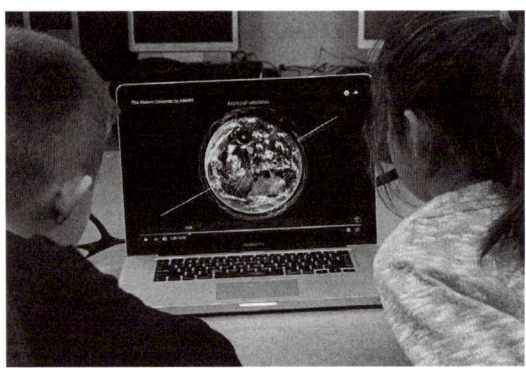

ENGAGING WITH DIGITAL MEDIA

EXPLAINING WITH A NON-FICTION READ-ALOUD

explore

In the *explore* phase, you provide students with cooperative exploration activities, giving them common, concrete experiences that help them begin constructing concepts and developing skills. Students can build models, collect data, make and test predictions, or design and test technological solutions. The purpose is to provide hands-on experiences you can use later to formally introduce a concept, process, or skill.

EXPLORING WITH A DESIGN CHALLENGE

explain

In the *explain* phase, learners articulate their ideas in their own words and listen critically to one another. You clarify their concepts, support learners as they develop scientifically accurate conceptions, and introduce scientific or engineering terminology. It is important that you clearly connect the students' explanations to experiences they had in the engage and explore phases.

elaborate

At the beginning of the *elaborate* phase, some students may still have misconceptions or may understand the concepts only in the context of the previous exploration. Elaboration activities can help students correct their remaining misconceptions and generalize the concepts in a broader context. These activities also challenge students to apply, extend, or elaborate on concepts and skills in a new situation, resulting in deeper understanding. Often, the elaborate phase is where students can apply what they have learned to meet a design challenge.

ELABORATING WITH A GAME

Chapter 3

EVALUATING WITH A POSTER SESSION

evaluate

In the *evaluate* phase, you evaluate students' understanding of concepts and their proficiency with various skills. You can use a variety of formal and informal procedures to assess conceptual understanding and progress toward learning outcomes. The evaluation phase also provides an opportunity for students to test their own understanding and skills.

Although the fifth phase is devoted to evaluation, a skillful teacher evaluates throughout the 5E Model, continually checking to see if students need more time or instruction to learn the key points in a lesson. Some ways to do this include informal questioning, teacher checkpoints, and class discussions. Each lesson in *Picture-Perfect STEM Lessons* also includes a formal evaluation such as extended-response questions or a poster session. These formal evaluations take place at the end of the lesson. A good resource for more information and practical suggestions for evaluating student understanding throughout the 5Es is *Seamless Assessment in Science: A Guide for Elementary and Middle School Teachers* by Abell and Volkmann (2006).

Cycle of Learning

In the discussion of the phases of the 5E Model, the 5Es are listed in a linear order. However, the model is most effective when you use it as a cycle of learning as in Figure 3.1 (p. 17). Each lesson begins with an engagement activity, but students can re-enter the 5E Model at other points in the cycle. For example, in Chapter 9, "Light It Up!," students explore by observing what is inside an incandescent lightbulb. Then, they begin the explain phase with a nonfiction book that explains how lightbulbs work. Next, they re-enter the explore phase by comparing different kinds of lightbulbs (incandescent, CFL, and LED) and then re-enter the explain phase to learn about the advantages and disadvantages of each kind of bulb. Students elaborate on what they have learned using LED bulbs and circuit tape to make a nightlight. Finally, student understanding is evaluated using an "Instruction Manual" writing assignment in which they explain how their nightlight works. Moving from the explain phase back into the explore phase and again to the explain phase gives students the opportunity to add to the experiences and information they construct in the lesson.

Roles of the Teacher and Student

The traditional roles of the teacher and student are virtually reversed in the 5E Model. Students take on much of the responsibility for learning as they construct knowledge through discovery, whereas in traditional models the teacher is responsible for dispensing information to be learned by the students. Table 3.1 (p. 20) shows actions of the teacher that are consistent with the 5E Model and actions that are inconsistent with the model.

In the 5E Model, the teacher acts as a guide: raising questions, providing opportunities for exploration or problem solving, asking for evidence to support student explanations, referring students to existing explanations, correcting misconceptions, and coaching students as they apply new concepts. This model differs greatly from the traditional format of lecturing, leading students step by step to a solution, providing definite answers, and testing isolated facts. The 5E Model requires the students to take on much of the responsibility for their own learning. Table 3.2 (p. 21) shows the actions of the student that are consistent with the 5E Model and those that are inconsistent with the model.

Table 3.1. The BSCS 5Es Teacher

Stage of the Instructional Model	What the Teacher Does That Is …	
	Consistent With This Model	**Inconsistent With This Model**
Engage	<ul><li>Creates interest</li><li>Generates curiosity</li><li>Raises questions</li><li>Elicits responses that uncover what the students know or think about the concept or topic</li></ul>	<ul><li>Explains concepts</li><li>Provides definitions and answers</li><li>States conclusions</li><li>Provides closure</li><li>Involves lectures</li></ul>
Explore	<ul><li>Encourages the students to work together without direct instruction from the teacher</li><li>Observes and listens to the students as they interact</li><li>Asks probing questions to redirect the students' investigations when necessary</li><li>Provides time for the students to puzzle through problems</li><li>Acts as a consultant for students</li></ul>	<ul><li>Provides answers</li><li>Tells or explains how to work through the problem</li><li>Provides closure</li><li>Tells the students that they are wrong</li><li>Gives information or facts that solve the problem</li><li>Leads the students step by step to a solution</li></ul>
Explain	<ul><li>Encourages the students to explain concepts and definitions in their own words</li><li>Asks for justification (evidence) and clarification from students</li><li>Formally provides definitions, explanations, and new labels</li><li>Uses students' previous experiences as the basis for explaining concepts</li></ul>	<ul><li>Accepts explanations that have no justification</li><li>Neglects to solicit the students' explanations</li><li>Introduces unrelated concepts or skills</li></ul>
Elaborate	<ul><li>Expects the students to use formal labels, definitions, and explanations provided previously</li><li>Encourages the students to apply or extend the concepts and skills in new situations</li><li>Reminds the students of alternate explanations</li><li>Refers the students to existing data and evidence and asks, "What do you already know?" "Why do you think x?" (Strategies from Explore also apply here.)</li></ul>	<ul><li>Provides definitive answers</li><li>Tells the students that they are wrong</li><li>Involves lectures</li><li>Leads students step by step to a solution</li><li>Explains how to work through the problem</li></ul>
Evaluate	<ul><li>Observes the students as they apply new concepts and skills</li><li>Assesses students' knowledge and/or skills</li><li>Looks for evidence that the students have changed their thinking or behaviors</li><li>Allows students to assess their own learning and group-process skills</li><li>Asks open-ended questions such as "Why do you think … ?" "What evidence do you have?" "What do you know about x?" "How would you explain x?"</li></ul>	<ul><li>Tests vocabulary words, terms, and isolated facts</li><li>Introduces new ideas or concepts</li><li>Creates ambiguity</li><li>Promotes open-ended discussion unrelated to the concept or skill</li></ul>

Copyright 2013. BSCS. All right reserved.

Table 3.2. The BSCS 5Es Student

Stage of the Instructional Model	What the Student Does That Is …	
	Consistent With This Model	**Inconsistent With This Model**
Engage	• Asks questions such as "Why did this happen?" "What do I already know about this?" "What can I find out about this?" • Shows interest in the topic	• Asks for the "right" answer • Offers the "right" answer • Insists on answers or explanations • Seeks one solution
Explore	• Thinks freely, within the limits of the activity • Tests predictions and hypotheses • Forms new predictions and hypotheses • Tries alternatives and discusses them with others • Records observations and ideas • Suspends judgment	• Lets others do the thinking and exploring (passive involvement) • Works quietly with little or no interaction with others (only appropriate when exploring ideas or feelings) • "Plays around" indiscriminately with no goal in mind • Stops with one solution
Explain	• Explains possible solutions or answers to others • Listens critically to others' explanations • Questions others' explanations • Listens to and tries to comprehend explanations that the teacher offers • Refers to previous activities • Uses recorded observations in explanations	• Proposes explanations from "thin air" with no relationship to previous experiences • Brings up irrelevant experiences and examples • Accepts explanations without justification • Does not attend to other plausible explanations
Elaborate	• Applies new labels, definitions, explanations, and skills in new but similar situations • Uses previous information to ask questions, propose solutions, make decisions, and design experiments • Draws reasonable conclusions from evidence • Records observations and explanations • Checks for understanding among peers	• "Plays around" with no goal in mind • Ignores previous information or evidence • Draws conclusions from "thin air" • Uses only those labels that the teacher provided in discussions
Evaluate	• Answers open-ended questions using observations, evidence, and previously accepted explanations • Demonstrates an understanding or knowledge of the concept or skill • Evaluates his or her own progress and knowledge • Asks related questions that would encourage future investigations	• Draws conclusions without using evidence or previously accepted explanations • Offers only yes or no and memorized definitions or explanations as answers • Fails to express satisfactory explanations in his or her own words • Introduces new, irrelevant topics

Copyright 2013. BSCS. All right reserved

The *Next Generation Science Standards* and the 5Es

In his book *Translating the* NGSS *for Classroom Instruction,* Rodger Bybee (2013) suggests that teachers planning learning experiences with the *Next Generation Science Standards* (*NGSS;* NGSS Lead States 2013) use an integrated instructional sequence such as the 5E Model. In his book *The BSCS 5E Instructional Model: Creating Teachable Moments,* Bybee (2015) recommends working backward from the performance expectations outlined in the *NGSS* to plan the learning experiences:

> *Begin by identifying your desired learning outcomes—for example, the performance expectations from* NGSS. *Then determine what would count as acceptable evidence of student learning. You should formulate strategies that set forth what counts as evidence of learning for the instructional sequence. This should be followed by actually designing assessments that will provide the evidence that students have learned the competencies described in the performance expectations. Then, and only then, begin developing the activities that will provide students opportunities to learn the concepts and practices described in the three dimensions of the performance expectations. (p. 68)*

Picture-Perfect STEM Lessons are designed using this model. Although we recognize that some districts have not fully adopted the NGSS as standards for their science teaching, the NGSS performance expectations are written in a way that integrates all three dimensions of *A Framework for K–12 Science Education* (NRC 2012; see Chapter 4 in this book for more on the three dimensions). These expectations serve to guide the learning outcomes for our lessons. From these outcomes, we work backward to formulate the activities that compose each stage of the 5E Model. We select quality STEM-related picture books for the engage and explain phases, and develop student-centered explorations and design challenges for the explore and elaborate phases. A wide range of assessments are used in the evaluate phase, including writing tasks, poster and multimedia projects, and quizzes, with the goal that these lessons are a step toward preparing students for the *NGSS* performance expectations outlined for each grade.

Using Picture Books in the 5Es

Both fiction and nonfiction picture books can be valuable components of the 5E Model when placed strategically within the cycle. We often begin lessons with a fiction book to pique students' curiosity or motivate them to want to learn more about a science concept. For example, Chapter 15, "From Seed to Tree," begins with *If You Hold a Seed*, a charming story about a boy who plants a seed and watches it grow into a tree. This read-aloud during the engage phase inspires the question "How does a seed become a towering tree?" and is followed by activities and a nonfiction read-aloud to learn the answer to that question. A storybook, however, might not be appropriate to use during the explore phase of the 5Es, in which students are participating in concrete, hands-on experiences. Likewise, a storybook might not be appropriate to use to clarify scientific concepts and introduce vocabulary during the explain phase. Sometimes, a narrative nonfiction text can be used to engage students. In Chapter 17, we draw students into the STEM lesson by engaging them with *Solving the Puzzle Under the Sea,* the true story of Marie Tharp who created the first maps of the ocean floor. This inspirational message of how one person's hard work and determination changed the way we saw our planet serves as a powerful invitation to inquiry.

You should also avoid using books too early in the learning cycle that contain a lot of scientific terminology or "give away" information students could discover on their own. It is important for students to have opportunities to construct meaning and articulate ideas in their own words before being introduced to scientific vocabulary. Nonfiction books, therefore, are most appropriate to use in the explain phase only after students have had these opportunities. For example, in the explain phase of Chapter 19, "Star Stuff," students compare the results of an activity during which they explore how the distance of a flashlight affects the apparent size and brightness of the light with information presented in the nonfiction book *Jump Into Science: Stars.*

The 5Es provide an ideal format for a constructivist sequence of activities, allowing students to form their own ideas, collect evidence to confirm or discount their ideas, design solutions to real-world problems, apply what they have learned to new situations, and demonstrate what they have learned. Thoughtful placement of fiction and nonfiction picture books within the 5E Model can help you engage, motivate, and explain while immersing your students in meaningful, integrated STEM learning experiences.

References

Abell, S. K., and M. J. Volkmann. 2006. *Seamless assessment in science: A guide for elementary and middle school teachers.* Chicago: Heinemann; Arlington, VA: NSTA Press.

Bybee, R. W. 1997. *Achieving scientific literacy: From purposes to practices.* Portsmouth, NH: Heinemann.

Bybee, R. W. 2013. *Translating the NGSS for classroom instruction.* Arlington, VA: NSTA Press.

Bybee, R. W. 2015. *The BSCS 5E Instructional Model: Creating teachable moments.* Arlington, VA: NSTA Press.

Colburn, A. 2003. *The lingo of learning: 88 education terms every science teacher should know.* Arlington, VA: NSTA Press.

National Research Council (NRC). 2012. *A framework for K–12 science education: Practices, crosscutting concepts, and core ideas.* Washington, DC: National Academies Press.

NGSS Lead States. 2013. *Next Generation Science Standards: For states, by states.* Washington, DC: National Academies Press. www.nextgenscience.org/next-generation-science-standards.

Children's Books Cited

Tomecek, S. 2006. *Jump into science: Stars.* Washington, DC: National Geographic.

MacKay, E. 2013. *If you hold a seed.* Philadelphia, PA: Running Press Kids.

Chapter 4

Connecting to the Standards

A Framework for K–12 Science Education and the Common Core State Standards for English Language Arts and Mathematics

In this book, the science, language arts, and mathematics standards that are addressed in the activities for each lesson are clearly identified. On the first page of each chapter, you will find a box titled "Lesson Objectives Connecting to the *Framework*," which lists the science and engineering practices, disciplinary core ideas, and crosscutting concepts that the lesson addresses. Throughout the lessons, you will find boxes noting the *Common Core State Standards, English Language Arts* (*CCSS ELA*) that are used during read-alouds and writing assignments and the *Common Core State Standards, Mathematics* (*CCSS Mathematics*) that are addressed in various activities. This chapter provides some background information about *A Framework for K–12 Science Education* (*Framework;* NRC 2012) and the *CCSS* and how our lessons connect to them.

A Framework for K–12 Science Education

The *Framework* was developed by the National Research Council, whose overarching goal was "to ensure that by the end of 12th grade, all students have some appreciation of the beauty and wonder of science; possess sufficient knowledge of science and engineering to engage in public discussions on related issues; are careful consumers of scientific and technological information related to their everyday lives; are able to continue to learn about science outside school; and have the skills to enter careers of their choice, including (but not limited to) careers in science, engineering, and technology" (NRC 2012, p. 1).

The *Framework* was developed around three major dimensions: (1) scientific and engineering practices, (2) crosscutting concepts, and (3) disciplinary core ideas. These three dimensions are the key components of the *Next Generation Science Standards* (*NGSS;* NGSS Lead States 2013) and many other state standards.

Dimension 1: Scientific and Engineering Practices

This dimension describes eight fundamental practices that scientists use as they investigate and build models and theories about the world, as well as the engineering practices that engineers use as they design and build systems (NRC 2012, p. 42). These practices are as follows:

1. Asking questions (for science) and defining problems (for engineering)
2. Developing and using models
3. Planning and carrying out investigations
4. Analyzing and interpreting data
5. Using mathematics and computational thinking
6. Constructing explanations (for science) and designing solutions (for engineering)
7. Engaging in argument from evidence
8. Obtaining, evaluating, and communicating information

For detailed information on the integration of science and engineering practices into the lessons in this book, see Chapter 5.

Dimension 2: Crosscutting Concepts

The seven crosscutting concepts outlined in the *Framework* appear throughout the lessons in this book. Table 4.1 shows an outline of the crosscutting concepts as they pertain to grades 3–5.

As you implement the lessons in this book, you can incorporate these important themes that appear over and over again throughout the disciplines of science, technology, engineering, and mathematics. Recognizing these themes helps students make connections between disciplines and understand that the same concept is relevant across different contexts. For example, the crosscutting concept of cause and effect underlies several lessons. In Chapter 8, "Wind It Up," students learn how winding and releasing springs causes motion. In Chapter 10, "Burn," students discover that wax moving up the wick of a burning candle is caused by capillary action, and they uncover other cause-and-effect relationships that occur when materials burn. In Chapter 13, "Spider Science," students learn that some traits in organisms are caused by their genetics (inherited traits) and other traits are caused by factors in the environment (acquired traits). In Chapter 16, "Hurricane!," students learn that the breaching and overtopping of levees was the primary cause of the devastation in New Orleans during Hurricane Katrina. Recognizing the relationship between cause and effect in both the natural and designed world is a key concept that relates to science and engineering.

Table 4.1. The Framework's *Crosscutting Concepts for Grades 3–5*

Crosscutting Concepts for Grades 3–5	
Patterns	• Similarities and differences in patterns can be used to sort, classify, communicate, and analyze simple rates of change for natural phenomena and designed products. • Patterns of change can be used to make predictions. • Patterns can be used as evidence to support an explanation.
Cause and Effect: Mechanism and Prediction	• Cause-and-effect relationships are routinely identified, tested, and used to explain change. • Events that occur together with regularity might or might not be a cause-and-effect relationship
Scale, Proportion, and Quantity	• Observable phenomena exist from very short to very long time periods. • Standard units are used to measure and describe physical quantities such as weight, time, temperature, and volume.
Systems and System Models	• A system is a group of related parts that make up a whole and can carry out functions its individual parts cannot. • A system can be described in terms of its components and their interactions.
Energy and Matter: Flows, Cycles, and Conservation	• Energy can be transferred in various ways and between objects. • Matter is made of particles. • Matter flows and cycles can be tracked in terms of the weight of the substances before and after a process occurs. The total weight of the substances does not change. This is what is meant by conservation of matter. Matter is transported into, out of, and within systems..
Structure and Function	• Different materials have different substructures, which can sometimes be observed. • Substructures have shapes and parts that serve functions
Stability and Change	• Change is measured in terms of differences over time and may occur at different rates. • Some systems appear stable, but over long periods of time will eventually change.

Source: Willard 2015, p. 102.

Because students often do not make the connections on their own, teachers must make these seven crosscutting concepts explicit for students to help them connect knowledge from different science fields into a coherent and scientifically based view of the world.

Dimension 3: Disciplinary Core Ideas

Disciplinary core ideas are grouped in four domains: (1) physical science; (2) life science; (3) Earth and space science; and (4) engineering, technology, and applications of science. The *Framework* committee has identified these core ideas of science and engineering as meeting at least two of the following criteria (NRC 2012):

1. Have broad importance across multiple sciences or engineering disciplines or be a key organizing principle of a single discipline.
2. Provide a key tool for understanding or investigating more complex ideas and solving problems.
3. Relate to the interests and life experiences of students or be connected to societal or personal concerns that require scientific or technological knowledge.
4. Be teachable and learnable over multiple grades at increasing levels of depth and sophistication. That is, the idea can be made accessible to younger students but is broad enough to sustain continued investigation over years.

The disciplinary core ideas are listed in a quick-reference chart in Table 4.2 (p. 28).

The lesson objectives in this book are closely aligned to a variety of the disciplinary core ideas outlined in dimension three. At the beginning of each lesson, we provide the disciplinary core ideas that are targeted within the lesson.

Next Generation Science Standards

For districts that have adopted the *NGSS*, the appendix (p. 365) provides a detailed correlation among the lessons presented in *Picture-Perfect STEM Lessons*, the three dimensions of the *NGSS*, and the corresponding performance expectations. Even if your district has not adopted the *NGSS* as the standard for science teaching, we encourage you to read through the charts in the appendix. Understanding how these lessons align to specific *NGSS* disciplinary core ideas, crosscutting concepts, science and engineering practices, and performance expectations will be helpful as you select and implement *Picture-Perfect STEM Lessons* in your classroom.

A note about *NGSS* performance expectations: The *NGSS* provide performance expectations that depict what students must do to show proficiency at each grade level. Performance expectations integrate all three dimensions of the *NGSS* into one task and are to be offered after students have had multiple experiences with the topic. The lessons in this book are intended to help students move toward specific performance expectations, which are identified at the beginning of each lesson. *However, the lesson will not by itself be sufficient to reach the performance expectations; rather, the lesson is meant to be one in a series of lessons that work toward the performance expectations.*

Common Core State Standards

The Common Core State Standards Initiative (*www.corestandards.org*) is a state-led effort to define the knowledge and skills students should acquire in their K–12 mathematics and ELA courses. It is a result of an extended, broad-based effort to fulfill the charge issued by the states to craft the next generation of K–12 standards to ensure that all students are college and career ready by the end of high school. The standards are research and evidence based, aligned with college and work expectations, rigorous, and internationally benchmarked. The lessons in this book were not designed to teach the *CCSS ELA* and *CCSS Mathematics* standards, but to provide opportunities for students to apply the standards appropriate to their grade level in an authentic way and in a meaningful context.

Common Core State Standards, English Language Arts

The *Common Core* suggests that the ELA standards be taught in the context of history/social

Chapter 4

Table 4.2. *The Framework's Disciplinary Core Ideas*

Disciplinary Core Ideas in Physical Science	Disciplinary Core Ideas in Life Science	Disciplinary Core Ideas in Earth and Space Science	Disciplinary Core Ideas in Engineering, Technology, and Applications of Science
PS1: Matter and Its Interactions PS1.A: Structure and Properties of Matter PS1.B: Chemical Reactions PS1.C: Nuclear Processes **PS2: Motion and Stability: Forces and Interactions** PS2.A: Forces and Motion PS2.B: Types of Interactions PS2.C: Stability and Instability in Physical Systems **PS3: Energy** PS3.A: Definitions of Energy PS3.B: Conservation of Energy and Energy Transfer PS3.C: Relationship Between Energy and Forces PS3.D: Energy in Chemical Processes and Everyday Life **PS4: Waves and Their Applications in Technologies for Information Transfer** PS4.A: Wave Properties PS4.B: Electromagnetic Radiation PS4.C: Information Technologies and Instrumentation	**LS1: From Molecules to Organisms: Structures and Processes** LS1.A: Structure and Function LS1.B: Growth and Development of Organisms LS1.C: Organization for Matter and Energy Flow in Organisms LS1.D: Information Processing **LS2: Ecosystems: Interactions, Energy, and Dynamics** LS2.A: Interdependent Relationships in Ecosystems LS2.B: Cycles of Matter and Energy Transfer in Ecosystems LS2.C: Ecosystem Dynamics, Functioning, and Resilience LS2.D: Social Interactions and Group Behavior **LS3: Heredity: Inheritance and Variation of Traits** LS3.A: Inheritance of Traits LS3.B: Variation of Traits **LS4: Biological Evolution: Unity and Diversity** LS4.A: Evidence of Common Ancestry and Diversity LS4.B: Natural Selection LS4.C: Adaptation LS4.D: Biodiversity and Humans	**ESS1: Earth's Place in the Universe** ESS1.A: The Universe and Its Stars ESS1.B: Earth and the Solar System ESS1.C: The History of Planet Earth **ESS2: Earth's Systems** ESS2.A: Earth Materials and Systems ESS2.B: Plate Tectonics and Large-Scale System Interactions ESS2.C: The Roles of Water in Earth's Surface Processes ESS2.D: Weather and Climate ESS2.E: Biogeology **ESS3: Earth and Human Activity** ESS3.A: Natural Resources ESS3.B: Natural Hazards ESS3.C: Human Impacts on Earth Systems ESS3.D: Global Climate Change	**ETS1: Engineering Design** ETS1.A: Defining and Delimiting an Engineering Problem ETS1.B: Developing Possible Solutions ETS1.C: Optimizing the Design Solution **ETS2: Links Among Engineering, Technology, Science, and Society** ETS2.A: Interdependence of Science, Engineering, and Technology ETS2.B: Influence of Engineering, Technology, and Science on Society and the Natural World

Source: Willard 2015, p. 3.

studies, science, and technical subjects (NGAC and CCSSO 2010). Grade-specific K–12 standards in reading, writing, speaking, listening, and language are included. Many of these grade-specific standards are used in *Picture-Perfect STEM Lessons* through the use of high-quality children's fiction and nonfiction picture books, research-based reading strategies, poster presentations, vocabulary development activities, and various writing assignments. In the boxes titled "Connecting to the Common Core," you will find the *CCSS ELA* strand(s) and topic the activity addresses, the grade level(s), and the standard number(s) (see Figure 4.1).

Because the codes for the *CCSS ELA* Standards are listed in the lessons instead of the actual standards statements, we have included the *CCSS ELA* grade-level statements in Table 4.3 (p. 30). This table is not a complete version of the *CCSS ELA*. Rather, it includes only the standards we address in our lessons for grades 3–5: reading (literature and informational text), writing, speaking and listening, and language. You can access the complete *Common Core State Standards ELA* at *www.corestandards.org/ELA-Literacy*.

Common Core State Standards, Mathematics

The *CCSS Mathematics* (NGAC and CCSSO 2010) are divided into two parts: eight mathematical practices that apply to every grade level and grade-specific standards of mathematical content. Many of these grade-specific standards are used in *Picture-Perfect STEM Lessons* as students measure, graph, compare, and design. In the boxes titled "Connecting to the Common Core," you will find the *CCSS Mathematics* domain the activity addresses, the grade level(s), and standard number(s) (see Figure 4.2).

Because the *CCSS Mathematics* codes are listed in the lessons instead of the actual standards statements, we have included the *CCSS Mathematics* grade-level standards in Table 4.4 (p. 37). This table is not a complete version of the *CCSS Mathematics*. Rather, it includes only the domains we address in our lessons for grades 3–5. You can access the complete *CCSS Mathematics* at *www.corestandards.org/Math*.

Figure 4.2. Sample **CCSS Mathematics** *Box*

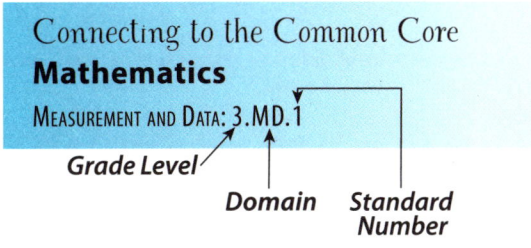

Figure 4.1. Sample **CCSS ELA** *Box*

Table 4.3. Common Core State Standards for English Language Arts and Literacy in History/Social Studies, Science, and Technical Subjects

Reading Standards for Literature 3–5

Grade 3 Students	Grade 4 Students	Grade 5 Students
Key Ideas and Details		
1. Ask and answer questions to demonstrate understanding of a text, referring explicitly to the text as the basis for the answers.	1. Refer to details and examples in a text when explaining what the text says explicitly and when drawing inferences from the text.	1. Quote accurately from a text when explaining what the text says explicitly and when drawing inferences from the text.
2. Recount stories, including fables, folktales, and myths from diverse cultures; determine the central message, lesson, or moral and explain how it is conveyed through key details in the text.	2. Determine a theme of a story, drama, or poem from details in the text; summarize the text.	2. Determine a theme of a story, drama, or poem from details in the text, including how characters in a story or drama respond to challenges or how the speaker in a poem reflects upon a topic; summarize the text.
3. Describe characters in a story (e.g., their traits, motivations, or feelings) and explain how their actions contribute to the sequence of events.	3. Describe in depth a character, setting, or event in a story or drama, drawing on specific details in the text (e.g., a character's thoughts, words, or actions).	3. Compare and contrast two or more characters, settings, or events in a story or drama, drawing on specific details in the text (e.g., how characters interact).
Craft and Structure		
4. Determine the meaning of words and phrases as they are used in a text, distinguishing literal from nonliteral language.	4. Determine the meaning of words and phrases as they are used in a text, including those that allude to significant characters found in mythology (e.g., Herculean).	4. Determine the meaning of words and phrases as they are used in a text, including figurative language such as metaphors and similes.
5. Refer to parts of stories, dramas, and poems when writing or speaking about a text, using terms such as chapter, scene, and stanza; describe how each successive part builds on earlier sections.	5. Explain major differences between poems, drama, and prose, and refer to the structural elements of poems (e.g., verse, rhythm, meter) and drama (e.g., casts of characters, settings, descriptions, dialogue, stage directions) when writing or speaking about a text.	5. Explain how a series of chapters, scenes, or stanzas fits together to provide the overall structure of a particular story, drama, or poem.
6. Distinguish their own point of view from that of the narrator or those of the characters.	6. Compare and contrast the point of view from which different stories are narrated, including the difference between first- and third-person narrations.	6. Describe how a narrator's or speaker's point of view influences how events are described.
Integration of Knowledge and Ideas		
7. Explain how specific aspects of a text's illustrations contribute to what is conveyed by the words in a story (e.g., create mood, emphasize aspects of a character or setting).	7. Make connections between the text of a story or drama and a visual or oral presentation of the text, identifying where each version reflects specific descriptions and directions in the text.	7. Analyze how visual and multimedia elements contribute to the meaning, tone, or beauty of a text (e.g., graphic novel, multimedia presentation of fiction, folktale, myth, poem).
8. (Not applicable to literature)	8. (Not applicable to literature)	8. (Not applicable to literature)
9. Compare and contrast the themes, settings, and plots of stories written by the same author about the same or similar characters (e.g., in books from a series).	9. Compare and contrast the treatment of similar themes and topics (e.g., opposition of good and evil) and patterns of events (e.g., the quest) in stories, myths, and traditional literature from different cultures.	9. Compare and contrast stories in the same genre (e.g., mysteries and adventure stories) on their approaches to similar themes and topics.
Range of Reading and Level of Text Complexity		
10. By the end of the year, read and comprehend literature, including stories, dramas, and poetry, at the high end of the grades 2–3 text complexity band independently and proficiently.	10. By the end of the year, read and comprehend literature, including stories, dramas, and poetry, in the grades 4–5 text complexity band proficiently, with scaffolding as needed at the high end of the range.	10. By the end of the year, read and comprehend literature, including stories, dramas, and poetry, at the high end of the grades 4–5 text complexity band independently and proficiently.

Copyright 2010. National Governors Association Center for Best Practices and Council of Chief State School Officers. All rights reserved.

Table 4.3. (continued)
Reading Standards for Informational Text 3–5

Grade 3 Students	Grade 4 Students	Grade 5 Students
Key Ideas and Details		
1. Ask and answer questions to demonstrate understanding of a text, referring explicitly to the text as the basis for the answers.	1. Refer to details and examples in a text when explaining what the text says explicitly and when drawing inferences from the text.	1. Quote accurately from a text when explaining what the text says explicitly and when drawing inferences from the text.
2. Determine the main idea of a text; recount the key details and explain how they support the main idea.	2. Determine the main idea of a text and explain how it is supported by key details; summarize the text.	2. Determine two or more main ideas of a text and explain how they are supported by key details; summarize the text.
3. Describe the relationship between a series of historical events, scientific ideas or concepts, or steps in technical procedures in a text, using language that pertains to time, sequence, and cause/effect.	3. Explain events, procedures, ideas, or concepts in a historical, scientific, or technical text, including what happened and why, based on specific information in the text.	3. Explain the relationships or interactions between two or more individuals, events, ideas, or concepts in a historical, scientific, or technical text based on specific information in the text.
Craft and Structure		
4. Determine the meaning of general academic and domain-specific words and phrases in a text relevant to a *grade 3 topic or subject area*.	4. Determine the meaning of general academic and domain-specific words or phrases in a text relevant to a *grade 4 topic or subject area*.	4. Determine the meaning of general academic and domain-specific words and phrases in a text relevant to a *grade 5 topic or subject area*.
5. Use text features and search tools (e.g., key words, sidebars, hyperlinks) to locate information relevant to a given topic efficiently.	5. Describe the overall structure (e.g., chronology, comparison, cause/effect, problem/solution) of events, ideas, concepts, or information in a text or part of a text.	5. Compare and contrast the overall structure (e.g., chronology, comparison, cause/effect, problem/solution) of events, ideas, concepts, or information in two or more texts.
6. Distinguish their own point of view from that of the author of a text.	6. Compare and contrast a firsthand and secondhand account of the same event or topic; describe the differences in focus and the information provided.	6. Analyze multiple accounts of the same event or topic, noting important similarities and differences in the point of view they represent.
Integration of Knowledge and Ideas		
7. Use information gained from illustrations (e.g., maps, photographs) and the words in a text to demonstrate understanding of the text (e.g., where, when, why, and how key events occur).	7. Interpret information presented visually, orally, or quantitatively (e.g., in charts, graphs, diagrams, time lines, animations, or interactive elements on Web pages) and explain how the information contributes to an understanding of the text in which it appears.	7. Draw on information from multiple print or digital sources, demonstrating the ability to locate an answer to a question quickly or to solve a problem efficiently.
8. Describe the logical connection between particular sentences and paragraphs in a text (e.g., comparison, cause/effect, first/second/third in a sequence).	8. Explain how an author uses reasons and evidence to support particular points in a text.	8. Explain how an author uses reasons and evidence to support particular points in a text, identifying which reasons and evidence support which point(s).
9. Compare and contrast the most important points and key details presented in two texts on the same topic.	9. Integrate information from two texts on the same topic in order to write or speak about the subject knowledgeably.	9. Integrate information from several texts on the same topic in order to write or speak about the subject knowledgeably.
Range of Reading and Level of Text Complexity		
10. By the end of the year, read and comprehend informational texts, including history/social studies, science, and technical texts, at the high end of the grades 2–3 text complexity band independently and proficiently.	10. By the end of year, read and comprehend informational texts, including history/social studies, science, and technical texts, in the grades 4–5 text complexity band proficiently, with scaffolding as needed at the high end of the range.	10. By the end of the year, read and comprehend informational texts, including history/social studies, science, and technical texts, at the high end of the grades 4–5 text complexity band independently and proficiently.

Copyright 2010. National Governors Association Center for Best Practices and Council of Chief State School Officers. All rights reserved.

Table 4.3. (continued)

Writing Standards 3–5

Grade 3 Students	Grade 4 Students	Grade 5 Students
Text Types and Purposes		
1. Write opinion pieces on topics or texts, supporting a point of view with reasons. 　a. Introduce the topic or text they are writing about, state an opinion, and create an organizational structure that lists reasons. 　b. Provide reasons that support the opinion. 　c. Use linking words and phrases (e.g., *because, therefore, since, for example*) to connect opinion and reasons. 　d. Provide a concluding statement or section.	1. Write opinion pieces on topics or texts, supporting a point of view with reasons and information. 　a. Introduce a topic or text clearly, state an opinion, and create an organizational structure in which related ideas are grouped to support the writer's purpose. 　b. Provide reasons that are supported by facts and details. Know final -e and common vowel team conventions for representing long vowel sounds. 　c. Link opinion and reasons using words and phrases (e.g., *for instance, in order to, in addition*). Decode two-syllable words following basic patterns by breaking the words into syllables. 　d. Provide a concluding statement or section related to the opinion presented.	1. Write opinion pieces on topics or texts, supporting a point of view with reasons and information. 　a. Introduce a topic or text clearly, state an opinion, and create an organizational structure in which ideas are logically grouped to support the writer's purpose. 　b. Provide logically ordered reasons that are supported by facts and details. 　c. Link opinion and reasons using words, phrases, and clauses (e.g., *consequently, specifically*). 　d. Provide a concluding statement or section related to the opinion presented.
2. Write informative/explanatory texts to examine a topic and convey ideas and information clearly. 　a. Introduce a topic and group related information together; include illustrations when useful to aiding comprehension. 　b. Develop the topic with facts, definitions, and details. 　c. Use linking words and phrases (e.g., *also, another, and, more, but*) to connect ideas within categories of information. 　d. Provide a concluding statement or section.	2. Write informative/explanatory texts to examine a topic and convey ideas and information clearly. 　a. Introduce a topic clearly and group related information in paragraphs and sections; include formatting (e.g., headings), illustrations, and multimedia when useful to aiding comprehension. 　b. Develop the topic with facts, definitions, concrete details, quotations, or other information and examples related to the topic. Use context to confirm or self-correct word recognition and understanding, rereading as necessary. 　c. Link ideas within categories of information using words and phrases (e.g., *another, for example, also, because*). 　d. Use precise language and domain-specific vocabulary to inform about or explain the topic.	2. Write informative/explanatory texts to examine a topic and convey ideas and information clearly. 　a. Introduce a topic clearly and group related information in paragraphs and sections; include formatting (e.g., headings), illustrations, and multimedia when useful to aiding comprehension. Read grade-level text orally with accuracy, appropriate rate, and expression on successive readings. 　b. Develop the topic with facts, definitions, concrete details, quotations, or other information and examples related to the topic. 　c. Link ideas within categories of information using words and phrases (e.g., *another, for example, also, because*). 　d. Use precise language and domain-specific vocabulary to inform about or explain the topic. 　e. Provide a concluding statement or section related to the information or explanation presented.
3. Write narratives to develop real or imagined experiences or events using effective technique, descriptive details, and clear event sequences. 　a. Establish a situation and introduce a narrator and/or characters; organize an event sequence that unfolds naturally. 　b. Use dialogue and descriptions of actions, thoughts, and feelings to develop experiences and events or show the response of characters to situations. 　c. Use temporal words and phrases to signal event order. 　d. Provide a sense of closure.	3. Write narratives to develop real or imagined experiences or events using effective technique, descriptive details, and clear event sequences. 　a. Orient the reader by establishing a situation and introducing a narrator and/or characters; organize an event sequence that unfolds naturally. 　b. Use dialogue and description to develop experiences and events or show the responses of characters to situations. 　c. Use a variety of transitional words and phrases to manage the sequence of events. 　d. Use concrete words and phrases and sensory details to convey experiences and events precisely. 　e. Provide a conclusion that follows from the narrated experiences or events.	3. Write narratives to develop real or imagined experiences or events using effective technique, descriptive details, and clear event sequences. 　a. Orient the reader by establishing a situation and introducing a narrator and/or characters; organize an event sequence that unfolds naturally. 　b. Use narrative techniques, such as dialogue, description, and pacing, to develop experiences and events or show the responses of characters to situations. 　c. Use a variety of transitional words, phrases, and clauses to manage the sequence of events. 　d. Use concrete words and phrases and sensory details to convey experiences and events precisely. 　e. Provide a conclusion that follows from the narrated experiences or events.

Table 4.3. (continued)
Writing Standards 3–5

Grade 3 Students	Grade 4 Students	Grade 5 Students
Production and Distribution of Writing		
4. With guidance and support from adults, produce writing in which the development and organization are appropriate to task and purpose. (Grade-specific expectations for writing types are defined in standards 1–3 above.)	4. Produce clear and coherent writing in which the development and organization are appropriate to task, purpose, and audience. (Grade-specific expectations for writing types are defined in standards 1–3 above.)	4. Produce clear and coherent writing in which the development and organization are appropriate to task, purpose, and audience. (Grade-specific expectations for writing types are defined in standards 1–3 above.)
5. With guidance and support from peers and adults, develop and strengthen writing as needed by planning, revising, and editing. (Editing for conventions should demonstrate command of Language standards 1–3 up to and including grade 3 on page 29.)	5. With guidance and support from peers and adults, develop and strengthen writing as needed by planning, revising, and editing. (Editing for conventions should demonstrate command of Language standards 1–3 up to and including grade 4 on page 29.)	5. With guidance and support from peers and adults, develop and strengthen writing as needed by planning, revising, editing, rewriting, or trying a new approach. (Editing for conventions should demonstrate command of Language standards 1–3 up to and including grade 5 on page 29.)
6. With guidance and support from adults, use technology to produce and publish writing (using keyboarding skills) as well as to interact and collaborate with others.	6. With some guidance and support from adults, use technology, including the Internet, to produce and publish writing as well as to interact and collaborate with others; demonstrate sufficient command of keyboarding skills to type a minimum of one page in a single sitting.	6. With some guidance and support from adults, use technology, including the Internet, to produce and publish writing as well as to interact and collaborate with others; demonstrate sufficient command of keyboarding skills to type a minimum of two pages in a single sitting.
Research to Build and Present Knowledge		
7. Conduct short research projects that build knowledge about a topic.	7. Conduct short research projects that build knowledge through investigation of different aspects of a topic.	7. Conduct short research projects that use several sources to build knowledge through investigation of different aspects of a topic.
8. Recall information from experiences or gather information from print and digital sources; take brief notes on sources and sort evidence into provided categories.	8. Recall relevant information from experiences or gather relevant information from print and digital sources; take notes and categorize information, and provide a list of sources.	8. Recall relevant information from experiences or gather relevant information from print and digital sources; summarize or paraphrase information in notes and finished work, and provide a list of sources.
9. (Begins in grade 4)	9. Draw evidence from literary or informational texts to support analysis, reflection, and research. a. Apply grade 4 Reading standards to literature (e.g., "Describe in depth a character, setting, or event in a story or drama, drawing on specific details in the text [e.g., a character's thoughts, words, or actions]."). b. Apply grade 4 Reading standards to informational texts (e.g., "Explain how an author uses reasons and evidence to support particular points in a text").	9. Draw evidence from literary or informational texts to support analysis, reflection, and research. a. Apply grade 5 Reading standards to literature (e.g., "Compare and contrast two or more characters, settings, or events in a story or a drama, drawing on specific details in the text [e.g., how characters interact]"). b. Apply grade 5 Reading standards to informational texts (e.g., "Explain how an author uses reasons and evidence to support particular points in a text, identifying which reasons and evidence support which point[s]").
Range of Writing		
10. Write routinely over extended time frames (time for research, reflection, and revision) and shorter time frames (a single sitting or a day or two) for a range of discipline-specific tasks, purposes, and audiences.	10. Write routinely over extended time frames (time for research, reflection, and revision) and shorter time frames (a single sitting or a day or two) for a range of discipline-specific tasks, purposes, and audiences.	10. Write routinely over extended time frames (time for research, reflection, and revision) and shorter time frames (a single sitting or a day or two) for a range of discipline-specific tasks, purposes, and audiences.

Copyright 2010. National Governors Association Center for Best Practices and Council of Chief State School Officers. All rights reserved.

Table 4.3. (continued)

Speaking and Listening Standards 3–5

Grade 3 Students	Grade 4 Students	Grade 5 Students
Comprehension and Collaboration		
1. Engage effectively in a range of collaborative discussions (one-on-one, in groups, and teacher-led) with diverse partners on grade *3 topics and texts,* building on others' ideas and expressing their own clearly. a. Come to discussions prepared, having read or studied required material; explicitly draw on that preparation and other information known about the topic to explore ideas under discussion. b. Follow agreed-upon rules for discussions (e.g., gaining the floor in respectful ways, listening to others with care, speaking one at a time about the topics and texts under discussion). c. Ask questions to check understanding of information presented, stay on topic, and link their comments to the remarks of others. d. Explain their own ideas and understanding in light of the discussion.	1. Engage effectively in a range of collaborative discussions (one-on-one, in groups, and teacher-led) with diverse partners on grade 4 topics and texts, building on others' ideas and expressing their own clearly. a. Come to discussions prepared, having read or studied required material; explicitly draw on that preparation and other information known about the topic to explore ideas under discussion. b. Follow agreed-upon rules for discussions and carry out assigned roles. c. Pose and respond to specific questions to clarify or follow up on information, and make comments that contribute to the discussion and link to the remarks of others. d. Review the key ideas expressed and explain their own ideas and understanding in light of the discussion	1. Engage effectively in a range of collaborative discussions (one-on-one, in groups, and teacher-led) with diverse partners on grade 5 topics and texts, building on others' ideas and expressing their own clearly. a. Come to discussions prepared, having read or studied required material; explicitly draw on that preparation and other information known about the topic to explore ideas under discussion. b. Follow agreed-upon rules for discussions and carry out assigned roles. c. Pose and respond to specific questions by making comments that contribute to the discussion and elaborate on the remarks of others. d. Review the key ideas expressed and draw conclusions in light of information and knowledge gained from the discussions.
2. Determine the main ideas and supporting details of a text read aloud or information presented in diverse media and formats, including visually, quantitatively, and orally.	2. Paraphrase portions of a text read aloud or information presented in diverse media and formats, including visually, quantitatively, and orally.	2. Summarize a written text read aloud or information presented in diverse media and formats, including visually, quantitatively, and orally.
3. Ask and answer questions about information from a speaker, offering appropriate elaboration and detail.	3. Identify the reasons and evidence a speaker provides to support particular points.	3. Summarize the points a speaker makes and explain how each claim is supported by reasons and evidence.
Presentation of Knowledge and Ideas		
4. Report on a topic or text, tell a story, or recount an experience with appropriate facts and relevant, descriptive details, speaking clearly at an understandable pace.	4. Report on a topic or text, tell a story, or recount an experience in an organized manner, using appropriate facts and relevant, descriptive details to support main ideas or themes; speak clearly at an understandable pace.	4. Report on a topic or text or present an opinion, sequencing ideas logically and using appropriate facts and relevant, descriptive details to support main ideas or themes; speak clearly at an understandable pace.
5. Add drawings or other visual displays to descriptions as desired to provide additional detail.	5. Add drawings or other visual displays to descriptions when appropriate to clarify ideas, thoughts, and feelings.	5. Include multimedia components (e.g., graphics, sound) and visual displays in presentations when appropriate to enhance the development of main ideas or themes.
6. Speak in complete sentences when appropriate to task and situation in order to provide requested detail or clarification. (See grade 3 Language standards 1 and 3 on page 28 for specific expectations.)	6. Differentiate between contexts that call for formal English (e.g., presenting ideas) and situations where informal discourse is appropriate (e.g., small-group discussion); use formal English when appropriate to task and situation. (See grade 4 Language standards 1 on page 28 for specific expectations.)	6. Adapt speech to a variety of contexts and tasks, using formal English when appropriate to task and situation. (See grade 5 Language standards 1 and 3 on page 28 for specific expectations.)

Copyright 2010. National Governors Association Center for Best Practices and Council of Chief State School Officers. All rights reserved.

Table 4.3. (continued)

Language Standards 3–5

Grade 3 Students	Grade 4 Students	Grade 5 Students
Conventions of Standard English		

1. Demonstrate command of the conventions of standard English grammar and usage when writing or speaking.
  a. Explain the function of nouns, pronouns, verbs, adjectives, and adverbs in general and their functions in particular sentences.
  b. Form and use regular and irregular plural nouns.
  c. Use abstract nouns (e.g., *childhood*).
  d. Form and use regular and irregular verbs.
  e. Form and use the simple (e.g., *I walked; I walk; I will walk*) verb tenses.
  f. Ensure subject-verb and pronoun-antecedent agreement.
  g. Form and use comparative and superlative adjectives and adverbs, and choose between them depending on what is to be modified.
  h. Use coordinating and subordinating conjunctions.
  i. Produce simple, compound, and complex sentences.
	1. Demonstrate command of the conventions of standard English grammar and usage when writing or speaking.
	  a. Use relative pronouns (*who, whose, whom, which, that*) and relative adverbs (*where, when, why*).
	  b. Form and use the progressive (e.g., *I was walking; I am walking; I will be walking*) verb tenses.
	  c. Use modal auxiliaries (e.g., *can, may, must*) to convey various conditions.
	  d. Order adjectives within sentences according to conventional patterns (e.g., a *small red bag* rather than a *red small bag*).
	  e. Form and use prepositional phrases.
	  f. Produce complete sentences, recognizing and correcting inappropriate fragments and run-ons.
	  g. Correctly use frequently confused words (e.g., *to, too, two; there, their*).
		1. Demonstrate command of the conventions of standard English grammar and usage when writing or speaking.
		  a. Explain the function of conjunctions, prepositions, and interjections in general and their function in particular sentences.
		  b. Form and use the perfect (e.g., *I had walked; I have walked; I will have walked*) verb tenses.
		  c. Use verb tense to convey various times, sequences, states, and conditions.
		  d. Recognize and correct inappropriate shifts in verb tense.
		  e. Use correlative conjunctions (e.g., *either/or, neither/nor*).

2. Demonstrate command of the conventions of standard English capitalization, punctuation, and spelling when writing.
  a. Capitalize appropriate words in titles.
  b. Use commas in addresses.
  c. Use commas and quotation marks in dialogue.
  d. Form and use possessives.
  e. Use conventional spelling for high-frequency and other studied words and for adding suffixes to base words (e.g., *sitting, smiled, cries, happiness*).
  f. Use spelling patterns and generalizations (e.g., word families, position-based spellings, syllable patterns, ending rules, meaningful word parts) in writing words.
  g. Consult reference materials, including beginning dictionaries, as needed to check and correct spellings.
	2. Demonstrate command of the conventions of standard English capitalization, punctuation, and spelling when writing.
	  a. Use correct capitalization.
	  b. Use commas and quotation marks to mark direct speech and quotations from a text.
	  c. Use a comma before a coordinating conjunction in a compound sentence.
	  d. Spell grade-appropriate words correctly, consulting references as needed.
		2. Demonstrate command of the conventions of standard English capitalization, punctuation, and spelling when writing.
		  a. Use punctuation to separate items in a series.
		  b. Use a comma to separate an introductory element from the rest of the sentence.
		  c. Use a comma to set off the words *yes* and *no* (e.g., *Yes, thank you*), to set off a tag question from the rest of the sentence (e.g., *It's true, isn't it?*), and to indicate direct address (e.g., *Is that you, Steve?*).
		  d. Use underlining, quotation marks, or italics to indicate titles of works.
		  e. Spell grade-appropriate words correctly, consulting references as needed.

Copyright 2010. National Governors Association Center for Best Practices and Council of Chief State School Officers. All rights reserved.

Table 4.3. (continued)

Language Standards 3–5 (continued)

Grade 3 Students	Grade 4 Students	Grade 5 Students
Knowledge of Language		
3. Use knowledge of language and its conventions when writing, speaking, reading, or listening. 　a. Choose words and phrases for effect.* 　b. Recognize and observe differences between the conventions of spoken and written standard English.	3. Use knowledge of language and its conventions when writing, speaking, reading, or listening. 　a. Choose words and phrases to convey ideas precisely.* 　b. Choose punctuation for effect.* 　c. Differentiate between contexts that call for formal English (e.g., presenting ideas) and situations where informal discourse is appropriate (e.g., small-group discussion).	3. Use knowledge of language and its conventions when writing, speaking, reading, or listening. 　a. Expand, combine, and reduce sentences for meaning, reader/listener interest, and style. 　b. Compare and contrast the varieties of English (e.g., dialects, registers) used in stories, dramas, or poems.
Vocabulary Acquisition and Use		
4. Determine or clarify the meaning of unknown and multiple-meaning word and phrases based on *grade 3 reading and content*, choosing flexibly from a range of strategies. 　a. Use sentence-level context as a clue to the meaning of a word or phrase. 　b. Determine the meaning of the new word formed when a known affix is added to a known word (e.g., *agreeable/disagreeable, comfortable/uncomfortable, care/careless, heat/preheat*). 　c. Use a known root word as a clue to the meaning of an unknown word with the same root (e.g., *company, companion*). 　d. Use glossaries or beginning dictionaries, both print and digital, to determine or clarify the precise meaning of key words and phrases.	4. Determine or clarify the meaning of unknown and multiple-meaning words and phrases based on *grade 4 reading and content*, choosing flexibly from a range of strategies. 　a. Use context (e.g., definitions, examples, or restatements in text) as a clue to the meaning of a word or phrase. 　b. Use common, grade-appropriate Greek and Latin affixes and roots as clues to the meaning of a word (e.g., *telegraph, photograph, autograph*). 　c. Consult reference materials (e.g., dictionaries, glossaries, thesauruses), both print and digital, to find the pronunciation and determine or clarify the precise meaning of key words and phrases.	4. Determine or clarify the meaning of unknown and multiple-meaning words and phrases based on *grade 5 reading and content*, choosing flexibly from a range of strategies. 　a. Use context (e.g., cause/effect relationships and comparisons in text) as a clue to the meaning of a word or phrase. 　b. Use common, grade-appropriate Greek and Latin affixes and roots as clues to the meaning of a word (e.g., *photograph, photosynthesis*). 　c. Consult reference materials (e.g., dictionaries, glossaries, thesauruses), both print and digital, to find the pronunciation and determine or clarify the precise meaning of key words and phrases.
5. Demonstrate understanding of word relationships and nuances in word meanings. 　a. Distinguish the literal and nonliteral meanings of words and phrases in context (e.g., *take steps*). 　b. Identify real-life connections between words and their use (e.g., describe people who are *friendly* or *helpful*). 　c. Distinguish shades of meaning among related words that describe states of mind or degrees of certainty (e.g., *knew, believed, suspected, heard, wondered*).	5. Demonstrate understanding of figurative language, word relationships, and nuances in word meanings. 　a. Explain the meaning of simple similes and metaphors (e.g., *as pretty as a picture*) in context. 　b. Recognize and explain the meaning of common idioms, adages, and proverbs. 　c. Demonstrate understanding of words by relating them to their opposites (antonyms) and to words with similar but not identical meanings (synonyms).	5. Demonstrate understanding of figurative language, word relationships, and nuances in word meanings. 　a. Interpret figurative language, including similes and metaphors, in context. 　b. Recognize and explain the meaning of common idioms, adages, and proverbs. 　c. Use the relationship between particular words (e.g., synonyms, antonyms, homographs) to better understand each of the words.
6. Acquire and use accurately grade-appropriate conversational, general academic, and domain-specific words and phrases, including those that signal spatial and temporal relationships (e.g., *After dinner that night we went looking for them*).	6. Acquire and use accurately grade-appropriate general academic and domain-specific words and phrases, including those that signal precise actions, emotions, or states of being (e.g., *quizzed, whined, stammered*) and that are basic to a particular topic (e.g., *wildlife, conservation,* and *endangered* when discussing animal preservation).	6. Acquire and use accurately grade-appropriate general academic and domain-specific words and phrases, including those that signal contrast, addition, and other logical relationships (e.g., *however, although, nevertheless, similarly, moreover, in addition*).

Copyright 2010. National Governors Association Center for Best Practices and Council of Chief State School Officers. All rights reserved.

Table 4.4. Common Core State Standards for Mathematics, 3–5

Operations and Algebraic Thinking	**3.OA**

Represent and solve problems involving multiplication and division.

1. Interpret products of whole numbers, e.g., interpret 5 × 7 as the total number of objects in 5 groups of 7 objects each. *For example, describe a context in which a total number of objects can be expressed as 5 × 7.*

2. Interpret whole-number quotients of whole numbers, e.g., interpret 56 ÷ 8 as the number of objects in each share when 56 objects are partitioned equally into 8 shares, or as a number of shares when 56 objects are partitioned into equal shares of 8 objects each. *For example, describe a context in which a number of shares or a number of groups can be expressed as 56 ÷ 8.*

3. Use multiplication and division within 100 to solve word problems in situations involving equal groups, arrays, and measurement quantities, e.g., by using drawings and equations with a symbol for the unknown number to represent the problem.[1]

4. Determine the unknown whole number in a multiplication or division equation relating three whole numbers. *For example, determine the unknown number that makes the equation true in each of the equations 8 × ? = 48, 5 = ☐ ÷ 3, 6 × 6 = ?*

Understand properties of multiplication and the relationship between multiplication and division.

5. Apply properties of operations as strategies to multiply and divide.[2] *Examples: If 6 × 4 = 24 is known, then 4 × 6 = 24 is also known. (Commutative property of multiplication.) 3 × 5 × 2 can be found by 3 × 5 = 15, then 15 × 2 = 30, or by 5 × 2 = 10, then 3 × 10 = 30. (Associative property of multiplication.) Knowing that 8 × 5 = 40 and 8 × 2 = 16, one can find 8 × 7 as 8 × (5 + 2) = (8 × 5) + (8 × 2) = 40 + 16 = 56. (Distributive property.)*

6. Understand division as an unknown-factor problem. *For example, find 32 ÷ 8 by finding the number that makes 32 when multiplied by 8.*

Multiply and divide within 100.

7. Fluently multiply and divide within 100, using strategies such as the relationship between multiplication and division (e.g., knowing that 8 × 5 = 40, one knows 40 ÷ 5 = 8) or properties of operations. By the end of Grade 3, know from memory all products of two one-digit numbers.

Solve problems involving the four operations, and identify and explain patterns in arithmetic.

8. Solve two-step word problems using the four operations. Represent these problems using equations with a letter standing for the unknown quantity. Assess the reasonableness of answers using mental computation and estimation strategies including rounding.[3]

9. Identify arithmetic patterns (including patterns in the addition table or multiplication table), and explain them using properties of operations. *For example, observe that 4 times a number is always even, and explain why 4 times a number can be decomposed into two equal addends.*

[1] See Glossary, Table 2.

[2] Students need not use formal terms for these properties.

[3] This standard is limited to problems posed with whole numbers and having whole-number answers; students should know how to perform operations in the conventional order when there are no parentheses to specify a particular order (Order of Operations).

Copyright 2010. National Governors Association Center for Best Practices and Council of Chief State School Officers. All rights reserved.

Table 4.4. (continued)

Number and Operations in Base Ten	3.NBT

Use place value understanding and properties of operations to perform multi-digit arithmetic.[4]

1. Use place value understanding to round whole numbers to the nearest 10 or 100.
2. Fluently add and subtract within 1,000 using strategies and algorithms based on place value, properties of operations, and/or the relationship between addition and subtraction.
3. Multiply one-digit whole numbers by multiples of 10 in the range 10–90 (e.g., 9 × 80, 5 × 60) using strategies based on place value and properties of operations.

Number and Operations—Fractions[5]	3.NF

Develop understanding of fractions as numbers.

1. Understand a fraction $1/b$ as the quantity formed by 1 part when a whole is partitioned into b equal parts; understand a fraction a/b as the quantity formed by a parts of size $1/b$.
2. Understand a fraction as a number on the number line; represent fractions on a number line diagram.
    a. Represent a fraction $1/b$ on a number line diagram by defining the interval from 0 to 1 as the whole and partitioning it into b equal parts. Recognize that each part has size $1/b$ and that the endpoint of the part based at 0 locates the number $1/b$ on the number line.
    b. Represent a fraction a/b on a number line diagram by marking off a lengths $1/b$ from 0. Recognize that the resulting interval has size a/b and that its endpoint locates the number a/b on the number line.
3. Explain equivalence of fractions in special cases, and compare fractions by reasoning about their size.
    a. Understand two fractions as equivalent (equal) if they are the same size, or the same point on a number line.
    b. Recognize and generate simple equivalent fractions, e.g., 1/2 = 2/4, 4/6 = 2/3. Explain why the fractions are equivalent, e.g., by using a visual fraction model.
    c. Express whole numbers as fractions, and recognize fractions that are equivalent to whole numbers. *Examples: Express 3 in the form 3 = 3/1; recognize that 6/1 = 6; locate 4/4 and 1 at the same point of a number line diagram.*
    d. Compare two fractions with the same numerator or the same denominator by reasoning about their size. Recognize that comparisons are valid only when the two fractions refer to the same whole. Record the results of the comparisons with the symbols >, =, or <, and justify the conclusions, e.g., by using a visual fraction model.

[4] A range of algorithms may be used.

[5] Grade 3 expectations in this domain are limited to fractions with denominators 2, 3, 4, 6, and 8.

Copyright 2010. National Governors Association Center for Best Practices and Council of Chief State School Officers. All rights reserved.

Table 4.4. (continued)

Measurement and Data	3.MD

Solve problems involving measurement and estimation of intervals of time, liquid volumes, and masses of objects.

1. Tell and write time to the nearest minute and measure time intervals in minutes. Solve word problems involving addition and subtraction of time intervals in minutes, e.g., by representing the problem on a number line diagram.
2. Measure and estimate liquid volumes and masses of objects using standard units of grams (g), kilograms (kg), and liters (l).[6] Add, subtract, multiply, or divide to solve one-step word problems involving masses or volumes that are given in the same units, e.g., by using drawings (such as a beaker with a measurement scale) to represent the problem.[7]

Represent and interpret data.

1. Draw a scaled picture graph and a scaled bar graph to represent a data set with several categories. Solve one- and two-step "how many more" and "how many less" problems using information presented in scaled bar graphs. *For example, draw a bar graph in which each square in the bar graph might represent 5 pets.*
2. Generate measurement data by measuring lengths using rulers marked with halves and fourths of an inch. Show the data by making a line plot, where the horizontal scale is marked off in appropriate units—whole numbers, halves, or quarters.

Geometric measurement: understand concepts of area and relate area to multiplication and to addition.

5. Recognize area as an attribute of plane figures and understand concepts of area measurement.
    a. A square with side length 1 unit, called "a unit square," is said to have "one square unit" of area, and can be used to measure area.
    b. A plane figure which can be covered without gaps or overlaps by n unit squares is said to have an area of n square units.
6. Measure areas by counting unit squares (square cm, square m, square in, square ft, and improvised units).
7. Relate area to the operations of multiplication and addition.
    a. Find the area of a rectangle with whole-number side lengths by tiling it, and show that the area is the same as would be found by multiplying the side lengths.
    b. Multiply side lengths to find areas of rectangles with whole-number side lengths in the context of solving real world and mathematical problems, and represent whole-number products as rectangular areas in mathematical reasoning.
    c. Use tiling to show in a concrete case that the area of a rectangle with whole-number side lengths a and b + c is the sum of a × b and a × c. Use area models to represent the distributive property in mathematical reasoning.
    d. Recognize area as additive. Find areas of rectilinear figures by decomposing them into non-overlapping rectangles and adding the areas of the non-overlapping parts, applying this technique to solve real world problems.

Geometric measurement: Recognize perimeter as an attribute of plane figures and distinguish between linear and area measures.

8. Solve real world and mathematical problems involving perimeters of polygons, including finding the perimeter given the side lengths, finding an unknown side length, and exhibiting rectangles with the same perimeter and different areas or with the same area and different perimeters.

Geometry	3.G

Reason with shapes and their attributes.

1. Understand that shapes in different categories (e.g., rhombuses, rectangles, and others) may share attributes (e.g., having four sides), and that the shared attributes can define a larger category (e.g., quadrilaterals). Recognize rhombuses, rectangles, and squares as examples of quadrilaterals, and draw examples of quadrilaterals that do not belong to any of these subcategories.
2. Partition shapes into parts with equal areas. Express the area of each part as a unit fraction of the whole. *For example, partition a shape into 4 parts with equal area, and describe the area of each part as 1/4 of the area of the shape.*

[6.] Excludes compound units such as cm^3 and finding the geometric volume of a container.

[7.] Excludes multiplicative comparison problems (problems involving notions of "times as much"; see Glossary, Table 2).

Copyright 2010. National Governors Association Center for Best Practices and Council of Chief State School Officers. All rights reserved.

Table 4.4. (continued)

| **Operations and Algebraic Thinking** | **4.OA** |

Use the four operations with whole numbers to solve problems.

1. Interpret a multiplication equation as a comparison, e.g., interpret 35 = 5 × 7 as a statement that 35 is 5 times as many as 7 and 7 times as many as 5. Represent verbal statements of multiplicative comparisons as multiplication equations.
2. Multiply or divide to solve word problems involving multiplicative comparison, e.g., by using drawings and equations with a symbol for the unknown number to represent the problem, distinguishing multiplicative comparison from additive comparison.[8]
3. Solve multistep word problems posed with whole numbers and having whole-number answers using the four operations, including problems in which remainders must be interpreted. Represent these problems using equations with a letter standing for the unknown quantity. Assess the reasonableness of answers using mental computation and estimation strategies including rounding.

Gain familiarity with factors and multiples.

4. Find all factor pairs for a whole number in the range 1–100. Recognize that a whole number is a multiple of each of its factors. Determine whether a given whole number in the range 1–100 is a multiple of a given one-digit number. Determine whether a given whole number in the range 1–100 is prime or composite.

Generate and analyze patterns.

5. Generate a number or shape pattern that follows a given rule. Identify apparent features of the pattern that were not explicit in the rule itself. *For example, given the rule "Add 3" and the starting number 1, generate terms in the resulting sequence and observe that the terms appear to alternate between odd and even numbers. Explain informally why the numbers will continue to alternate in this way.*

| **Number and Operations in Base Ten**[9] | **4.NBT** |

Generalize place value understanding for multi-digit whole numbers.

1. Recognize that in a multi-digit whole number, a digit in one place represents ten times what it represents in the place to its right. *For example, recognize that 700 ÷ 70 = 10 by applying concepts of place value and division.*
2. Read and write multi-digit whole numbers using base-ten numerals, number names, and expanded form. Compare two multi-digit numbers based on meanings of the digits in each place, using >, =, and < symbols to record the results of comparisons.
3. Use place value understanding to round multi-digit whole numbers to any place.

Use place value understanding and properties of operations to perform multi-digit arithmetic.

4. Fluently add and subtract multi-digit whole numbers using the standard algorithm.
5. Multiply a whole number of up to four digits by a one-digit whole number, and multiply two two-digit numbers, using strategies based on place value and the properties of operations. Illustrate and explain the calculation by using equations, rectangular arrays, and/or area models.
6. Find whole-number quotients and remainders with up to four-digit dividends and one-digit divisors, using strategies based on place value, the properties of operations, and/or the relationship between multiplication and division. Illustrate and explain the calculation by using equations, rectangular arrays, and/or area models.

[8] See Glossary, Table 2.

[9] Grade 4 expectations in this domain are limited to whole numbers less than or equal to 1,000,000.

Copyright 2010. National Governors Association Center for Best Practices and Council of Chief State School Officers. All rights reserved.

Table 4.4. (continued)

Number and Operations—Fractions[10]	4.NF

Extend understanding of fraction equivalence and ordering.

1. Explain why a fraction a/b is equivalent to a fraction $(n \times a)/(n \times b)$ by using visual fraction models, with attention to how the number and size of the parts differ even though the two fractions themselves are the same size. Use this principle to recognize and generate equivalent fractions.
2. Compare two fractions with different numerators and different denominators, e.g., by creating common denominators or numerators, or by comparing to a benchmark fraction such as 1/2. Recognize that comparisons are valid only when the two fractions refer to the same whole. Record the results of comparisons with symbols >, =, or <, and justify the conclusions, e.g., by using a visual fraction model.

Build fractions from unit fractions by applying and extending previous understandings of operations on whole numbers.

3. Understand a fraction a/b with $a > 1$ as a sum of fractions $1/b$.
   a. Understand addition and subtraction of fractions as joining and separating parts referring to the same whole.
   b. Decompose a fraction into a sum of fractions with the same denominator in more than one way, recording each decomposition by an equation. Justify decompositions, e.g., by using a visual fraction model. *Examples: 3/8 = 1/8 + 1/8 + 1/8; 3/8 = 1/8 + 2/8 ; 2 1/8 = 1 + 1 + 1/8 = 8/8 + 8/8 + 1/8.*
   c. Add and subtract mixed numbers with like denominators, e.g., by replacing each mixed number with an equivalent fraction, and/or by using properties of operations and the relationship between addition and subtraction.
   d. Solve word problems involving addition and subtraction of fractions referring to the same whole and having like denominators, e.g., by using visual fraction models and equations to represent the problem.
4. Apply and extend previous understandings of multiplication to multiply a fraction by a whole number.
   a. Understand a fraction a/b as a multiple of 1/b. *For example, use a visual fraction model to represent 5/4 as the product $5 \times (1/4)$, recording the conclusion by the equation $5/4 = 5 \times (1/4)$.*
   b. Understand a multiple of a/b as a multiple of $1/b$, and use this understanding to multiply a fraction by a whole number. *For example, use a visual fraction model to express $3 \times (2/5)$ as $6 \times (1/5)$, recognizing this product as 6/5. (In general, $n \times (a/b) = (n \times a)/b$.)*
   c. Solve word problems involving multiplication of a fraction by a whole number, e.g., by using visual fraction models and equations to represent the problem. *For example, if each person at a party will eat 3/8 of a pound of roast beef, and there will be 5 people at the party, how many pounds of roast beef will be needed? Between what two whole numbers does your answer lie?*

Understand decimal notation for fractions, and compare decimal fractions.

5. Express a fraction with denominator 10 as an equivalent fraction with denominator 100, and use this technique to add two fractions with respective denominators 10 and 100.[11] *For example, express 3/10 as 30/100, and add 3/10 + 4/100 = 34/100.*
6. Use decimal notation for fractions with denominators 10 or 100. *For example, rewrite 0.62 as 62/100; describe a length as 0.62 meters; locate 0.62 on a number line diagram.*
7. Compare two decimals to hundredths by reasoning about their size. Recognize that comparisons are valid only when the two decimals refer to the same whole. Record the results of comparisons with the symbols >, =, or <, and justify the conclusions, e.g., by using a visual model.

[10] Grade 3 expectations in this domain are limited to fractions with denominators 2, 3, 4, 6, 8, 10, 12, and 100.

[11] Students who can generate equivalent fractions can develop strategies for adding fractions with unlike denominators in general. But addition and subtraction with unlike denominators in general is not a requirement at this grade.

Copyright 2010. National Governors Association Center for Best Practices and Council of Chief State School Officers. All rights reserved.

Table 4.4. (continued)

Measurement and Data	4.MD

Solve problems involving measurement and conversion of measurements from a larger unit to a smaller unit.

1. Know relative sizes of measurement units within one system of units including km, m, cm; kg, g; lb, oz.; l, ml; hr, min, sec. Within a single system of measurement, express measurements in a larger unit in terms of a smaller unit. Record measurement equivalents in a two-column table. *For example, know that 1 ft is 12 times as long as 1 in. Express the length of a 4 ft snake as 48 in. Generate a conversion table for feet and inches listing the number pairs (1, 12), (2, 24), (3, 36), ...*

2. Use the four operations to solve word problems involving distances, intervals of time, liquid volumes, masses of objects, and money, including problems involving simple fractions or decimals, and problems that require expressing measurements given in a larger unit in terms of a smaller unit. Represent measurement quantities using diagrams such as number line diagrams that feature a measurement scale.

3. Apply the area and perimeter formulas for rectangles in real world and mathematical problems. *For example, find the width of a rectangular room given the area of the flooring and the length, by viewing the area formula as a multiplication equation with an unknown factor.*

Represent and interpret data.

4. Make a line plot to display a data set of measurements in fractions of a unit (1/2, 1/4, 1/8). Solve problems involving addition and subtraction of fractions by using information presented in line plots. *For example, from a line plot find and interpret the difference in length between the longest and shortest specimens in an insect collection.*

Geometric measurement: Understand concepts of angle and measure angles.

5. Recognize angles as geometric shapes that are formed wherever two rays share a common endpoint, and understand concepts of angle measurement:

   a. An angle is measured with reference to a circle with its center at the common endpoint of the rays, by considering the fraction of the circular arc between the points where the two rays intersect the circle. An angle that turns through 1/360 of a circle is called a "one-degree angle," and can be used to measure angles.

   b. An angle that turns through n one-degree angles is said to have an angle measure of n degrees.

6. Measure angles in whole-number degrees using a protractor. Sketch angles of specified measure.

7. Recognize angle measure as additive. When an angle is decomposed into non-overlapping parts, the angle measure of the whole is the sum of the angle measures of the parts. Solve addition and subtraction problems to find unknown angles on a diagram in real world and mathematical problems, e.g., by using an equation with a symbol for the unknown angle measure.

Geometry	4.G

Draw and identify lines and angles, and classify shapes by properties of their lines and angles.

1. Draw points, lines, line segments, rays, angles (right, acute, obtuse), and perpendicular and parallel lines. Identify these in two-dimensional figures.

2. Classify two-dimensional figures based on the presence or absence of parallel or perpendicular lines, or the presence or absence of angles of a specified size. Recognize right triangles as a category, and identify right triangles.

3. Recognize a line of symmetry for a two-dimensional figure as a line across the figure such that the figure can be folded along the line into matching parts. Identify line-symmetric figures and draw lines of symmetry.

Copyright 2010. National Governors Association Center for Best Practices and Council of Chief State School Officers. All rights reserved.

Table 4.4. (continued)

Operations and Algebraic Thinking	5.OA

Write and interpret numerical expressions.

1. Use parentheses, brackets, or braces in numerical expressions, and evaluate expressions with these symbols.
2. Write simple expressions that record calculations with numbers, and interpret numerical expressions without evaluating them. *For example, express the calculation "add 8 and 7, then multiply by 2" as 2 × (8 + 7). Recognize that 3 × (18932 + 921) is three times as large as 18932 + 921, without having to calculate the indicated sum or product.*

Analyze patterns and relationships.

3. Generate two numerical patterns using two given rules. Identify apparent relationships between corresponding terms. Form ordered pairs consisting of corresponding terms from the two patterns, and graph the ordered pairs on a coordinate plane. *For example, given the rule "Add 3" and the starting number 0, and given the rule "Add 6" and the starting number 0, generate terms in the resulting sequences, and observe that the terms in one sequence are twice the corresponding terms in the other sequence. Explain informally why this is so.*

Number and Operations in Base Ten	5.NBT

Understand the place value system.

1. Recognize that in a multi-digit number, a digit in one place represents 10 times as much as it represents in the place to its right and 1/10 of what it represents in the place to its left.
2. Explain patterns in the number of zeros of the product when multiplying a number by powers of 10, and explain patterns in the placement of the decimal point when a decimal is multiplied or divided by a power of 10. Use whole-number exponents to denote powers of 10.
3. Read, write, and compare decimals to thousandths.
    a. Read and write decimals to thousandths using base-ten numerals, number names, and expanded form, e.g., 347.392 = 3 × 100 + 4 × 10 + 7 × 1 + 3 × (1/10) + 9 × (1/100) + 2 × (1/1000).
    b. Compare two decimals to thousandths based on meanings of the digits in each place, using >, =, and < symbols to record the results of comparisons.
4. Use place value understanding to round decimals to any place.

Perform operations with multi-digit whole numbers and with decimals to hundredths.

5. Fluently multiply multi-digit whole numbers using the standard algorithm.
6. Find whole-number quotients of whole numbers with up to four-digit dividends and two-digit divisors, using strategies based on place value, the properties of operations, and/or the relationship between multiplication and division. Illustrate and explain the calculation by using equations, rectangular arrays, and/or area models.
7. Add, subtract, multiply, and divide decimals to hundredths, using concrete models or drawings and strategies based on place value, properties of operations, and/or the relationship between addition and subtraction; relate the strategy to a written method and explain the reasoning used.

Copyright 2010. National Governors Association Center for Best Practices and Council of Chief State School Officers. All rights reserved.

Table 4.4. (continued)

Number and Operations—Fractions	5.NF

Use equivalent fractions as a strategy to add and subtract fractions.

1. Add and subtract fractions with unlike denominators (including mixed numbers) by replacing given fractions with equivalent fractions in such a way as to produce an equivalent sum or difference of fractions with like denominators. *For example, 2/3 + 5/4 = 8/12 + 15/12 = 23/12. (In general, a/b + c/d = (ad + bc)/bd.)*

2. Solve word problems involving addition and subtraction of fractions referring to the same whole, including cases of unlike denominators, e.g., by using visual fraction models or equations to represent the problem. Use benchmark fractions and number sense of fractions to estimate mentally and assess the reasonableness of answers. *For example, recognize an incorrect result 2/5 + 1/2 = 3/7, by observing that 3/7 < 1/2.*

Apply and extend previous understandings of multiplication and division to multiply and divide fractions.

3. Interpret a fraction as division of the numerator by the denominator ($a/b = a \div b$). Solve word problems involving division of whole numbers leading to answers in the form of fractions or mixed numbers, e.g., by using visual fraction models or equations to represent the problem. *For example, interpret 3/4 as the result of dividing 3 by 4, noting that 3/4 multiplied by 4 equals 3, and that when 3 wholes are shared equally among 4 people each person has a share of size 3/4. If 9 people want to share a 50-pound sack of rice equally by weight, how many pounds of rice should each person get? Between what two whole numbers does your answer lie?*

4. Apply and extend previous understandings of multiplication to multiply a fraction or whole number by a fraction.
    a. Interpret the product $(a/b) \times q$ as a parts of a partition of q into b equal parts; equivalently, as the result of a sequence of operations a × q ÷ b. *For example, use a visual fraction model to show (2/3) × 4 = 8/3, and create a story context for this equation. Do the same with (2/3) × (4/5) = 8/15. (In general, (a/b) × (c/d) = ac/bd.)*
    b. Find the area of a rectangle with fractional side lengths by tiling it with unit squares of the appropriate unit fraction side lengths, and show that the area is the same as would be found by multiplying the side lengths. Multiply fractional side lengths to find areas of rectangles, and represent fraction products as rectangular areas.

5. Interpret multiplication as scaling (resizing), by:
    a. Comparing the size of a product to the size of one factor on the basis of the size of the other factor, without performing the indicated multiplication.
    b. Explaining why multiplying a given number by a fraction greater than 1 results in a product greater than the given number (recognizing multiplication by whole numbers greater than 1 as a familiar case); explaining why multiplying a given number by a fraction less than 1 results in a product smaller than the given number; and relating the principle of fraction equivalence $a/b = (n \times a)/(n \times b)$ to the effect of multiplying a/b by 1.

6. Solve real world problems involving multiplication of fractions and mixed numbers, e.g., by using visual fraction models or equations to represent the problem.

7. Apply and extend previous understandings of division to divide unit fractions by whole numbers and whole numbers by unit fractions.[12]
    a. Interpret division of a unit fraction by a non-zero whole number, and compute such quotients. *For example, create a story context for (1/3) ÷ 4, and use a visual fraction model to show the quotient. Use the relationship between multiplication and division to explain that (1/3) ÷ 4 = 1/12 because (1/12) × 4 = 1/3.*
    b. Interpret division of a whole number by a unit fraction, and compute such quotients. *For example, create a story context for 4 ÷ (1/5), and use a visual fraction model to show the quotient. Use the relationship between multiplication and division to explain that 4 ÷ (1/5) = 20 because 20 × (1/5) = 4.*
    c. Solve real world problems involving division of unit fractions by non-zero whole numbers and division of whole numbers by unit fractions, e.g., by using visual fraction models and equations to represent the problem. *For example, how much chocolate will each person get if 3 people share 1/2 lb of chocolate equally? How many 1/3-cup servings are in 2 cups of raisins?*

[12] Students able to multiply fractions in general can develop strategies to divide fractions in general, by reasoning about the relationship between multiplication and division. But division of a fraction by a fraction is not a requirement at this grade.

Copyright 2010. National Governors Association Center for Best Practices and Council of Chief State School Officers. All rights reserved.

Table 4.4. (continued)

Measurement and Data	**5.MD**

Convert like measurement units within a given measurement system.

1. Convert among different-sized standard measurement units within a given measurement system (e.g., convert 5 cm to 0.05 m), and use these conversions in solving multi-step, real world problems.

Represent and interpret data.

2. Make a line plot to display a data set of measurements in fractions of a unit (1/2, 1/4, 1/8). Use operations on fractions for this grade to solve problems involving information presented in line plots. *For example, given different measurements of liquid in identical beakers, find the amount of liquid each beaker would contain if the total amount in all the beakers were redistributed equally.*

Geometric measurement: understand concepts of volume and relate volume to multiplication and to addition.

3. Recognize volume as an attribute of solid figures and understand concepts of volume measurement.
    a. A cube with side length 1 unit, called a "unit cube," is said to have "one cubic unit" of volume, and can be used to measure volume.
    b. A solid figure which can be packed without gaps or overlaps using n unit cubes is said to have a volume of n cubic units.
4. Measure volumes by counting unit cubes, using cubic cm, cubic in, cubic ft, and improvised units.
5. Relate volume to the operations of multiplication and addition and solve real world and mathematical problems involving volume.
    a. Find the volume of a right rectangular prism with whole-number side lengths by packing it with unit cubes, and show that the volume is the same as would be found by multiplying the edge lengths, equivalently by multiplying the height by the area of the base. Represent threefold whole-number products as volumes, e.g., to represent the associative property of multiplication.
    b. Apply the formulas $V = l \times w \times h$ and $V = b \times h$ for rectangular prisms to find volumes of right rectangular prisms with whole-number edge lengths in the context of solving real world and mathematical problems.
    c. Recognize volume as additive. Find volumes of solid figures composed of two non-overlapping right rectangular prisms by adding the volumes of the non-overlapping parts, applying this technique to solve real world problems.

Geometry	**5.G**

Graph points on the coordinate plane to solve real-world and mathematical problems.

1. Use a pair of perpendicular number lines, called axes, to define a coordinate system, with the intersection of the lines (the origin) arranged to coincide with the 0 on each line and a given point in the plane located by using an ordered pair of numbers, called its coordinates. Understand that the first number indicates how far to travel from the origin in the direction of one axis, and the second number indicates how far to travel in the direction of the second axis, with the convention that the names of the two axes and the coordinates correspond (e.g., *x*-axis and *x*-coordinate, *y*-axis and *y*-coordinate).
2. Represent real world and mathematical problems by graphing points in the first quadrant of the coordinate plane, and interpret coordinate values of points in the context of the situation.

Classify two-dimensional figures into categories based on their properties.

3. Understand that attributes belonging to a category of two-dimensional figures also belong to all subcategories of that category. *For example, all rectangles have four right angles and squares are rectangles, so all squares have four right angles.*
4. Classify two-dimensional figures in a hierarchy based on properties.

Copyright 2010. National Governors Association Center for Best Practices and Council of Chief State School Officers. All rights reserved.

References

National Governors Association Center for Best Practices and Council of Chief State School Officers (NGAC and CCSSO). 2010. *Common core state standards.* Washington, DC: NGAC and CCSSO.

National Research Council (NRC). 2012. *A framework for K–12 science education: Practices, crosscutting concepts, and core ideas.* Washington DC: National Academies Press.

NGSS Lead States. 2013. *Next Generation Science Standards: For states, by states.* Washington, DC: National Academies Press. *www.nextgenscience.org/next-generation-science-standards.*

Willard, T., ed. 2015. *The NSTA quick-reference guide to the NGSS: Elementary school.* Arlington, VA: NSTA Press.

Science and Engineering Practices

Picture-Perfect STEM Lessons incorporates all three dimensions of *A Framework for K–12 Science Education* (*Framework*; NRC 2012). However, one dimension stands out as key to developing a successful STEM classroom—the science and engineering practices. Our previous *Picture-Perfect Science* books emphasized scientific inquiry as a key component, but we think the science and engineering practices allow for a broader implementation of inquiry by expanding the inquiry process into the world of engineering. We like how Mariel Milano, a member of the *Next Generation Science Standards* (*NGSS*; NGSS Lead States 2013) writing team, describes the shift from inquiry to the practices. Milano (2013, pp. 15–16) says, "The perspective presented in the Framework is not one of replacing inquiry; rather it is one of expanding and enriching the teaching and learning of science and engineering. The practices are in some ways inquiry unpacked."

The practices go well beyond the facts of science to help students understand how scientific knowledge develops and how it can be applied to solving real-world problems through engineering. The *Framework* states, "Any education that focuses predominantly on the detailed products of scientific labor—the facts of science—without developing an understanding of how those facts were established or that ignores the many important applications of science in the world misrepresents science and marginalizes the importance of engineering" (NRC 2012, p. 43). Simply put, the practices embody two aspects: (1) how we know what we know in science and (2) how to apply what we know in science.

Engineering is receiving more attention than ever before in elementary school. It is sometimes referred to as the "stealth" profession because, although we use thousands of designed objects each day, we seldom think about the engineering practices involved in the creation and production of those objects. From the pen you write with, to the window you look through, to the cell phone in your pocket, all of these objects were most likely designed by engineers. For the first time ever, we are working with national standards that include engineering as a disciplinary core idea and the practices of engineers as a key component. Milano (2013, pp. 10–11) says, "The NGSS demonstrates a commitment to fully integrating engineering and technology into the structure of science education by elevating engineering design to the same level as scientific inquiry in classroom instruction."

In this chapter, we describe the scientific and engineering practices, show how they apply to all of the components of STEM, and explain how students in grades 3–5 can use these practices in the classroom. Focusing on practices rather than discrete skills or methods helps teachers and students understand that there is more than one approach to doing science and engineering. In other words, there is no set "scientific method" or "design process" that all scientists and engineers use in lockstep. Instead, scientists and engineers engage in a wide variety of practices as part of their work. The *Framework* identifies eight science and engineering practices that students should engage in throughout grades K–12. These practices are not meant to be a linear sequence of steps to be taken in a specific order; rather, they are to be used iteratively and in combination with each other (see p. 25 for a list of the practices).

All four components of STEM—science, technology, engineering, and mathematics—are included in these practices. Although in name the science and engineering practices seem to include only the *S* and *E* components of STEM, through reading these practices one can see that technology and mathematics are integral parts. The science and engineering practices are closely tied to science content but go well beyond knowing scientific facts; they are based on the idea that "students cannot understand scientific and engineering ideas without engaging in the practices of inquiry and the discourses by which such ideas are developed and refined" (NRC 2012, p. 218).

The *NGSS* provide the following guiding principles for implementing the science and engineering practices (see pages 49–50 of NGSS Lead States 2013):

1. Students in grades K–12 should engage in all eight practices over each grade band.
2. Practices grow in complexity and sophistication across the grades.
3. Each practice may reflect science or engineering.
4. Practices represent what students are expected to do and are not teaching methods or curriculum.
5. The eight practices are not separate; they intentionally overlap and interconnect.
6. Performance expectations focus on some, but not all, capabilities associated with a practice.
7. Engagement in practices is language intensive and requires students to participate in science discourse.

Picture-Perfect STEM Lessons upholds these principles by paying attention to the progression of the practices over the grades, overlapping several practices within each lesson, and incorporating science discourse throughout the activities. The rationale that these practices be incorporated throughout grades K–12 comes from Chapter 3 of the *Framework,* which the *NGSS* are based on.

Science and Engineering Practices in the 3–5 Classroom

So what do the science and engineering practices look like in the 3–5 classroom? Table 5.1 shows the components of the science and engineering practices specific to grades 3–5.

On the second page of each lesson in this book, the science and engineering practice(s) addressed are identified. Although several of the practices might be addressed in a particular lesson, we emphasize between two and four practices in each lesson. Next we discuss a few examples of how these practices are incorporated into the grades 3–5 lessons in this book.

In Chapter 6, "The Inventor's Secret," students learn how Thomas Edison encouraged Henry Ford to "keep at it" in his quest to make an affordable, gas-powered automobile. Then, students have the opportunity to design a balloon car that meets certain criteria and constraints. This design activity incorporates practice 6, constructing explanations and designing solutions. Specifically, students generate and compare multiple solutions to a problem based on how well the solutions meet the criteria and constraints of the design problem. After students build their balloon cars, they test and refine their designs. This engages students in practice 3, planning and carrying out investigations, as they test, compare, and improve their cars.

Testing balloon car designs

Table 5.1. Science and Engineering Practices for Grades 3–5

Asking Questions and Defining Problems for Grades 3–5

Asking questions and defining problems in 3–5 builds on K–2 experiences and progresses to specifying qualitative relationships.

- Ask questions about what would happen if a variable is changed.
- Identify scientific (testable) and non-scientific (non-testable) questions.
- Ask questions that can be investigated and predict reasonable outcomes based on patterns such as cause-and-effect relationships.
- Use prior knowledge to describe problems that can be solved.
- Define a simple design problem that can be solved through the development of an object, tool, process, or system and includes several criteria for success and constraints on materials, time, or cost.

Developing and Using Models for Grades 3–5

Modeling in 3–5 builds on K–2 experiences and progresses to building and revising simple models and using models to represent events and design solutions.

- Identify limitations of models.
- Collaboratively develop and/or revise a model based on evidence that shows the relationships among variables for frequent and regularly occurring events.
- Develop a model using an analogy, example, or abstract representation to describe a scientific principle or design solution.
- Develop and/or use models to describe and/or predict phenomena.
- Develop a diagram or simple physical prototype to convey a proposed object, tool, or process.
- Use a model to test cause-and-effect relationships or interactions concerning the functioning of a natural or designed system.

Planning and Carrying Out Investigations for Grades 3–5

Planning and carrying out investigations to answer questions or test solutions to problems in 3–5 builds on K–2 experiences and progresses to include investigations that control variables and provide evidence to support explanations or design solutions.

- Plan and conduct an investigation collaboratively to produce data to serve as the basis for evidence, using fair tests in which variables are controlled and the number of trials considered.
- Evaluate appropriate methods and/or tools for collecting data.
- Make observations and/or measurements to produce data to serve as the basis for evidence for an explanation of a phenomenon or test a design solution.
- Make predictions about what would happen if a variable changes.
- Test two different models of the same proposed object, tool, or process to determine which better meets criteria for success.

Analyzing and Interpreting Data for Grades 3–5

Analyzing data in 3–5 builds on K–2 experiences and progresses to introducing quantitative approaches to collecting data and conducting multiple trials of qualitative observations. When possible and feasible, digital tools should be used.

- Represent data in tables and/or various graphical displays (bar graphs, pictographs, and/or pie charts) to reveal patterns that indicate relationships.
- Analyze and interpret data to make sense of phenomena, using logical reasoning, mathematics, and/or computation.
- Compare and contrast data collected by different groups in order to discuss similarities and differences in their findings.
- Analyze data to refine a problem statement or the design of a proposed object, tool, or process.
- Use data to evaluate and refine design solutions.

Source: Willard 2015, pp. 100–101.

Table 5.1. Science and Engineering Practices for Grades 3–5 (continued)

Using Mathematics and Computational Thinking for Grades 3–5

Mathematical and computational thinking in 3–5 builds on K–2 experiences and progresses to extending quantitative measurements to a variety of physical properties and using computation and mathematics to analyze data and compare alternative design solutions.

- Organize simple data sets to reveal patterns that suggest relationships.
- Describe, measure, estimate, and/or graph quantities such as area, volume, weight, and time to address scientific and engineering questions and problems.
- Create and/or use graphs and/or charts generated from simple algorithms to compare alternative solutions to an engineering problem.

Constructing Explanations and Designing Solutions for Grades 3–5

Constructing explanations and designing solutions in 3–5 builds on K–2 experiences and progresses to the use of evidence in constructing explanations that specify variables that describe and predict phenomena and in designing multiple solutions to design problems.

- Construct an explanation of observed relationships (e.g., the distribution of plants in the backyard).
- Use evidence (e.g., measurements, observations, patterns) to construct or support an explanation or design a solution to a problem.
- Identify the evidence that supports particular points in an explanation.
- Apply scientific ideas to solve design problems.
- Generate and compare multiple solutions to a problem based on how well they meet the criteria and constraints of the design solution.

Engaging in Argument From Evidence for Grades 3–5

Engaging in argument from evidence in 3–5 builds on K–2 experiences and progresses to critiquing the scientific explanations or solutions proposed by peers by citing relevant evidence about the natural and designed world(s).

- Compare and refine arguments based on an evaluation of the evidence presented.
- Distinguish among facts, reasoned judgment based on research findings, and speculation in an explanation.
- Respectfully provide and receive critiques from peers about a proposed procedure, explanation, or model by citing relevant evidence and posing specific questions.
- Construct and/or support an argument with evidence, data, and/or a model.
- Use data to evaluate claims about cause and effect.
- Make a claim about the merit of a solution to a problem by citing relevant evidence about how it meets the criteria and constraints of the problem.

Obtaining, Evaluating, and Communicating Information for Grades 3–5

Obtaining, evaluating, and communicating information in 3–5 builds on K–2 experiences and progresses to evaluating the merit and accuracy of ideas and methods.

- Read and comprehend grade-appropriate complex texts and/or other reliable media to summarize and obtain scientific and technical ideas and describe how they are supported by evidence.
- Compare and/or combine across complex texts and/or other reliable media to support the engagement in other scientific and/or engineering practices.
- Combine information in written text with that contained in corresponding tables, diagrams, and/or charts to support the engagement in other scientific and/or engineering practices.
- Obtain and combine information from books and/or other reliable media to explain phenomena or solutions to a design problem.
- Communicate scientific and/or technical information orally and/or in written formats, including various forms of media as well as tables, diagrams, and charts.

Source: Willard 2015, pp. 100–101.

Chapter 5

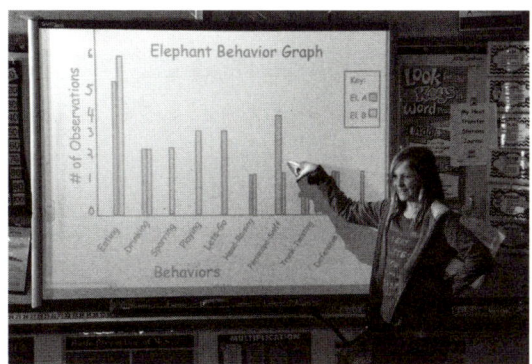

GRAPHING ELEPHANT BEHAVIORS

USING GOOGLE EARTH

In Chapter 12, "Better Together," students engage in practice 4, analyzing and interpreting data, and practice 5, using mathematics and computational thinking, as they analyze data on elephant behaviors using an ethogram. Students graph the data and use their graphs to make inferences about the elephants. Then, they incorporate practice 7, engaging in argument from evidence, as they use evidence to support the idea that living in groups helps some animals survive.

We do not think that every STEM lesson needs to include a specific design challenge. In some lessons in this book, students can engage in the science and engineering practices by using technologies that apply scientific concepts and developing an understanding of how the technologies work. For example, in Chapter 17, "Solving the Puzzle Under the Sea," students engage in practice 8, obtaining, evaluating, and communicating information, as they use the Google Earth tool to explore Earth's mountains. Students also learn how sonar technology helps us create maps of the ocean floor. They are involved in practice 4, analyzing and interpreting data, as they use information from simulated sonar readings to create a profile of the ocean floor. So although students are not designing a solution to a specific problem, they are learning about technologies that have been developed to help us make models of our planet.

The Differences Between Science and Engineering

Science and engineering are closely related and interdependent. As stated in the *Framework*, "It is impossible to do engineering today without applying science in the process and, in many areas of science, designing and building new experiments requires scientists to engage in some engineering practices" (NRC 2012, p. 32). However, it is helpful to understand the difference between what scientists do and what engineers do. You can begin to see a key difference between science and engineering by the way practice 1 is written: "Asking questions (for science) and defining problems (for engineering)." Science begins with a question and engineering begins with a problem. The rest of the practices follow accordingly. Table 5.2 (p. 52) shows how the eight science and engineering practices apply differently to the fields of science and engineering.

Some lessons in this book emphasize the engineering side of the practices. For example, in Chapter 6, "The Inventor's Secret," students are deeply immersed in the practices of engineers as they use the design process to design, test, and refine a balloon car. In Chapter 11, "From Edison to the iPod," students learn how music players have advanced over the years (from the phonograph to the iPod) and design a prototype of a music player of the future. These lessons not only give students an awareness of the work of engineers but also provide

Picture-Perfect STEM Lessons, 3–5

Table 5.2. Science Practices and Engineering Practices

Science Practices	Engineering Practices
1. **Asking questions**	1. **Defining problems**
2. **Developing and using models** to understand natural phenomena	2. **Developing and using models** to analyze systems and test solutions
3. **Planning and carrying out investigations** to answer a question about the natural world	3. **Planning and carrying out investigations** to collect data to specify design criteria or test a solution
4. **Analyzing and interpreting data** from an investigation to identify patterns and derive meaning	4. **Analyzing and interpreting data** from an investigation to compare different solutions to see which best solves the problem
5. **Using mathematics and computational thinking** to represent variables and their relationships and to predict natural phenomena	5. **Using mathematics and computational thinking** to calculate and simulate different designs to see which are best
6. **Constructing explanations** of natural phenomena	6. **Designing solutions** to solve problems
7. **Engaging in argument from evidence** to evaluate different lines of reasoning in searching for an explanation of a natural phenomenon	7. **Engaging in argument from evidence,** using a systematic method to find the best solution for a problem
8. **Obtaining, evaluating, and communicating information** by listing and reading others' ideas critically and communicating one's own ideas clearly	8. **Obtaining, evaluating, and communicating information** to learn how others have solved a problem and to present persuasive arguments in favor of a given solution

Source: Adapted from Milano 2013.
Note: A more detailed version of this table can be found on pages 50–53 in *A Framework for K–12 Science Education* (NRC 2012).

them with opportunities to think like engineers. However, lessons such as Chapter 7, "Mesmerized," and Chapter 13, "Spider Science," focus more on the science side of the practices, where students are carrying out investigations, identifying patterns in the natural world, and constructing explanations of natural phenomena.

In most of the lessons, however, science and engineering are closely tied together. For example, in Chapter 9, "Light it Up," students explore how different kinds of lightbulbs work and apply that scientific knowledge to design a nightlight using LED lights and circuit tape. In Chapter 14, "Bionic Animals," students observe the structures different animals use for locomotion and design a prosthetic device to mimic the structure and function of an animal body part. In both of these lessons, scientific understanding is essential to the design challenge. These experiences can help students see the interdependent relationship between science and engineering.

The Engineering Design Process

There are many approaches to the engineering design process, but all of them follow the same basic pattern—a series of steps done iteratively to solve a problem. In these lessons, we use the simple, three-component model described in the *NGSS* as our framework (see Figure 5.1).

The *NGSS* make it clear that these component ideas do not always follow in order and that "at any stage a problem solver can redefine the problem or generate new solutions to replace an idea that is just not working out" (NGSS Lead States 2013, Appendix I).

The *NGSS* also provide the following explanation of engineering design in grades 3–5:

Figure 5.1. Engineering Design in Grades 3–5

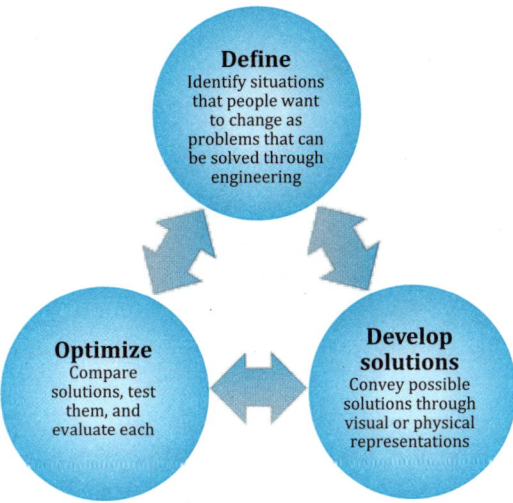

Source: NGSS Lead States 2013, Appendix I.

At the upper elementary grades, engineering design engages students in more formalized problem solving. Students define a problem using criteria for success and constraints or limits of possible solutions. Students research and consider multiple possible solutions to a given problem. Generating and testing solutions also becomes more rigorous as students learn to optimize solutions by revising them several times to obtain the best possible design. (NGSS Lead States 2013, Appendix I).

Picture-Perfect STEM Lessons integrates the three component ideas of engineering throughout the book. Students are provided with opportunities to define problems, develop solutions, and optimize solutions. For example, in Chapter 16, "Hurricane!," after students learn about how the failed levees contributed to the devastation in New Orleans during Hurricane Katrina, they have to design a model of a levee and test its ability to hold back a storm surge. They are given certain criteria that must be met and constraints that must be considered. In Chapter 20, "From Trash to Treasure," students hear an inspiring story about a woman in The Gambia who turned the problem of plastic bag pollution into a solution by designing a new use for the bags. Then, students design their own upcycled product out of discarded plastics.

UPCYCLED PLASTIC PRODUCT

Student-Directed Exploration With the Practices

In *Picture-Perfect STEM Lessons,* students are engaged in science and engineering practices in a guided manner throughout the lessons. At the end of each lesson, a "For Further Exploration" box is provided to help you encourage your students to use the science and engineering practices in a more student-directed format. This box lists questions and challenges related to the lesson that students may select to research, investigate, or innovate. Students may also use the questions as examples to help them generate their own questions. After selecting one of the questions in the box or formulating their own questions, students can individually or collaboratively make predictions, design investigations or surveys to test their predictions, collect evidence, devise explanations, design solutions, and/or examine related resources. They can communicate their findings through a science notebook, at a poster session or gallery walk, or by producing a media project.

For example, in Chapter 9, "Light It Up!," students learn to implement various science and engineering practices as they investigate how different kinds of lightbulbs work and design a nightlight using a paper circuit made with circuit tape and an LED lightbulb. After this guided investigation, they research information about common uses for LED lights, plan and carry out their own investigations to find out what kinds of objects can be used as a switch in a paper circuit, or apply what they have learned to design another practical use of a paper circuit. Students can choose a question from the box or brainstorm some questions of their own. We believe that once students have participated in the guided format of *Picture-Perfect STEM Lessons,* they are more likely to be able to apply the science and engineering practices on their own. The "For Further Exploration" box can help you provide those opportunities.

At the end of the chapter, we have provided some "think sheets" to guide students as they approach the "For Further Exploration" questions (see pp. 56–59). The Innovation Think Sheet (pp. 58–59) is based on the design process offered by PBS's *Design Squad Global* (see Figure 5.2). The *Design Squad Global* website suggests that "when you see the design process in action, you'll notice that it's rarely the smooth succession of steps that the diagram implies. The steps often overlap and blur, and their order is sometimes reversed—it's a creative, fluid way of working that has to be adapted to each individual situation. As you guide kids through the design process, you'll want to be flexible and receptive to the different approaches your kids may try" (PBS 2016).

We suggest that students share the results of their explorations with each other through poster sessions. Scientists, engineers, and researchers routinely hold poster sessions to communicate their findings. Here are some suggestions for poster sessions:

- Posters should include a title, the researchers' names, a brief description of the investigation, and a summary of the main findings.

A GALLERY WALK

- Observations, data tables, and/or graphs should be included as evidence to justify conclusions.
- The print should be large enough that people can read it from a distance.
- Students should have the opportunity to present their posters to the class.
- The audience in a poster session should examine the evidence, ask thoughtful questions, identify faulty reasoning, and suggest alternative explanations to presenters in a polite, respectful manner.

Poster sessions not only mirror the work of real scientists but also provide excellent opportunities for authentic assessment. Another way to share students' posters is during a gallery walk. In a gallery walk, students put their posters on display for their classmates to view and critique. Students taking the gallery walk use sticky notes to post suggestions, questions, and praise directly on their classmates' posters. Writing on sticky notes encourages interaction, and the comments provide immediate feedback for the "exhibitors." Here are some guidelines for a gallery walk:

- All necessary information about the investigation should be included on the poster because students will not be giving an oral presentation.
- Like a visit to an art gallery, the gallery walk should be done quietly. Students should be respectful of their classmates' poster displays.
- Students should have the opportunity to read the comments about their own posters and make changes if necessary.

Figure 5.2. Design Process From Design Squad Global

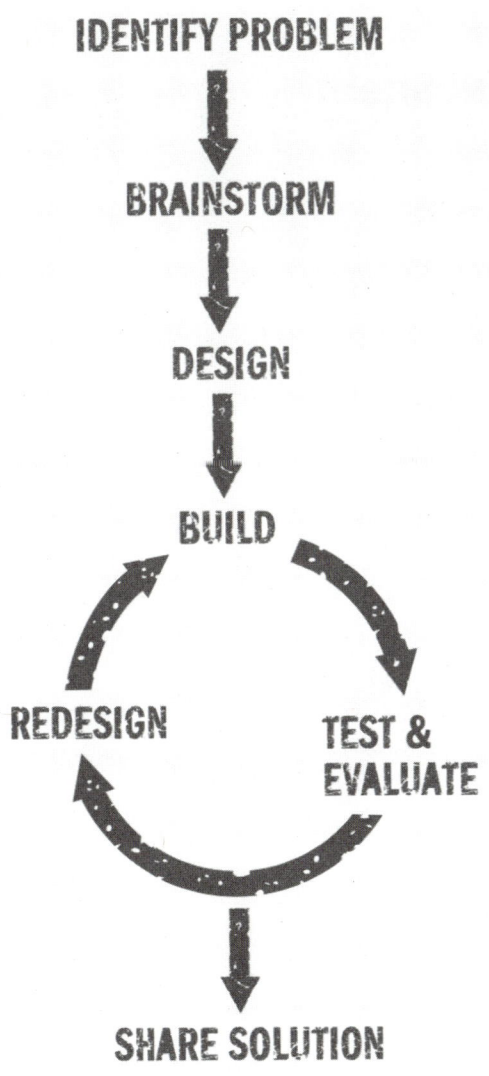

Source: PBS 2016.

Implementing the guided lessons in this book along with the "For Further Exploration" suggestions at the end of each lesson provides a framework for moving from teacher-guided to self-directed learning.

In summary, *Picture-Perfect STEM Lessons* engages students in these scientific and engineering practices to capture their interest, motivate their continued study, and above all instill in them a sense of wonder about the natural and designed world. The end result is that by actually doing science and engineering rather than merely learning about it, students will recognize that the work of scientists and engineers is creative and rewarding and deeply impacts their world.

References

Milano, M. 2013. The *Next Generation Science Standards* and engineering for young learners: Beyond bridges and egg drops. *Science and Children* 51 (2): 10–16.

National Research Council (NRC). 2012. *A framework for K–12 science education: Practices, crosscutting concepts, and core ideas.* Washington, DC: National Academies Press.

NGSS Lead States. 2013. *Next Generation Science Standards: For states, by states.* Washington, DC: National Academies Press. *www.nextgenscience. org/next-generation-science-standards.*

PBS. 2016. Design squad global. Parents and educators: Online workshop. *http://pbskids.org/designsquad/ parentseducators/workshop/process.html.*

Willard, T., ed. 2015. *The NSTA quick-reference guide to the NGSS: Elementary school.* Arlington, VA: NSTA Press.

Name: _____   Date: _____

Research Think Sheet

1. My question: _____

2. My prediction: _____

3. How I will find information on the question:

   _____

   _____

   _____

   _____

4. My answer:

   _____

   _____

   _____

   _____

   _____

5. My sources:

   _____

   _____

   _____

   _____

   _____

   _____

Name: _____   Date: _____

Investigation Think Sheet

1. My question: _____

2. My prediction: _____

3. My procedure and materials: _____

_____

_____

_____

4. My data (observations, measurements, graphs, etc.):

```
┌─────────────────────────────────────────────────────────────┐
│                                                             │
│                                                             │
│                                                             │
│                                                             │
│                                                             │
│                                                             │
│                                                             │
└─────────────────────────────────────────────────────────────┘
```

5. My conclusion:

_____

_____

_____

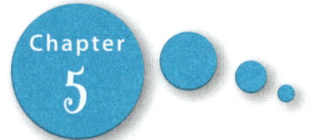

Name: _____   Date: _____

Innovation Think Sheet

What is the problem or need?

_____

_____

Brainstorm! What are some ways to solve the problem or meet the need?

_____

_____

_____

Design it! Sketch your best design idea:

Build it! Sketch your solution:

Test it! How well did it solve the problem or meet the need?

_____

_____

_____

Redesign it! What can you change to improve your design?

_____

_____

_____

Share it! Describe your final solution in words and pictures:

_____

_____

_____

Chapter 6

The Inventor's Secret

Description

Two books that emphasize the power of perseverance are paired, engaging students in a challenge to design a toy car using everyday materials. Students are introduced to the design process and use it to improve their car design.

Suggested Grade Levels: 3–5

LESSON OBJECTIVES Connecting to the *Framework*		
Science and Engineering Practices	**Disciplinary Core Ideas**	**Crosscutting Concept**
Planning and Carrying Out Investigations Constructing Explanations and Designing Solutions	**ETS1.C:** Optimizing the Design Solution **ETS2.B:** Influence of Engineering, Technology, and Science on Society and the Natural World	Structure and Function

Featured Picture Books

TITLE: ***Rosie Revere, Engineer***
AUTHOR: **Andrea Beaty**
ILLUSTRATOR: **David Roberts**
PUBLISHER: **Abrams Books for Young Readers**
YEAR: **2013**
GENRE: **Story**
SUMMARY: *Young Rosie dreams of being an engineer. Alone in her room at night, she constructs great inventions from odds and ends. Afraid of failure, Rosie hides her creations under her bed until a fateful visit from her great-great-aunt Rose, who shows her that a first flop isn't something to fear—it's something to celebrate.*

TITLE: ***The Inventor's Secret: What Thomas Edison Told Henry Ford***
AUTHOR: **Suzanne Slade**
ILLUSTRATOR: **Jennifer Black Reinhardt**
PUBLISHER: **Charlesbridge**
YEAR: **2015**
GENRE: **Narrative Information**
SUMMARY: *This delightful book about Henry Ford's quest to make an affordable car also portrays the friendship between him and Thomas Edison, including the fateful moment when Ford learned Edison's secret to inventing—keep at it!*

Picture-Perfect STEM Lessons, 3–5

Time Needed

This lesson will take several class periods. Suggested scheduling is as follows:

Day 1: Engage with *Rosie Revere, Engineer* Read-Aloud and **Explore** with Balloon Car Challenge, Part 1, and PBS *Design Squad Global* Video

Day 2: Explore with Balloon Car Challenge, Part 2

Day 3: Explain with *The Inventor's Secret* Read-Aloud and The Design Process and **Elaborate** with Edison Quotes

Day 4: Evaluate with Balloon Car Challenge, Part 3

Materials

For Build and Test a Balloon Car (per class)

You will need a variety of materials that can be used for different parts of the balloon cars. Organize the materials into five bins labeled *power source, body, wheels, axles,* and *materials for attaching*. The materials needed for each bin are as follows:

- Power source
  - Straws
  - Balloons, all the same size
  - Tape
  - Rubber bands
- Body
  - Empty water bottle
  - Clean, empty juice box or other small box
  - Disposable cup
  - Cardboard
- Wheels
  - Plastic bottle caps
  - Candy mints (with hole in the middle)
  - CDs
  - Cardboard
- Axles
  - Straws
  - Wooden skewers
  - Chopsticks
  - Cotton swabs
- Materials for attaching
  - Tape
  - Foam

SAFETY

- Before using balloons in the classroom, be sure that no one is allergic to latex.
- Have students wear safety glasses or goggles during this activity.
- Remind students not to eat any food used in this activity.
- Wash hands with soap and water after completing this activity.

- Modeling clay
- Glue
- Pieces of dry sponge

For Build and Test a Balloon Car (per group of 3–4 students)

- Air pump for blowing up balloons
- Metersticks or tape measures

Additional class materials

- The Design Process poster (enlarged version of p. 74; full-color version available on the Extras page at *www.nsta.org/PicturePerfectSTEM3-5*)
- Edison Quotes (1 set cut into strips)

Student Pages

- Balloon Car Design Challenge
- Redesign, Build, Test & Evaluate
- Prototype Display Card
- STEM at Home

Background for Teachers

Thomas Edison and Henry Ford are two of the most famous inventors in history. Edison's inventions are too numerous to list here (2,332 patents worldwide), but many of them have changed the way we live. His most famous inventions were the phonograph, kinetoscope, dictaphone, and, of course, the incandescent light bulb. Although Ford did not invent the automobile, he is credited with developing the first car that most working people could afford, the Model T. His introduction of the Model T revolutionized transportation, and the moving assembly line he designed to manufacture the Model T transformed American industry.

Edison was Ford's hero. Sixteen years Edison's junior, a young and ambitious Henry Ford admired Edison's imagination and seemingly limitless talent for inventing new technologies. As Ford struggled with designing an affordable, gas-powered car, Edison was inventing at an incredible pace. In 1896, Ford had the chance to meet Edison at a convention in New York. After seeing Ford's designs, Edison banged his fist on the table and told Ford to "Keep at it!" Ford later said of the fortuitous meeting, "That bang on the table was worth worlds to me! No man up to then had given me any encouragement … and out of the clear blue sky the greatest inventive genius of the world had given me a complete approval." (p. 40 of *The Inventor's Secret*) Ford went on to experience great success with his Model T and Ford Motor Company.

Edison and Ford eventually became good friends. They even went on camping trips together and discussed their ideas. Ford eventually built a house right next door to Edison's winter home in Fort Meyers, Florida. The two friends installed a gate between the two houses so that they could visit each other. The gate was later nicknamed the *friendship gate*.

The story of *The Inventor's Secret* is used to engage students in a design challenge that supports the "Keep at it!" theme as they build toy cars out of everyday materials. This lesson features a version of the design process used by inventors and engineers. Several versions of this process exist, but they all have the same basic components. The version used by the PBS Kids television show *Design Squad Global* emphasizes the iterative component of the design process: to keep building, testing, evaluating, and redesigning until you are satisfied with the solution (in other words, "Keep at it!"). See the "Websites" section for more information about *Design Squad Global*. Both of the picture books featured in this lesson highlight the Build, Test & Evaluate, Redesign Cycle in this model.

A Framework for K–12 Science Education represents the cycle in ETS1.C: Optimizing the Design Solution (NRC 2012). This engineering design disciplinary core idea states that, by the end of grade 5, students should understand that "different solutions need to be tested in order to determine which of them best solves the problem, given the criteria and the constraints" (NRC 2012, p. 209). The lesson in this chapter also addresses the practices of planning and carrying out investigations and constructing explanations and designing solutions, as well as the crosscutting concept of structure and function.

Note: The balloon car design challenge introduced in the explore phase of this lesson was chosen because Henry Ford designed automobiles. However, this lesson could be used to frame other design challenges.

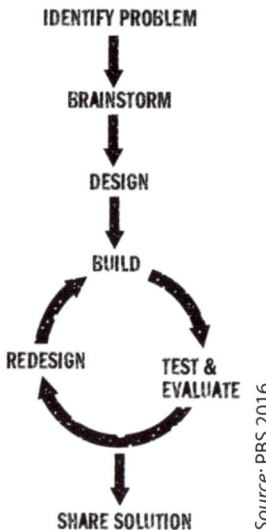

engage

Rosie Revere, Engineer Read-Aloud

Connecting to the Common Core
Reading: Literature
Key Ideas and Details: 3.3, 4.3

 Questioning

Show students the cover of *Rosie Revere, Engineer*. Introduce the author, Andrea Beaty, and the illustrator, David Roberts. Students may recognize this author–illustrator team from the book *Iggy Peck, Architect* or *Ada Twist, Scientist*. Open the book so students can see the picture on the front and back covers. *Ask*

? What do you think this book is about?

As you read, stop periodically to ask the following questions:

? Page 7—Why do you think Rosie hides her inventions? (Answers will vary.)

? Page 13—Why *did* she hide her inventions? (She was laughed at when they didn't work. She was embarrassed.)

? Page 32 (show the illustration of the flying cheese-copter on the copyright page)—Was Rosie's invention ultimately a success? (yes) Why? (It flew!)

After reading, *ask*

? What happened that encouraged Rosie to not be ashamed of her invention attempts? (Her great-great-aunt Rose visited her and told her that it was the perfect first try.)

? How is Rosie an engineer? (Students should recognize that Rosie takes things apart, builds things to solve problems, and tests her inventions. All of these are things that engineers do.)

? What would you say is the moral of the story? (Never give up. True failure comes only if you quit.)

> Connecting to the Common Core
> **Reading: Literature**
> KEY IDEAS AND DETAILS: 3.2, 4.2, 5.2

After reading, students may be interested in viewing a 6 min. video that features the author of the book discussing what inspired her to write the book and how science and art are related (see "Websites" section).

Next, open the book to pages 6–7, which show Rosie's attic. *Ask*

? Where does Rosie get all of the materials she uses to create new gadgets and gizmos? (They should recognize from the illustrations that Rosie gets her parts from machines, appliances, and discarded toys.)

explore

Balloon Car Challenge, Part 1: Identify Problem, Brainstorm, and Design

Tell students that, as did Rosie, they are going to make a vehicle out of some everyday materials. Give students the Balloon Car Design Challenge student page and present the problem, or design challenge, to students.

Identify Problem

Explain that inventors and engineers always have desired features or outcomes in mind when designing solutions to problems or design challenges, whether those solutions are projects, products, systems, or technologies. These desired features and outcomes are known as *criteria*. The criteria for this design challenge are as follows:

- Must be powered by a balloon
- Must travel in a straight line
- Must travel at least 30 cm on one full balloon of air

Explain that inventors and engineers also have to work within constraints when designing solutions. *Constraints* are typically limits on time, materials, and money. The constraints for the Balloon Car Design Challenge are as follows:

- Materials: You may use only materials provided by or approved by your teacher.
- Time: You must build your balloon car within the time limit set by your teacher.

Brainstorm

Show students the materials they can use. Have them brainstorm how they could assemble the materials to make a car. Be sure to explain that there is no perfect design or one "right" way of making the car. Different solutions can be made to solve the same problem. The best designs meet the criteria within the given constraints. Stimulate creative thinking by asking questions such as the following:

? What are the choices to use for the axles?
? What do you think would work best for the wheels?
? How could you attach the wheels to the axles?
? What could you use to make the body of the car?
? How could you attach the balloon to the car?

PBS *Design Squad Global* Video

After students have brainstormed several ideas, show them the "4-Wheel Balloon Car" video from the PBS Show *Design Squad Global* for more ideas (see "Websites" section). This website also features step-by-step instructions for making the cars.

Design

Have students sketch their balloon car idea. Next, have them write how they will test their car to see if it meets the criteria. When students have completed the student page, sign off at the teacher checkpoint. Tell students that they will be able to build and test their car during the next class period.

Chapter 6

BUILDING A BALLOON CAR

Balloon Car Challenge, Part 2: Build and Test & Evaluate

Remind students of the criteria and constraints defined in the challenge. Show them where to find their supplies. Set a realistic time limit appropriate for the age of your students. Tell students that they may test their cars during that time, but they will need to stop as soon as they hear the timer or hear you say, "Stop!" Give students updates on the time they have remaining as they build and test. When time is up, have students stop and put their cars aside. Tell them that they will have a chance to revise their original designs during another class period. Then, *ask*

? Did your car meet the criteria (powered by a balloon, travels in a straight line, and moves at least 30 cm on one full balloon of air)?

? Did you stay within the constraints of the challenge (built with approved materials and built within the time limit)?

? What is something that didn't work?

? What is something that worked well?

? What ideas do you have for improving your car?

? Did you ever feel like giving up?

? What would you say is the "secret" to inventing or engineering?

explain

The Inventor's Secret Read-Aloud

 Questioning

Tell students that you have a book that might help them with their car design process. Show students the cover of *The Inventor's Secret,* and introduce the author and illustrator. Share the subtitle, *What Thomas Edison Told Henry Ford,* and *ask*

? Who was Thomas Edison? (Answers will vary, but students will likely know Edison invented the light bulb.)

? Who was Henry Ford? (Answers will vary, but students will likely associate Ford with cars.)

? Did Thomas Edison and Henry Ford know each other? (Answers will vary.)

? What do you think the "inventor's secret" might be? (Answers will vary.)

> **Connecting to the Common Core**
> **Reading: Informational Text**
> KEY IDEAS AND DETAILS: 3.1, 4.1, 5.1

 Determining Importance

Tell students that, as you read, you would like them to listen for the "inventor's secret" and signal (raise their hands) when they hear it.

National Science Teachers Association

 Questioning

Read the book aloud, stopping to ask the following questions (you may want to write the questions on sticky notes and place them within the book as reminders for you before reading):

? Page 8—Who was older, Thomas Edison or Henry Ford? (Edison) By how much? (16 years)

? Page 9—What did young Thomas and young Henry have in common? (They both wondered how things work, they built things, and sometimes they got into trouble for building things.)

? Page 24—What criteria (desired features or outcomes) did Henry Ford want his car to meet? (To run on gas, to be easy to drive, to be big enough for families, and to be affordable)

? Page 25—How do you think Henry Ford was feeling at this point in the story? (discouraged, jealous, frustrated) What makes you think that? (The book says he "couldn't stand it any longer," and he looks discouraged in the illustrations.)

? Page 29—Why do you think Thomas Edison banged his fist on the table? (to emphasize his point that Ford should "Keep at it!")

? Page 37—Why did Henry Ford have so many different Models—Model A, Model C, Model F, Model K, and so on? (Each one failed to meet Henry Ford's criteria for success, so he kept improving on them. He used the alphabet for the order of his revised prototypes.)

? Page 37—Which model was the one that finally met the criteria? (the Model T)

? Page 39—So what is the inventor's secret? ("Keep at it!")

Tell students that this story is based on true events. Read the quote from Henry Ford on page 40:

> "That bang on the table was worth worlds to me! No man up to then had given me any encouragement...and out of a clear sky the greatest inventive genius in the world have given me complete approval."—Henry Ford after meeting Thomas Edison

Ask

? Why do you think that moment was so important to Henry Ford? (He was feeling discouraged and needed encouragement to carry on his work. Thomas Edison was someone he admired.)

Read the information on pages 40–41 about Edison and Ford's friendship, and read the author's and illustrator's notes, which give the reader insight on their purpose and background.

The Design Process

Explain to students that inventors and engineers (like Thomas Edison and Henry Ford) use the Design Process when approaching a problem. There are several versions of the design process, but they all have the same basic ideas. Make a poster-sized version of The Design Process (see p. 74), which features the model used on the PBS show *Design Squad Global*. Using the poster and Table 6.1 (p. 68) as a guide, discuss each step of the process and what it means. Then, ask students to find evidence in the text that shows how Ford used each step of the design process while working on his car. Refer to the book when giving examples (page numbers are listed).

Ask

? What part of the design process do you think was highlighted in this book? (the Build, Test & Evaluate, Redesign Cycle)

? What evidence in the text makes you think so? (The author spends a lot of time writing about Ford's different attempts, tests, and redesigns.)

Explain that the Build, Test & Evaluate, Redesign Cycle can go on and on for a long time. *Ask*

? How long after Edison and Ford's meeting (depicted in the book) do you think it was before Ford introduced the Model T? (Students can use the time line in the back of the book to find the dates of these two events. The meeting

Picture-Perfect STEM Lessons, 3–5

Chapter 6

Table 6.1. Design Process Discussion Guide

Step	What It Means	How Henry Ford Used It

Identify a Problem	Find a problem to solve and identify the criteria for success.	• (p. 14) Ford's problem was to invent a new kind of vehicle that did not need a horse to pull it.
		• (p. 24) Ford identified the following criteria for success for his car:
		  • Runs on gas
		  • Easy to drive
		  • Big enough for families
		  • Affordable

Brainstorm	Come up with lots of ideas that might work.	• (pp. 18–23) Ford had many ideas—using a steam engine, hooking up a homemade engine to a mower, using a two-cylinder gas engine, and using a four-cylinder gas engine.
Design	Make drawings, sketches, or plans for creating the solution.	• (p. 28) Ford shared a sketch of his four-cylinder engine with Edison. • (pp. 38–39) Some of Ford's and Edison's sketches for patents are shown.
Build, Test & Evaluate, Redesign	Build the solution, see if it works, make changes, test it again, and so on. This step is a cycle that can continue until the solution meets the criteria.	• (pp. 22–37) Throughout the book, Ford was building vehicles, testing them, evaluating them based on his criteria, redesigning, building again, testing, and so on.
Share Solution	Tell others about your solution.	Ford introduced the Model T (Tin Lizzie) to the world in 1908 and eventually sold millions of them.

was in 1896, and the Model T was released in 1908. That is 12 years!)

? In those 12 years, do you think Henry Ford had moments when he felt like giving up? (yes)

 Making Connections: Text to Text

? Think back to *Rosie Revere, Engineer*. Where was the Build, Test & Evaluate, Redesign Cycle evident in the story? (She built the cheese-copter, she tested it, it crashed, she thought it was a failure, she worked the rest of the day redesigning it, and finally it flew.)

? Do you think other inventors and engineers ever feel like giving up? (yes)

elaborate

Edison Quotes

Many would consider Edison the greatest inventor (or engineer) of all time. His words inspired Ford not to give up, and his words have also encouraged many others as well. People all over the world have long been inspired by Edison's ingenuity, work ethic, appreciation of his team, and, of course, perseverance.

Form student groups of three or four and give each group a strip of paper with one of the following quotes from Edison:

- *Genius is 1% inspiration and 99% perspiration.*
- *I have not failed. I have just found 10,000 ways that do not work.*
- *Our greatest weakness lies in giving up. The most certain way to succeed is to try just one more time.*
- *Just because something doesn't do what you planned it to do, doesn't mean it's useless.*
- *I never did a day's work in my life. It was all fun.*
- *I have friends in overalls that I would not swap for the favor of the kings of the world.*
- *To invent, you need a good imagination and a pile of junk.*
- *If we did all the things we were capable of, we would literally astound ourselves.*

Ask each group to read their quote, look up any words they don't know, discuss the meaning with their group members, and consider what the quote tells us about Edison as a person and an inventor. Then, have each group share their Edison quote and interpretation with the rest of the class. After everyone is done sharing, *ask*

? What do these quotes tell us about Thomas Edison? (He didn't give up, he did not let failure make him quit, he loved the work he did, he valued his team, etc.)

? What does your group's quote tell you about being an inventor or engineer? (Answers will vary.)

 Making Connections: Text to Text

Connecting to the Common Core
Reading: Informational Text
INTEGRATION OF KNOWLEDGE AND IDEAS: 3.9, 4.9, 5.9

Next, *ask*

? What do the two books we read in this lesson, *Rosie Revere, Engineer* and *The Inventor's Secret*, have in common? (They both have the same message: Don't give up.)

? How are the two books different? (*Rosie Revere, Engineer* is fiction and *The Inventor's Secret* is nonfiction.)

Refer to the part of the story where Rosie feels discouraged because her flying machine didn't work (pp. 18–21), but her great-great-aunt Rose tells her, "Your brilliant first flop was a raging success! Come on, let's get busy and on to the next!" Explain that an important part of engineering is dealing with failures and learning from them. Then, read the last line of page 27, "Life might have its failures, but this was not it. The only true failure can come if you quit." Explain that engineers often try many designs before they find one that works best and that to be a good engineer you must persevere through failed attempts. Failures give engineers a chance to go back and improve on their original idea until they solve the problem.

evaluate

Balloon Car Challenge, Part 3: Redesign, Build, Test & Evaluate, and Share Solution

Tell students that you would like them to apply the design process they learned about in the explain phase of this lesson to their balloon car design challenge. Refer to the poster, point out each step, and ask students how they have addressed each step so far. Here are the answers:

- Identify Problem: The problem was provided for us—design a balloon car.
- Brainstorm: We brainstormed as a class and watched a video for more ideas.
- Design: We sketched our ideas.
- Build: We built our cars with the supplies.
- Test & Evaluate: We tested them to see if they met the criteria in the design challenge.

Tell students that now it is time to complete the remaining steps of the process—Redesign and Share Solution. *Ask*

? Did your first car meet the criteria for success presented in the design challenge?

Chapter 6

TESTING BALLOON CARS

> **?** Can you think of ways to make it go farther on one balloonful of air?
>
> **?** Can you think of any other ways to improve it?

Tell students that you would like them to think of the first version of the car they built as their "Model A." Encourage students to redesign, rebuild, and test their car to make it work better. Point out on the poster that Redesign, Build, and Test & Evaluate are parts of a cycle and they will need to repeat all of those steps until they have a prototype they are satisfied with and ready to share. Explain that a *prototype* is a working model of a product, usually the last model before the design goes into production. Explain that it is important to keep track of each change they make and the affect the change has on the motion of the car. Give each student a copy of the Redesign, Build, Test & Evaluate student page. On the table, they can keep track of how each change affected the motion of the design. Tell students it is important to test one change at a time to know how that change affected the motion of the car. If they change more than one thing at time, then it will be difficult to determine what change made the difference. Remind students of the inventor's secret—"Keep at it!" You may even want to create a sign to hang in the classroom with Edison's encouraging words on it.

> **Connecting to the Common Core**
> **Reading: Informational Text**
> INTEGRATION OF KNOWLEDGE AND IDEAS: 3.9, 4.9, 5.9

Finally, when students are satisfied with their designs, have them share their balloon car prototype with the class by demonstrating how it works and then displaying it. Create display cards by copying the Display Card student page on cardstock. Have students fill out the information on the card (name, model letter, distance traveled, criteria met and constraints taken into account checked off) and fold it in half to create a table tent to put next to their models.

STEM at Home

Have students complete the "I learned that …" and "My favorite part of the lesson was …" portions of the STEM at Home student page as a reflection on their learning. They may choose to do the following at-home activity with an adult helper and share their results with the class. If students do not have access to the internet or these materials at home, you may choose to have them complete this activity at school.

"At home, we can watch a short video on the *Design Squad Global* website from PBS Kids."

Visit http://pbskids.org/designsquad/video *and choose a design challenge video to watch.*

"After we watch the video, we can discuss these questions:

1. What problem was the design squad trying to solve?
2. What parts of the design process did you observe in the video?
3. Which idea do you think would work best?"

For Further Exploration

This section is provided to help you encourage your students to use the science and engineering practices in a more student-directed format. This box lists questions and challenges related to the lesson that students may select to research, investigate, or innovate. Students may also use the questions as examples to help them generate their own questions. After selecting one of the questions in the box or formulating their own questions, students can individually or collaboratively make predictions, design investigations or surveys to test their predictions, collect evidence, devise explanations, design solutions, or examine related resources. They can communicate their findings through a science notebook, at a poster session or gallery walk, or by producing a media project.

Research

Have students brainstorm researchable questions:

? How many patents does Thomas Edison have? What are some of his lesser known inventions?

? How did Henry Ford's assembly line change American industry?

? What kinds of engineers are involved in car design?

Investigate

Have students brainstorm testable questions to be solved through science or math:

? Which balloon car in your class goes the farthest?

? Which balloon car in your class goes the fastest?

? What do your classmates think the best invention of all time is? Take a survey! Graph the results, then analyze your graph. What can you conclude?

Innovate

Have students brainstorm problems to be solved through engineering:

? Can you design a way to mass produce your best model car?

? Can you invent something to solve an everyday problem?

? Can you design a reminder (poster, bookmark, etc.) for your classmates to "Keep at it!" when they get discouraged with a task or challenge?

Picture-Perfect STEM Lessons, 3–5

Reference

National Research Council (NRC). 2012. *A framework for K–12 science education: Practices, crosscutting concepts, and core ideas.* Washington, DC: National Academies Press.

Websites

Design Squad Global
   http://pbskids.org/designsquad

Design Squad Global "4-Wheel Balloon Car" (video)
   http://pbskids.org/designsquad/build/4-wheel-balloon-car

Rosie Revere, Engineer: Engineering and Discovery With Author Andrea Beaty (video)
   www.youtube.com/watch?v=EymQZsv9Me8

More Books to Read

Barretta, G. 2012. *Timeless Thomas: How Thomas Edison changed our lives.* New York: Henry Holt and Company.
   Summary: This clever book shows modern-day devices that had their beginnings in Edison's lab. Colorful, and at times humorous, illustrations depict Edison and his team of employees working in the lab, while the opposite side of each page shows present-day versions of his inventions. End matter includes a time line of Edison's most famous inventions and short bios of some of his employees.

Brown, D. 2010. *A wizard from the start: The incredible boyhood and amazing inventions of Thomas Edison.* New York: Houghton Mifflin.
   Summary: This "storyography" depicts Thomas Edison's life from his humble beginnings as a farmer's son selling newspapers on trains and reading through public libraries shelf by shelf, to his inventing career, to eventually his becoming a world-renowned legend.

Editors at Time for Kids. 2005. *Thomas Edison: A brilliant inventor.* Time for Kids Biographies. New York: HarperCollins.
   Summary: This book gives readers an up-close look at the life and work of Thomas Edison and is illustrated with historical and contemporary photographs.

Editors at TIME for Kids. 2008. *Henry Ford: Putting the world on wheels.* Time for Kids Biographies. New York: HarperColins.
   Summary: This book gives readers an up-close look at the life and work of Henry Ford and is illustrated with both historical and contemporary photographs.

Mortensen, L. 2007. *Thomas Edison: Inventor, scientist, and genius.* Minneapolis, MN: Picture Window Books.
   Summary: Simple text and whimsical illustrations depict the life of Thomas Edison from his childhood antics to the inventions that changed the world.

PBS. 2016. Design squad global. Parents and educators: Online workshop. *http://pbskids.org/designsquad/parentseducators/workshop/process.html.*

Spires, A. 2014. *The most magnificent thing.* Tonawanda, NY: Kids Can Press.
   Summary: One day, a little girl decides she is going to make the most magnificent thing! She knows just how it will look. She knows just how it will work. But making the most magnificent thing is harder than she thinks.

Name: _____

Balloon Car Design Challenge

Problem: Design a balloon-powered car.

Criteria (desired features or outcomes):

1. Must be powered by a balloon
2. Must travel in a straight line
3. Must travel at least 30 cm on one full balloon of air

Constraints (limits on available resources and time):

1. Materials: You may use only materials provided by or approved by your teacher.
2. Time: You must build your balloon car within the time limit set by your teacher.

Sketch your ideas.

How will you test your design? _____
_____
_____

Teacher Checkpoint: ☐

The Design Process

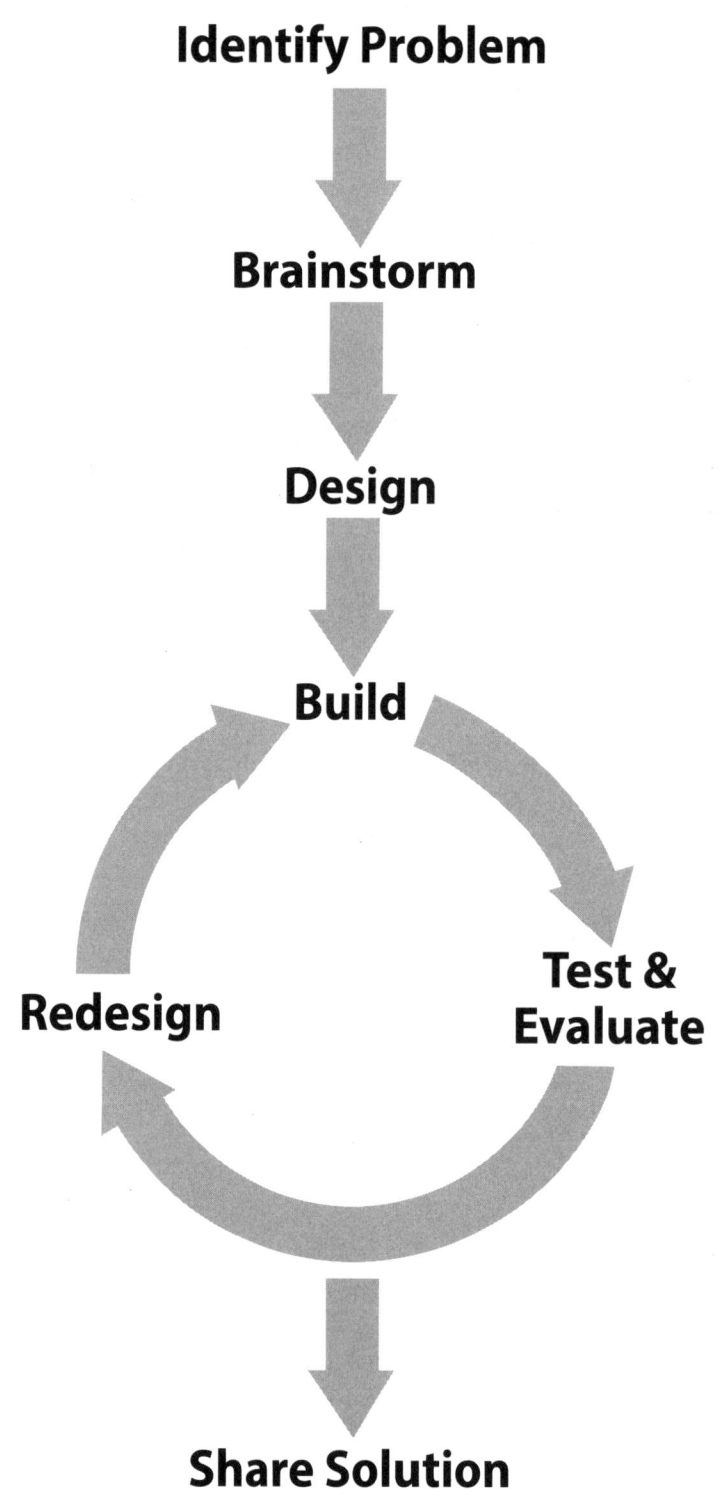

Edison Quotes

Genius is 1% inspiration and 99% perspiration. —Thomas Edison
I have not failed. I have just found 10,000 ways that do not work. —Thomas Edison
Our greatest weakness lies in giving up. The most certain way to succeed is to try just one more time. —Thomas Edison
Just because something doesn't do what you planned it to do, doesn't mean it's useless. —Thomas Edison
I never did a day's work in my life. It was all fun. —Thomas Edison
I have friends in overalls that I would not swap for the favor of the kings of the world. —Thomas Edison
To invent, you need a good imagination and a pile of junk. —Thomas Edison
If we did all the things we were capable of, we would literally astound ourselves. —Thomas Edison

Name: _____

Chapter 6

Redesign, Build, Test & Evaluate

Use the Redesign, Build, Test & Evaluate Cycle to improve your car. Record each change you make in the table below and then record the distance each model travels and any observations about its motion.

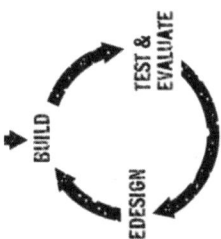

Model	Change to Design (Example: Changed the wheels to a different material)	Distance Traveled (cm)	Observations
A	No change, original design		

Name : _____

- -

Prototype Display Card

Engineer _____

Model _____ Distance Traveled _____

Criteria Met (✓)

☐ Powered by a balloon

☐ Moves in a straight line

☐ Travels a distance of at least 30 cm on one full balloon of air

Within Constraints (✓)

☐ Materials: Built with only approved materials

☐ Time: Built within the time limit

Name: _____

STEM at Home

Dear _____,

At school, we have been learning about the **design process.**

I learned that: _____
_____

My favorite part of the lesson was: _____
_____
_____

At home, we can watch a short video on the *Design Squad Global* website from PBS Kids.

Visit *http://pbskids.org/designsquad/video* and choose a Design Challenge Video to watch.

THE DESIGN PROCESS

IDENTIFY PROBLEM → BRAINSTORM → DESIGN → BUILD → TEST & EVALUATE → REDESIGN → SHARE SOLUTION

After we watch the video, we can discuss these questions:

What problem was the design squad trying to solve?
_____
_____

What parts of the design process did you observe in the video?
_____
_____

Which idea do you think would work best? _____
_____
_____

National Science Teachers Association

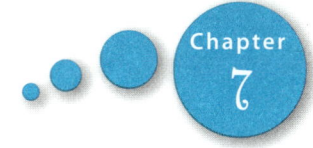

Mesmerized

Description

After hearing the true story of how Ben Franklin challenged Franz Mesmer's "magic force" with a blind test, students take part in a blind taste test of two different beverages. They learn that even though advertisers try to "mesmerize" us with their ads, they must have evidence to support the claims they make. Then, students design and perform their own product comparison tests and share their results at a Consumer Fair.

Suggested Grade Levels: 3–5

LESSON OBJECTIVES Connecting to the *Framework*		
Science and Engineering Practices	**Disciplinary Core Ideas**	**Crosscutting Concept**
Asking Questions and Defining Problems Planning and Carrying Out Investigations Engaging in Argument From Evidence	**ETS1.C:** Optimizing the Design Solution **EST2.B:** Influence of Engineering, Technology, and Science on Society and the Natural World	Cause and Effect

Featured Picture Books

TITLE: **Mesmerized: How Ben Franklin Solved a Mystery That Baffled All of France**
AUTHOR: **Mara Rockliff**
ILLUSTRATOR: **Iacopo Bruno**
PUBLISHER: **Candlewick**
YEAR: **2015**
GENRE: **Story**
SUMMARY: *This peek into a fascinating moment in history shows the early development and practice of the scientific method and how one of American's most beloved founding fathers boldly challenged superstition with applied common sense.*

TITLE: **Let's Think About the Power of Advertising**
AUTHOR: **Elizabeth Raum**
PUBLISHER: **Heinemann**
YEAR: **2015**
GENRE: **Non-Narrative Information**
SUMMARY: *From the Let's Think About series, this book looks at the power of advertising—how it works, the pros and cons, the impact of consumerism, and the effect of advertising on our daily lives.*

Time Needed

This lesson will take several class periods. Suggested scheduling is as follows:

Day 1: **Engage** with *Mesmerized* Read-Aloud and **Explore** with Blind Taste Test

Day 2: **Explore** with Looking at Ads and **Explain** with *Let's Think About the Power of Advertising* Read-Aloud and "Food Ad Tricks" Video

Day 3: **Explain** with Don't Be Fooled Article

Day 4: **Elaborate** with Detergent Comparison Test Design

Day 5 and Beyond: **Evaluate** with Product Testing and Consumer Fair

Materials

For the Blind Taste Test (per class)

- Two 2-liter bottles of drinks (such as non-caffeinated soda, punch, or juice), one a popular name brand and the other a generic or store brand—same flavor, same color, and same temperature
- 2 paper bags (large enough to cover the 2-liter bottles)
- Small plastic cups or condiment containers (2 per student)
- Permanent marker

SAFETY
This activity should not be done in a science laboratory where hazardous chemicals have been used.

For Looking at Ads (per group of 4 students)

- 4 print advertisements from newspapers, magazines, circulars, and so on (pre-selected to ensure they are age-appropriate!)

For Product Testing

- Posterboard, tri-fold display board, or large file folder for the display
- Markers

  Note: Students will provide their own products and supplies for the product testing.

Student Pages

- Looking at Ads
- Don't Be Fooled
- Detergent Comparison Test Design
- Product Testing
- Consumer Fair Rubric
- STEM at Home

Chapter 7

Background for Teachers

We are bombarded with advertisements each and every day—from the billboards we drive past, to the commercials we hear on the radio, to the ads that pop up on our computers, to the product placements in our favorite TV shows. Advertisements claim their products and services will make us happier, healthier, solve a problem, or make life easier. But are these ads true? The Federal Trade Commission states that "when consumers see or hear an advertisement, whether it's on the Internet, radio or television, or anywhere else, federal law says that ad must be truthful, not misleading, and, when appropriate, backed by scientific evidence. The Federal Trade Commission enforces these truth-in-advertising laws, and it applies the same standards no matter where an ad appears—in newspapers and magazines, online, in the mail, or on billboards or buses." (Federal Trade Commission, "Truth in Advertising").

When companies make claims about the products they sell, they must support them with scientific evidence. They begin with a *claim* stating what their product can do or stating how their product outperforms other products. Companies typically conduct two different types of tests depending on the type of data they need to collect: *product claim tests* to gather evidence to support their claims or *product comparison tests* to determine how their product compares with other brands. The tests they perform must be *reproducible* by other researchers. In product comparison tests, variables must be controlled. In this lesson, students learn that the *experimental variable,* the one that is being tested, is the only variable that can change in a *fair test.* The *controlled variables* are the ones that must be kept the same. For example, when testing to see if one brand of detergent works better than another, the experimental variable would be the brand of detergent. The controlled variables would be the amount of detergent, the temperature of the water, the washing machine cycle, the same types and color of clothes, and so on. Companies run these kinds of studies often so they can make claims about their products that are based on evidence.

This lesson begins with an example of how the scientific method can be used to challenge false advertising. In the fascinating true story *Mesmerized: How Ben Franklin Solved a Mystery That Baffled All of France,* students learn how Benjamin Franklin used the scientific method to prove that the mysterious Dr. Franz Mesmer was making false claims about his "magic force" that could cure multiple ailments. Dr. Mesmer was actually putting his subjects in a hypnotic state, which is where the word *mesmerized* comes from. Through this read aloud, students learn how a version of the scientific method can be used to invalidate a claim. This idea is then applied to product testing. Students learn how to conduct a fair test and then design their own product comparison test.

Through these experiences, students are involved in the scientific practices of asking questions and defining problems, planning and carrying out investigations, and engaging in argument from evidence, as well as the crosscutting concept of testing cause-and-effect relationships.

engage

Mesmerized Read-Aloud

Connecting to the Common Core
Reading: Literature
KEY IDEAS AND DETAILS: 3.1, 4.1, 5.1

Questioning

Show students the cover of *Mesmerized* and introduce the author, Mara Rockliff, and the illustrator, Iacopo Bruno. *Ask*

? What does the word *mesmerized* mean? (to be so fascinated by something that you can't look away or notice anything else around you)

Picture-Perfect STEM Lessons, 3–5

Chapter 7

Read the subtitle, *How Ben Franklin Solved a Mystery That Baffled All of France,* and *ask*

? This book is about Benjamin Franklin, also called Ben Franklin. Who was Ben Franklin? What was he known for? (Answers will vary, but students may recognize him as a founding father of the United States, an inventor, or the face on the $100 bill.)

Read the book aloud, stopping at the following pages to *ask*

? Page 10—Why were the people of Paris "absolutely gaga" about Ben Franklin? (Everyone had heard about his famous kite experiment. People in Paris were excited about new discoveries in science at the time.)

? Page 13—How were Dr. Mesmer and Ben Franklin different? (Dr. Mesmer was fancy and dramatic. Ben Franklin was plain and simple.)

? Page 16—How much was Dr. Mesmer charging for his treatments? (100 gold louis; a louis was a coin used in France at the time)

? Page 17—Do you think Dr. Mesmer was really curing people? (Answers will vary.)

? Page 21—What was Ben Franklin's hypothesis? (that what the patients felt was caused by their own minds, not by an invisible force as Dr. Mesmer claimed)

? Page 21—How do you think Ben Franklin is going to test this hypothesis? (Answers will vary.)

? Page 27—How did Ben Franklin test his hypothesis? (He blindfolded the patients.)

? Page 29—Why do you think Ben Franklin tested his hypothesis on so many different patients? (so he could compare the results of each test to see if they were consistent)

? Page 39—Why did some of Dr. Mesmer's patients feel better after his "treatments"? (They felt better because of the Placebo Effect; they believed they would feel better, so they did.)

? Page 41—How did Ben Franklin use the scientific method to find the truth? (Go back through the book and find the insets that describe how Ben Franklin used the scientific method:

? Page 19—He **observed** the difference between the patients' reactions and his own.

? Page 21—He **hypothesized** that what the patients felt was caused by their own minds, not by an invisible force.

? Page 27—He **tested** his hypothesis by blindfolding the patients.

? Page 29—The test results **supported** his hypothesis that the so-called force did not exist.)

Tell students that real scientists rarely use a single "scientific method." In fact, the so-called scientific method has many variations and may include different steps. However, these various methods all have one thing in common—the idea that when searching for an answer, you should formulate a testable question and then collect evidence to find the answer. This evidence might be in the form of observations (like Ben Franklin's) or numerical data such as measurements or counts. It is important to note that not every investigation requires a hypothesis (i.e., a tentative explanation based on observations and research). Scientists and researchers sometimes make observations without a hypothesis in mind.

After reading, *ask*

? Why do you think so many people were fooled by Dr. Mesmer? (They wanted his claim to be true. He was dramatic, bold, and good at getting people to believe him.)

 Making Connections: Text to Self

Ask

? Have you ever been fooled by an advertisement that was too good to be true? (Answers will vary.)

Table 7.1. Blind Taste Test Preferences Chart

Drink	Predicted Preference (Number of Votes)	Actual Preference (Number of Votes)
Name brand		
Store brand		
No preference		

explore

Blind Taste Test

Make the chart in Table 7.1 on the board. Then, show students two 2-liter bottles of drinks, one a popular name-brand and the other a generic or store brand. They should have the original labels on them and be the same flavor, color, and temperature. Before doing the "blind" taste test, ask students to vote on which one they think will taste the best. They can vote for the name brand or store brand, or they can indicate that they have no preference. Write the totals of their votes on the board below the words "Predicted Preference."

Next, out of view of the class, place the two bottles in paper bags. Use a marker to label one bag "Drink X" and the other "Drink Y." Tell students that because they no longer know which drink is which, this is considered a *blind test*. Although they are not actually blindfolded as in the book *Mesmerized*, they are not able to see which drink they taste.

Give each student a sample of Drink X and Drink Y in small disposable cups or condiment containers labeled *X* and *Y*. Then, have students vote for the one they like best (or indicate no preference). Write the total number of votes for each on two different sticky notes so that students can't see them. Then, dramatically reveal which drink was which by writing *X* or *Y* next to "Name Brand" and "Store Brand" and placing the corresponding sticky notes with the vote tallies on the chart. Compare the number of votes before and after tasting, and have students discuss possible reasons for the outcome. Then, *ask*

? How many of you correctly predicted your preferred drink?

? How many of you did not?

? How many of you were surprised by the results?

Tell students that you followed the same basic procedure that Ben Franklin followed in the book *Mesmerized*:

- Observation: I **observed** that there are lots of advertisements for the name brand drink.
- Hypothesis: I **hypothesized** that students would think the name brand drink would taste the best because of its advertising.
- Test: I **tested** to see which one they prefer before and after tasting and recorded the votes.
- Results and Conclusions: I tallied the votes and compared the totals to see if my hypothesis was supported or not supported by the evidence.

 Making Connections: Text to Self

Ask

? Have you ever seen the advertisements for Drink X?

Picture-Perfect STEM Lessons, 3–5

Chapter 7

BLIND TASTE TEST

LOOKING AT ADS

? Have you ever seen the advertisements for Drink Y?
? Have you ever been "mesmerized" by an advertisement?
? Do you think advertisements influenced your preference?

explore

Looking at Ads

 Making Connections

Ask

? Do you have a favorite commercial?
? Are there any jingles or slogans that you know by heart?
? Have you ever been persuaded by a commercial, print advertisement, or online ad to buy something?

Give each student a Looking at Ads student page and each group of four students four different print advertisements (pre-selected from a variety of magazines, newspapers, mailings, etc.). Have students complete Part 1 by reading through each of their group's advertisements and recording the name of the product or service, what it claims to do or its slogan, what evidence is provided (if any) to support its claims, and whether the ad makes them want to buy the product or service.

explain

Let's Think About the Power of Advertising Read-Aloud

 **Questioning**

After students have had a chance to look through each of their group's four ads, have them compare their responses to the ads with each other. Then, *ask*

? What was similar about the ads you looked at?
? What was different about the ads you looked at?
? How many of the ads gave evidence to back up their claims?
? What is the purpose of advertising?
? Do you think companies spend more money on print ads like the ones you just looked at or other forms of advertising such as TV ads?

National Science Teachers Association

? What are some techniques that companies use in their advertisements to get you to want to buy their products or services?

? Is advertising important? Why or why not?

Chunking

Then, tell students you have a very interesting nonfiction book that can help answer those questions. Show students the cover of *Let's Think About the Power of Advertising*. Because the book is nonfiction and has some information that doesn't pertain to the questions, you can skip some pages and "chunk" the information into manageable bits as you read. You may want to use sticky notes in advance to mark the pages you will be reading. Read aloud the following pages, stopping to ask questions or discuss the reading as noted.

Questioning

After reading pages 4–9, *ask*

? What is the purpose of advertising? (It is a form of communication used to inform people about a particular event, idea, product, or service.)

? Is all advertising designed to sell a product or service? (No, politicians, causes, and charities, etc. also advertise.)

? On what kind of ads do companies today spend the most money? (TV ads)

? Where have you noticed *ambient* ads, like the ones on billboards, buses, and so on? For example, have you ever seen the Wienermobile parked somewhere? (Answers will vary.)

Turn and Talk

After reading pages 10–11, have students turn to a partner and discuss the "What Do You Think?" inset about product placement.

Questioning

After reading pages 12–17, *ask*

? So is advertising important? (yes) Why? (Advertising helps businesses make their products known and compete for customers, and advertising helps us decide what to buy when there are so many choices to make.)

? Are there any products you can think of that you associate with certain celebrities? (Answers will vary.)

After reading pages 18–21 and pages 32–33, *ask*

? What are some techniques that companies use in their advertisements to persuade you to want to buy their products or services? (creating positive feelings, using jingles or music, repetition, setting watches in ads at 10:10 to form a "smiling face," using facts and figures, using "weasel words," the bandwagon technique, "glittering generalities," "puffery," making food look good in photographs, etc.)

Next, have students complete Looking at Ads: Part 2 by recording examples of weasel words, celebrities, glittering generalities, and puffery from the four ads they looked at earlier. Have students share examples with other groups.

"Food Ad Tricks" Video

Then, tell students that you have a very entertaining video to show them about a common advertising technique. Show the "Food Ad Tricks: Helping Students Understand Food Advertisements" video (see "Website" section).

> **Connecting to the Common Core**
> **Reading: Informational Text**
> INTEGRATION OF KNOWLEDGE AND IDEAS: 3.9, 4.9, 5.9

Turn and Talk

After viewing, have students turn to a partner and discuss the following question:

? What new information did you learn about advertising techniques from the video that wasn't in the book? (There is a job in the advertising industry called a *food stylist*, which means the person is sort of a make-up artist for food. A food stylist has many techniques

Chapter 7

for making hamburgers and French fries look appealing in photographs, but you wouldn't want to eat that good-looking food!)

Don't Be Fooled Article

Revisit page 32 of *Let's Think About the Power of Advertising*, which says, "Making false claims or lying in ads is illegal. Companies cannot say that a medicine cures an illness unless there is scientific proof that it does so." Explain that this applies to not only companies that make medicine but also any company that makes a claim about a product. *Ask*

? How do you think companies test their products?

? Think back to the advertisements you observed. What kinds of claims did they make? How do you think they tested those claims?

 **Pairs Read**

> **Connecting to the Common Core**
> **Reading: Informational Text**
> Range of Reading and Level of Text Complexity 3.10, 4.10, 5.10

Tell students that you have an article that can help them understand more about how companies test their products. Give each student a copy of the Don't Be Fooled article and have students read it independently or as a pairs read. In a pairs read, one student reads a paragraph while the other listens and then makes comments ("I think …"), asks questions ("I wonder …"), or shares new learning ("I didn't know …"). Alternately, you may want to use this article as a homework assignment. After reading, have them answer the questions on the last page.

> **Connecting to the Common Core**
> **Reading: Informational Text**
> Key Ideas and Details: 3.1, 4.1, 5.1

Discuss the answers together:

1. What product or service is the Zip Ultra advertisement selling? (powdered laundry detergent)
2. Do you see any "weasel words" in the Zip Ultra advertisement that suggest a positive meaning without giving a guarantee that can be measured, such as "people say" or "studies show?" ("Most folks agree …")
3. What claim does the Zip Ultra advertisement make about how well the product works? (Zip Ultra gently cleans away tough stains better than the leading brand of powdered laundry detergent … without scrubbing!)
4. What do companies need to have in order to make a claim in an advertisement? (evidence in the form of observations or measurements)
5. What would be the experimental variable in a product comparison test of Zip Ultra and another brand? (the brand of detergent)
6. What variables would need to be controlled, or kept the same, in a product comparison test of Zip Ultra and another brand? (the amount of detergent, the temperature of the water, the washing machine cycle, the same types and color of clothes, the same stains, etc.)

elaborate

Detergent Comparison Test Design

Next, tell students that they are going to be consumer product researchers, designing a way to test the claim made in the Zip Ultra ad. Pass out the Detergent Comparison Test Design and read the question "Which powdered laundry detergent cleans stains better without scrubbing, Zip Ultra or Skid-Daddle?"

Tell students that they must design a test that is *fair*—all of the variables except for brand of detergent are controlled, or kept the same—and *reproducible*, meaning another researcher could

National Science Teachers Association

follow their written procedure for the test. Have students work in pairs to come up with a test that is both fair and reproducible. For example, you might take two of the same size square of cotton, rub the same amount of chocolate syrup or other staining substance on them with a cotton swab, let the stains set for the same amount of time, add the same amount of water to the same amount of each detergent to make a solution, drop the same amount of detergent solution on the stains with an eyedropper, let each detergent work for 10 min., then rinse each cotton square 10 times in clean water. To determine which removed more stains, you might observe the difference between the cleaned cotton squares and an unstained cotton square. In this example, the experimental variable is the type of detergent. The controlled variables include the following:

- Type of fabric
- Type of stain
- Time the stain sets
- Amount of detergent
- Amount of water
- Temperature of the water
- Time the detergent is on the fabric

Students could also design a test that involves a variety of stains or a variety of clothing or design one that involves using a washing machine. The most important thing is that students understand that the test must be fair (the variables controlled) and the procedure written so specifically that another researcher could reproduce the test. After students have come up with a test that they think is fair and reproducible, have them share their procedure with another pair who can then critique it. For fun, you can have students create an advertisement for the "leading brand,"—Skid-Daddle Detergent—on the back of the student page.

evaluate

Product Testing

Tell students that they are going to be consumer product researchers again, but this time they will get to choose which products to test, and they will actually be testing them! You may decide to have students work individually or in teams. Give students a copy of the Product Testing student page. First, they must choose the products they will compare and then come up with a testable question. Some sample product comparison tests are as follows:

- Does a name brand diaper work as well as a store brand diaper?
- Which brand of paper towels are the strongest?
- Which brand of marker draws the longest line before drying out?
- Which brand of orange juice do people prefer?

After students have selected a testable question, they can identify the experimental variable and the controlled variables. Next, have students develop a procedure for their test on the student page. Refer to Ben Franklin's blind test in the book *Mesmerized*. Remind students that if they are testing people's reactions or preferences, it is best to do so without the people knowing which product they are using. When students have completed the Product Testing student page, check to make sure the question is testable, the variables are correctly identified, and the procedure is safe, reasonable, and reproducible. Then, sign the teacher checkpoint, and give students the Consumer Fair Rubric and a file folder or display board on which they can create a display to show their results. Students can perform their product tests at home. Allow students ample time to complete their tests and create their displays.

Consumer Fair

> Connecting to the Common Core
> **Writing**
> TEXT TYPES AND PURPOSES: 3.2, 4.2, 5.2

Chapter 7

Have students display their results in a Consumer Fair. You may want to set up the Consumer Fair as a gallery walk where students put their posters on display for their classmates to view. Students taking the gallery walk use sticky notes to post comments, questions, or praise on or near their classmates' posters. Encourage students to include any packaging or advertisements for their products along with their posters. You can use the Consumer Fair Rubric to score student displays and provide comments.

STEM at Home

Have students complete the "I learned that …" and "My favorite part of the lesson was …" portions of the STEM at Home student page as a reflection on their learning. They may choose to do the following at-home activity with an adult helper and share their results with the class. If students do not have access to the internet or these materials at home, you may choose to have them complete this activity at school.

"Reading product reviews is a good way to figure out if a product is worth buying. Product reviews are written by people who have purchased the product, not by advertisers. At home, we can look at product reviews on Amazon."

 *Go to Amazon's website at* www.amazon.com.

"With an adult helper, think of something you would like to buy (a toy, game, household item, etc.). Type the name of the item in the search box. Scroll down to read the customer reviews."

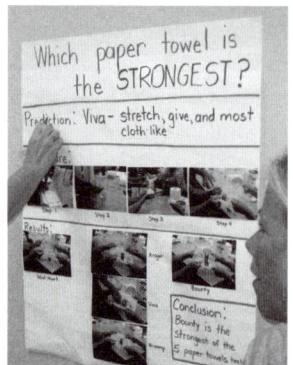 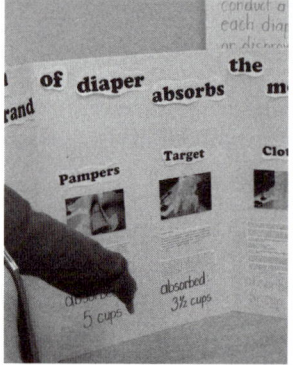

Consumer Fair

National Science Teachers Association

For Further Exploration

This section is provided to help you encourage your students to use the science and engineering practices in a more student-directed format. This box lists questions and challenges related to the lesson that students may select to research, investigate, or innovate. Students may also use the questions as examples to help them generate their own questions. After selecting one of the questions in the box or formulating their own questions, students can individually or collaboratively make predictions, design investigations or surveys to test their predictions, collect evidence, devise explanations, design solutions, or examine related resources. They can communicate their findings through a science notebook, at a poster session or gallery walk, or by producing a media project.

Research
Have students brainstorm researchable questions:

- What is an advertising agency? What kinds of jobs do people do there?
- Research a favorite product. How much does that company spend on advertising in a year?
- In what other ways do professional food stylists make food look more appetizing?

Investigate
Have students brainstorm testable questions to be solved through science or math:

- How many advertisements do you encounter in one day? Keep a log.
- Next time you watch a TV show or movie, look for product placements. How many can you find?
- Compare the cost of a giant box of cereal with the cost of the smallest box of the same cereal. Which one costs less per serving?

Innovate
Have students brainstorm problems to be solved through engineering:

- Can you develop an advertising campaign for a cause you believe in?
- Can you design a new-and-improved version of a product?
- Can you develop a blind test to test your classmates' product preferences?

Reference
Federal Trade Commission. Truth in advertising. *www.ftc.gov/news-events/media-resources/truth-advertising*.

Website
"Food Ad Tricks: Helping Kids Understand Food Ads on TV" (video)
www.youtube.com/watch?v=fUjz_eilX8k

Another Book to Read
Rosenstock, B. 2014. *Ben Franklin's big splash: The mostly true story of his first invention.* Honesdale, PA: Calkins Creek.
Summary: Delightful ink-and-watercolor illustrations and lively text tell the story of Ben Franklin's first invention at age 11—swim fins.

Chapter 7

Name: _____

Looking at Ads: Part 1

Read through four print advertisements for a product or service. Then, fill out the information for each advertisement.

Product or Service	What is the claim or slogan? What does the ad say the product or service will do?	Does the ad provide any evidence to support its claims? If so, list it.	Does this ad make you want to buy the product or service? Why or why not?
1.			
2.			
3.			
4.			

Name : _____

Looking at Ads: Part 2

1. Do you see any **weasel words** in the ads that suggest a positive meaning without giving a guarantee that can be measured, such as, "people say," "most people agree," or "studies show"? If so, list them.

   _____

   _____

2. Do the ads use **celebrities**? If so, whom? _____

3. Do you see any **glittering generalities** that cannot be proven. For example, the product will "change your life" or make you a "cool kid"? If so, list them.

   _____

   _____

4. Do any of the ads use **puffery** such as "World's Best" or "Kid's Favorite"? If so, list them.

   _____

   _____

Don't Be Fooled

Advertisements are everywhere! From newspaper and magazine ads to billboards and online ads—you may not realize it, but companies are working hard to sell you products and services every day. When you see an ad that states a product will "last 50% longer" than its competitors or will make your teeth whiter "in just 6 weeks," have you ever wondered if the claims are true? As a smart **consumer** (a person who purchases goods and services), it is important to know what happens in order for companies to legally make these kinds of statements.

A **claim** is a statement about what a product can do. In the United States, it is illegal for a company to make a false claim about a product. If a company claims that its product does a certain thing or gives specific results, the company must have **evidence** to prove it. Companies collect this evidence by doing studies, or tests, that are **reproducible** by other researchers and that support their product claims. The evidence could be in the form of observations or measurements.

A **product claim test** is a study that either proves or disproves a product claim. For example, the claim that a brand of toothpaste will make teeth whiter in 6 weeks might be tested by measuring tooth whiteness (using a color chart) before and after 6 weeks of using the product. In a **product comparison test,** a product is compared to other products that are similar in order to compare their effectiveness. For example, a company that makes paper towels might compare their product to other companies' paper towels by measuring and comparing the amount of liquid each brand absorbs.

When doing a product test, or any kind of scientific test, be sure that it is a **fair test.** A fair test is one in which the variables are controlled. There are two types of variables in a scientific test: **experimental variables** and **controlled variables**. An experimental variable is the one thing that is being tested. For example, if you want to see which brand of sports drink people prefer, the experimental variable would be the brand of drink. The other variables—such as the flavor, temperature, amount, and type of container—must be controlled, or kept exactly the same, to keep the test fair.

So what happens if a company makes a claim about a product that another company or a consumer thinks might be false? The other company or consumer may file a complaint with the **Federal Trade Commission (FTC),** a government organization that enforces truth-in-advertising laws. The company in question will be asked to provide evidence that the claim is true. If the FTC decides that there is not enough evidence to prove the claim is true, the company could be fined and ordered to remove the advertisements that include that claim. How would you test the claim of the "Zip Ultra" ad below?

Be "Kleen" like me and use Zip Ultra!

Hey sports fans! Have you tried new **Zip Ultra** on your dirty clothes? Most folks agree Zip Ultra gently cleans away tough stains better than the leading brand of powdered laundry detergent … without scrubbing! Just ask celebrity racecar driver Thunder McKleen. He washes his uniform with Zip Ultra after every race!

Zip on down to your nearest Zip Ultra retailer and buy some today!

Picture-Perfect STEM Lessons, 3–5

Name: _____

First, read the article, "Don't Be Fooled." Next, study the Zip Ultra advertisement and answer the following questions:

1. What product or service is the Zip Ultra advertisement selling?

   _____

2. What "weasel words" do you see in the Zip Ultra advertisement that suggest a positive meaning without giving a guarantee that can be measured, such as "people say" or "studies show?"

   _____

   _____

3. What claim does the Zip Ultra advertisement make about how well the product works?

   _____

4. What do companies need to have in order to make a claim in an advertisement?

   _____

5. What would be the experimental variable in a product comparison test of Zip Ultra and another brand?

   _____

6. What variables would need to be controlled, or kept the same, in a product comparison test of Zip Ultra and another brand?

   _____

   _____

Name: _____

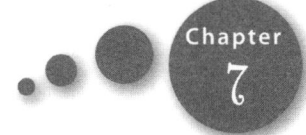

Detergent Comparison Test Design

New Zip Ultra vs. Skid-Daddle

Question: Which powdered laundry detergent cleans stains better without scrubbing, Zip Ultra or Skid-Daddle?

Directions: Design a test to find out which of these two laundry detergents cleans stains better without scrubbing.

Experimental variable: _____

Controlled variables: _____

_____

_____

Procedure: _____

_____

_____

_____

_____

_____

_____

_____

Picture-Perfect STEM Lessons, 3–5

Name: _____

Product Testing
Design your own product comparison test!

Products: _____

Question: _____

_____

Experimental variable: _____

Controlled variables: _____

_____

Procedure: _____

_____

_____

_____

_____

_____

_____

Teacher Checkpoint: ☐

Name: _____

Consumer Fair Rubric

Share the findings of your product test. Include the criteria listed below on your display.

Score	Criteria
____ 4 ____ 3 ____ 2 ____ 1	Display the names of the products and the question you investigated.
____ 4 ____ 3 ____ 2 ____ 1	Identify the experimental variable and controlled variables in your investigation.
____ 4 ____ 3 ____ 2 ____ 1	List the step-by-step procedure you followed to answer the question you investigated.
____ 4 ____ 3 ____ 2 ____ 1	Write a conclusion and provide the evidence that supports your conclusion (using tables, graphs, photos, and/or other visual aids).
____ 4 ____ 3 ____ 2 ____ 1	Include product costs, packaging, and/or advertisements that make claims about what the products do.

4—*Excellent*   3—*Above Average*   2—*Average*   1—*Below Average*

_____ Total Points/20

Picture-Perfect STEM Lessons, 3–5

Name: _____

STEM at Home

Dear _____,

At school, we have been learning about **product testing.**

I learned that: _____

_____

_____

My favorite part of the lesson was: _____

_____

Reading product reviews is a good way to figure out if a product is worth buying. Product reviews are written by people who have purchased the product, not by advertisers. At home we can look at product reviews on Amazon.

 Go to Amazon's website at *www.amazon.com.*

With an adult helper, think of something you would like to buy (a toy, game, household item, etc.). Type the name of the item in the search box. Scroll down to read the customer reviews. Then, fill in the blanks and answer the questions below.

Name of the Product: _____

Average review score? _____ Number of reviews? _____

Do you still want to buy this product? Why or why not? _____

_____

How are reviews helpful in making decisions about what to buy?

_____

_____

_____

Chapter 8

Wind It Up

Description

After reading a story about a loveable toy robot, students explore wind-up toys, observe their patterns of motion, graph the distance they travel, and predict their future motion. Then, they take the toys apart to see how they work. Through a nonfiction read-aloud, students learn how springs inside wind-up toys store energy that is released when the spring unwinds, causing the toy to move. Finally, students build their own wind-up toy and design an instruction manual that explains how it works.

Suggested Grade Levels: 3–5

LESSON OBJECTIVES Connecting to the *Framework*		
Science and Engineering Practices	**Disciplinary Core Ideas**	**Crosscutting Concept**
Asking Questions and Defining Problems Analyzing and Interpreting Data Constructing Explanations and Designing Solutions Using Mathematics and Computational Thinking	**PS2.A:** Forces and Motion **ETS1.C:** Optimizing the Design Solution	Patterns Cause and Effect

Featured Picture Books

TITLE: **Clink**
AUTHOR: **Kelly DiPucchio**
ILLUSTRATOR: **Matthew Myers**
PUBLISHER: **Balzer + Bray**
YEAR: **2011**
GENRE: **Story**
SUMMARY: *While newer, fancier robots are quickly purchased, Clink, an old-fashioned robot who can make only toast and music, gathers dust and feels downhearted until a young boy enters the shop looking for something special.*

TITLE: **Making Machines With Springs**
AUTHOR: **Chris Oxlade**
PUBLISHER: **Heinemann Raintree**
YEAR: **2015**
GENRE: **Non-Narrative Information**
SUMMARY: *From the Simple Machine Projects series, this book explains how springs work and their many practical uses. It includes kid-friendly activities to try and projects to build.*

Picture-Perfect STEM Lessons, 3–5

Time Needed

This lesson will take several class periods. Suggested scheduling is as follows:

Day 1: **Engage** with *Clink!* Read-Aloud and **Explore** with Wind-Up Robots

Day 2: **Explain** with *Making Machines With Springs* Read-Aloud

Day 3: **Elaborate** with "Wind Up Racer" Video and Designing Spool Car Racers

Day 4: **Evaluate** with Spool Car Racer Instruction Manual

Materials

For Wind-Up Toys (per pair)

- Wind-up toy robot (Fun Express Wind-up Toy Robots [1 dozen] are available on Amazon.com.)
- Tape measure or ruler

For Wind-Up Toys (for teacher use)

- Small screwdriver set
- Protective gloves
- Document camera and projector (optional)

For Making Machines With Springs *Read-Aloud (per pair)*

- Spring from a retractable ballpoint pen
- Clothes pin or chip clip
- Thick rubber band
- Thin rubber band
- Pencil
- Plastic ruler
- Book

For Spool Car Racer

- Wooden spool
- Rubber band
- Small washer
- Masking tape
- Pencil
- Wooden toothpick
- Rubber bands of various sizes (for testing different designs after the spool car racers are built)
- Wooden spools of various sizes (for testing different designs after the spool car racers are built)
- Markers and stickers to decorate spool car racers (optional)

SAFETY

- If some students are allergic to the latex found in rubber bands, be sure to substitute that item with latex-free elastic bands.
- Safety glasses or goggles and protective gloves are required for this activity.
- Use caution when handling screw drivers, springs, and clips.

Chapter 8

Student Pages

- Wind It Up
- Spool Car Racer Building Instructions
- Spool Car Racer Instruction Manual
- STEM at Home

Background for Teachers

Springs are everywhere! When you wind a watch, stretch a rubber band, turn a door handle, sit on a couch, or type on a computer keyboard, springs help you! Springs are not simple machines, but they often work with simple machines to do many useful jobs. Springs come in many shapes and sizes, from the tiny springs in ballpoint pens to the huge suspension springs on trucks. The most recognizable type of spring is the *coil spring,* a metal wire wrapped around a circle of a fixed size. Another common type of spring is the *clock spring.* These are the springs used in clocks and wind-up toys. They are made of a strip of metal that is wound up into a tight circle. When you wind the toy, the spring gets tighter. This stores energy; when you let go, the spring begins unwinding. The toy moves until the spring is unwound. Until about 100 years ago, most moving toys were wind-up toys.

COIL SPRING

When you push or pull on a spring, you change the arrangement of the coils. This force stores energy in the spring. That energy is released when the force pushing or pulling it is removed, and the spring returns back to its original shape. In 1660, British scientist Robert Hooke discovered that if you double the pull on a spring, it stretches twice as much. Springs are usually made of *spring steel,* a special alloy that

CLOCK SPRING

can endure stretching and compressing and return to its original shape. Other materials can be used as springs as well, as long as they have a certain amount of rigidity and flexibility. In fact, anything that is elastic, meaning its arrangement can be changed and then it restores itself, can be used to store energy in the same way.

A rubber band can be stretched and will return to its original configuration. Energy is stored when the rubber band is stretched, and energy is released when the forces stretching the rubber band are removed. Some trucks and trains have springs made of big blocks of rubber that are squeezed when the truck or train goes over a bump. Springs can even be made of plastic. Of course, plastic springs are usually not as strong as steel springs, but they are strong enough to do the job for which they are intended. Sometimes, metal springs are not ideal because they can interfere with medical imaging or other iron-sensitive devices, so plastic springs are used.

In this lesson, students observe the motion of wind-up toys, and then take one apart to see how it works. Inside the toy, students find a clock spring and learn how winding the spring stores energy that is released when the spring becomes unwound. The force of the spring pushes against the gears of the toy, which makes the toy move. In this activity, students use mathematics and computational thinking to graph the distance the wind-up toy moves versus how many times the toy is wound. They then predict the wind-up toy's future motion based on these observations. Students are also engaged in the science and engineering practice of constructing explanations and designing solutions as they learn how springs work and eventually make their own wind-up toy—a spool car racer. The crosscutting concepts of patterns and cause and effect are explored as students observe patterns of motion and learn how winding and releasing springs causes motion.

Picture-Perfect STEM Lessons, 3–5

Chapter 8

engage

Clink Read-Aloud

> **Connecting to the Common Core**
> **Reading: Literature**
> KEY IDEAS AND DETAILS: 3.1, 4.1, 5.1

Show students the cover of *Clink* and introduce the author, Kelly DiPucchio, and the illustrator, Matthew Myers. Read the back inside flap of the book jacket to students, which explains that DiPucchio was inspired to write this book when she saw one of Myer's paintings of a sad-looking robot. Read *Clink* aloud. After reading, show students the drawings inside the covers of the book. These are patent illustrations for a robot like Clink. An inventor must submit complete drawings, showing every detail of his or her invention, in order to apply for a patent. The drawings must be done in a special format, similar to the ones pictured in the book. For fun, read the "General Notes" aloud!

 **Inferring**

Ask

- ? Why doesn't anyone want to buy Clink? (He is an old toy.)
- ? Why does the boy think Clink is perfect? (The boy likes to fix things, likes burnt toast, and likes to dance.)
- ? How do you think Clink works? What are his parts? Are there any clues in the text or illustrations? (His body is a radio and his head is a toaster. Several illustrations show springs flying out of him. In fact, a rusty spring hits the boy square in the forehead! Tell students that most toasters use springs to pop up the toast, so it is possible that the springs came from Clink's head. Other clues as to how he works are included in the patent illustrations inside the covers.)

Then, show students a wind-up robot toy and *ask*

? How do you think this toy robot works? (Answers will vary.)

explore

Wind-Up Robots

Give each pair of students a wind-up robot (or other wind-up toy that walks or rolls.) Give them a few minutes to play with the toy and observe how it moves. Caution them not to twist the knob too much! *Ask*

- ? How do you get the toy to move? (wind it up by twisting the knob)
- ? How do you think the number of times you wind it up affects the distance the toy travels? (Answers will vary.)
- ? How could you find out? (We could count the number of winds and measure how far it goes.)
- ? What would be a good way to keep track of your data? (by recording the number of winds and the distance traveled on a data table)

> **Connecting to the Common Core**
> **Mathematics**
> MEASUREMENT AND DATA: 3.MD.3, 3.MD.4

EXPLORING WITH WIND-UP TOYS

Give each student a copy of the Wind It Up student page and a tape measure or ruler. Tell students that the best way to keep track of the number of times you wind the toy is to hold the knob still and spin the toy until it makes one complete rotation. Have students wind their toy one time, place it on the table, and record how far it travels in centimeters. Call on pairs to share their data and compare the data for each group. (You may choose to calculate and record classroom averages for each number of turns.) *Ask*

? How far do you think the toy will go with two turns? (Answers will vary.)

Have students repeat the procedure with two, three, and four turns, each time predicting future motion based on the pattern of the toy's past motion. *Ask*

? How does the number of turns affect the distance? (The more turns, the greater the distance traveled. Students may notice specific patterns based on mathematical relationships, e.g., 1 turn = 2 cm, 2 turns = 4 cm, and so on.)

? Were you surprised by any of the results? (Answers will vary.)

Tell students that a good way to compare data and look for patterns is to make a graph. Help students create their graph on the Wind It Up student page by asking these guiding questions,

? What kind of graph is best for comparing totals? (bar graph)

? What label should you put on the *x*-axis? (number of turns)

? How will you divide the *x*-axis? (from 0 to the highest number of turns recorded)

? What label should you put on the *y*-axis? (distance traveled in centimeters)

? How will you divide the *y*-axis? (from zero to the longest distance)

? What would be a good title for the graph? ("Number of Wind-Up Toy Turns Versus Distance Traveled," for example)

Have students create their graphs. Then, have them share their graph with another group and compare the similarities and differences. *Ask*

? How does your graph compare with the other group's graph? (Answers will vary.)

? When did your toy move the shortest distance after winding? (with one turn)

? When did your toy move the greatest distance? (Answers will vary based on the toys used.)

? Is it possible to wind up the toy too many times? (yes)

? Supposing you *could* wind the toy one more turn without breaking it, how far do you predict your toy would travel? (Answers will vary, but students should predict a distance that is reasonable, based on the regular patterns of the toy's past motion.)

Ask

? What do you think is inside the wind-up toy? What makes it go? (Answers will vary.)

? How can we find out? (We can take the toy apart.)

Have students observe you taking apart one of the wind-up toys with a small screwdriver. If you have one, use a document camera and projector so students can see all of the parts in detail as you take the toy apart. Try to show that the gears are in contact with the part that contains the spring.

SAFETY
You will need a screwdriver and protective gloves to disassemble the toy.

OBSERVING A DISASSEMBLED WIND-UP TOY

Picture-Perfect STEM Lessons, 3–5

After you have taken apart the toy, *ask*

? Which part do you think is the one that makes the toy go? (the wind-up piece)

? What do you think is inside? (Answers will vary.)

Carefully open up the wind-up piece and show students the clock spring inside. Demonstrate that by turning the knob, a force from your hand winds the spring up tight and when you let go, the spring unwinds and a force from the spring pushes against the gears, causing the toy to move.

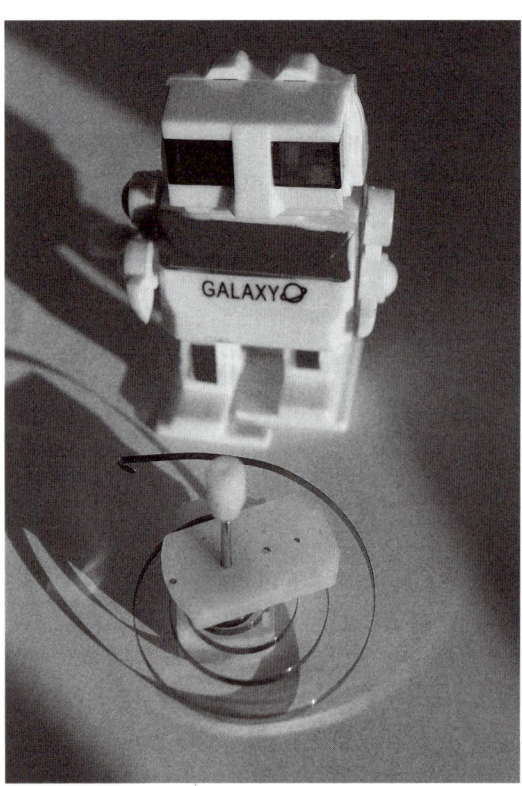

WIND-UP TOY SPRING

explain

Making Machines With Springs Read-Aloud

Connecting to the Common Core
Reading: Informational Text
KEY IDEAS AND DETAILS: 3.1, 4.1, 5.1

Show students the cover of *Making Machines With Springs*. Tell them that this book can help them understand how the spring inside the wind-up robot made it move.

 Turn and Talk

Ask

? Other than wind-up toys, where are springs used? (Answers will vary.)

 Stop and Try It

Read aloud pages 4–7. Then, stop and have students try the activities on pages 8–11, or do the activities as demonstrations. Next, *ask*

? The inset of page 6 reads, "You need two or more forces to make a spring change shape." What forces do you need to make a spring stretch? (two pulling forces, one from each end)

? What forces do you need to compress a spring? (two pushing forces, one on each end; students might push the spring against something, like their desk, and think that requires a force from only one end, so explain that the desk is actually pushing back against the spring as well)

? What forces did you need to make the other springs work? (Chip clip—two pushing forces; rubber band—two pulling forces; plastic ruler—one push holding the ruler down and another force pushing on it to make it bend.)

? What kinds of materials were the springs you tried made of? (metal, plastic, and rubber)

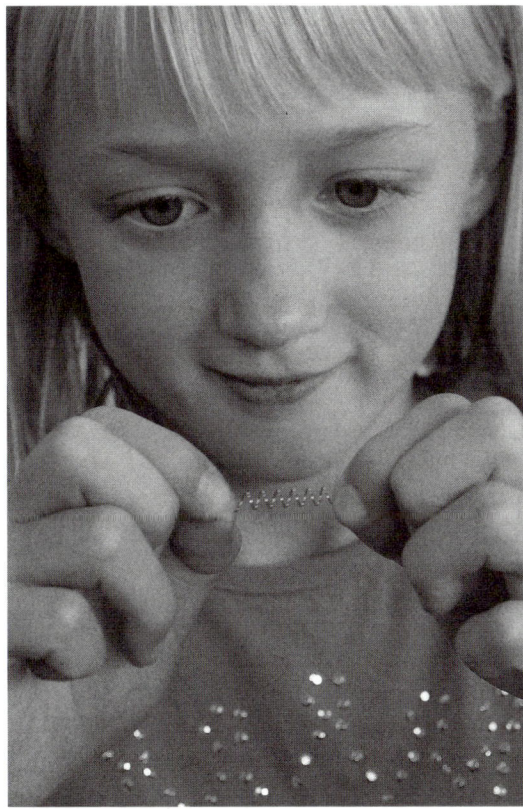

LOOKING AT SPRINGS

? What properties of these materials make them work as springs? (They are flexible, but they will snap back to their original shape.)

Read pages 12–13 about the different jobs that springs do. Skip page 14, and read page 15 about clock springs. Point out the picture of the spring on page 15 and *ask*

? What does this spring remind you of? (The spring in the toy robot.)
? What is this kind of spring called? (a clock spring)

Skip the activity on pages 16–19, and read pages 20–21. *Ask*

? So how does our wind-up toy robot work? (When we wind the toy, the spring inside gets tighter. This stores the energy we used to do the winding. When we let go, energy is released as the spring unwinds. The toy moves until the spring is completely unwound.)
? Where does the force come from that winds the spring? (our hand turning the knob)
? Where does the force come from that turns the gears? (the unwinding spring)

Show students the activity on pages 22–25 and tell them that, during the next class period, they will be making something similar that is powered by a rubber band. If time permits, read the rest of the book aloud.

 ### Turn and Talk

Ask

? Where are springs used? (wind-up toys, door handles, bicycle brakes, rubber bands, slinkys, trampolines, ballpoint pens, chip clips, clothespins, video game controllers, computer keyboards, flashlights, bungee cords, clocks, and so on.)

 ### Synthesizing

Give each student an index card, and have them write their name and an answer to the following question, including a labeled sketch:

? How does a wind-up toy robot work?

Collect the cards and use them as an informal assessment of student learning. Students should be moving toward an understanding that when they turn the knob, they apply a force that winds the spring, thereby storing energy in the spring. When they let go of the knob, the spring releases energy, applying a force that turns the gears and produces the walking motion of the toy robot.

elaborate

"Wind Up Racer" Video

Tell students that they are going to design and build a toy powered by a rubber band! Show students the video "Wind Up Racer" from Steve Spangler Science (see "Website" section). After the video, *ask*

Chapter 8

? How is the rubber band–powered toy like a wind up toy? (You wind it up to make it work. It stores energy in the rubber band, which acts like the spring in the wind-up toy.)

Designing Spool Car Racers

Give each student a wooden spool, rubber band, paper clip, washer, pencil, and a copy of the Spool Car Racer Building Instructions. Give them time to build and test their spool car racers.

After students have all successfully built their spool car racers, *ask*

A SPOOL CAR RACER

? How do you get to your spool car to go faster and farther? (wind it up more times)

? Is there a limit to how many times you can wind it? (Yes, if the rubber band gets too tight, it gets stuck. The rubber band can also break.)

? How could you modify your spool car to get it to go farther, faster, and straighter? (Use a different size or thickness of rubber band or use a different size of spool.)

Provide students with some spools and rubber bands of various sizes. Encourage them to modify, test, and evaluate their designs. A table is provided on the student page to help students keep track of their spool car modifications and the distance the spool car travels. They can use markers or stickers to decorate their final design, or *prototype*.

TESTING SPOOL CAR RACERS

evaluate

Spool Car Instruction Manual

Connecting to the Common Core
Writing
PRODUCTION AND DISTRIBUTION OF WRITING: 3.4, 4.4, 5.4

 Synthesizing

Give students a copy of the Spool Car Racer Instruction Manual. (Pages 111–112 should be printed front to back and folded in half.) Tell them that you would like them to create an instruction manual like the ones that come with most toys. The instruction manual should include the following:

- Cover: A catchy name for the toy (on the line above the words *Instruction Manual*), a sketch

of the toy, an advertising slogan, and the toy designer's name
- Page 1: A list of parts and a labeled diagram, similar to a patent illustration (refer them to the inside covers of *Clink* for inspiration)
- Page 2: Instructions for using the toy, an explanation of how to make it go farther, and troubleshooting (what to do if the toy does not work)
- Back Cover: An explanation of how the toy works using the words *turn, force, rubber band, energy,* and *motion*

Students should be able to correctly place the terms in the cloze paragraph, which explains, "When you turn the pencil, you apply a *force* that winds the rubber band, storing *energy* in the rubber band. When you let go of the pencil, the rubber band releases *energy,* applying the force that causes the rolling *motion* of the spool."

STEM at Home

Have students complete the "I learned that …" and "My favorite part of the lesson was …" portions of the STEM at Home student page as a reflection on their learning. They may choose to do the following at-home activity with an adult helper and share their results with the class. If students do not have access to the internet or these materials at home, you may choose to have them complete this activity at school.

"At home, we can watch a short video called 'Scientist Profile: Toyologist,' which is about an engineer who designs toys."

Search "Scientist Profile: Toyologist" at www.pbslearningmedia.org *to find the video at* www.pbslearningmedia.org/resource/67a59215-806a-4d6c-9810-da3b1d-3d30e0/67a59215-806a-4d6c-9810-da3b1d3d30e0.

"After we watch the video, we can design our own wind-up toy and sketch it here. We can also draw a simple version of a patent illustration for our toy on the back of this page (similar to the ones we can find by doing a Google images search for 'patent illustration wind-up toy')."

For Further Exploration

This section is provided to help you encourage your students to use the science and engineering practices in a more student-directed format. This box lists questions and challenges related to the lesson that students may select to research, investigate, or innovate. Students may also use the questions as examples to help them generate their own questions. After selecting one of the questions in the box or formulating their own questions, students can individually or collaboratively make predictions, design investigations or surveys to test their predictions, collect evidence, devise explanations, design solutions, or examine related resources. They can communicate their findings through a science notebook, at a poster session or gallery walk, or by producing a media project.

Research

Have students brainstorm researchable questions:

? What is an automaton?

? How does a spring scale work, and what does it measure?

? How are torsion springs used in watches?

Investigate

Have students brainstorm testable questions to be solved through science or math:

? What is the relationship between the distance you pull back a toy "pull-back" car and the distance it moves? Graph the results, then analyze your graph. What can you conclude?

? How does a pull-back car work? Take one apart!

? How does a jack-in-the-box toy work? Take one apart!

Innovate

Have students brainstorm problems to be solved through engineering:

? Can you invent your own wind-up toy that is powered by a rubber band?

? Can you make a coil spring out of a paper clip, twist tie, or other material?

? Can you design your own jack-in-the-box toy using a coil spring?

Website

"Wind Up Racer"(video)
www.stevespanglerscience.com/lab/experiments/wind-up-racer

More Books to Read

Greathouse, L. 2010. *How toys work*. Huntington Beach, CA: Teacher Created Materials.
Summary: This book explores the design and function of electric, magnetic, and motion-powered toys.

Kalman, B. 2014. *Toys and games: Long ago and today*. New York: Crabtree Publishing.
Summary: Part of the *From Olden Days to Modern Ways* series, this book discusses how toys and games have changed over the years and how some have stood the test of time.

Sohn, E. 2016. *Experiments in forces and motion with toys and everyday stuff*. Mankato, MN: Capstone Press.
Summary: From the *Fun in Science* series, this book explains some basic physics concepts with activities that use toys and household items.

Wulffson, D. 2014. *Toys! Amazing stories behind some great inventions*. New York: Square Fish.
Summary: This book describes the creation of a variety of toys and games—from seesaws to Silly Putty and toy soldiers to Trivial Pursuit.

Name : _____

Chapter 8

Wind It Up

How does the number of times you wind the toy affect the distance the toy travels? Let's investigate to find out!

1. Record your data in the table. Be careful not to twist the knob too much!

Number of Winds	Predicted Distance Traveled	Actual Distance Traveled
0	0 cm	0 cm
1		
2		
3		
4		

2. Make a graph to represent your data.

   Title: _____

Picture-Perfect STEM Lessons, 3–5

109

Spool Car Racer
Building Instructions

Materials:

- Wooden spool
- Rubber band
- Pencil
- Small washer
- Masking tape
- Toothpick

Instructions:

1. Break the toothpick so that it is a little shorter than the diameter of the spool.
2. Run the rubber band through the spool.
3. Insert the toothpick piece through one end of the rubber band, and tape the toothpick flat against the end of the spool (making sure none of it hangs over the edges).
4. Run the rubber band on the other side of the spool through the washer.
5. Put a pencil through the open loop of the rubber band.
6. Wind the rubber band by spinning the pencil around and around.
7. Put your car down on the table or floor and watch it go!

Can you make the car go farther?

Try different-size spools or rubber bands! Change one thing at a time, and keep track of the distance traveled for each change below.

Change to Design	Distance Traveled

National Science Teachers Association

Instruction Manual

Toy Designer: _____

How It Works

> Word Bank (words may be used more than once):
> *force, energy, and motion*

When you turn the pencil, you apply a _____ that winds the rubber band, storing _____ in the rubber band. When you let go of the pencil, the rubber band releases _____, applying the _____ that causes the rolling _____ of the spool.

Picture-Perfect STEM Lessons, 3–5

Parts

_____

_____

_____

_____

Labeled Diagram

[blank box]

Instructions

To make it go, you need to:

_____

_____

_____

To make it go farther, you need to:

_____

_____

Troubleshooting

If the toy does not work, try:

_____

_____

_____

_____

National Science Teachers Association

Name : _____

STEM at Home

Dear _____,

At school, we have been learning about how **wind-up toys** work.

I learned that: _____

_____

_____

My favorite part of the lesson was:

_____

_____

At home, we can watch a short video called "Scientist Profile: Toyologist," which is about an engineer who designs toys.

    Search "Scientist Profile: Toyologist" at *www.pbslearningmedia.org to find the video at www.pbslearningmedia.org/resource/67a59215-806a-4d6c-9810-da3b1d3d30e0/67a59215-806a-4d6c-9810-da3b1d3d30e0.*

After we watch the video, we can design our own wind-up toy and sketch it here. ➡

We can also draw a simple version of a patent illustration for our toy on the back of this page (similar to the ones we can find by doing a Google images search for "patent illustration wind-up toy").

Picture-Perfect STEM Lessons, 3–5

Light It Up!

Description

Students explore how incandescent lightbulbs, compact fluorescent lamps (CFLs), and light-emitting diodes (LEDs) transform electricity into light and some heat. Then, they use mathematics to compare the energy efficiency of these light sources. After reading a story about a child who is afraid of the dark, they apply their learning about energy transformations by designing a nightlight using an LED bulb, circuit tape, and a battery.

Suggested Grade Levels: 3–5

LESSON OBJECTIVES Connecting to the *Framework*		
Science and Engineering Practices	**Disciplinary Core Ideas**	**Crosscutting Concept**
Constructing Explanations and Designing Solutions Analyzing and Interpreting Data	**PS3.B:** Conservation of Energy and Energy Transfer **ETS1.A:** Defining Engineering Problems **ETS2.B:** Influence of Engineering, Technology, and Science on Society and the Natural World	Energy and Matter

Featured Picture Books

TITLE: **How Things Work: Lightbulbs**
AUTHOR: **Joanne Mattern**
PUBLISHER: **Children's Press/Franklin Watts**
YEAR: **2016**
GENRE: **Non-Narrative Information**
SUMMARY: *This book introduces the science of the lightbulb and explains how innovators have continued to improve on the little glass bulb that brightens our world.*

TITLE: **Orion and the Dark**
AUTHOR: **Emma Yarlett**
ILLUSTRATOR: **Emma Yarlett**
PUBLISHER: **Templar**
YEAR: **2014**
GENRE: **Story**
SUMMARY: *Orion is scared of wasps, monsters, girls, storms, and dogs, but—most of all—he's scared of the dark!*

Picture-Perfect STEM Lessons, 3–5

Time Needed

This lesson will take several class periods. Suggested scheduling is as follows:

Day 1: **Engage** with Mystery Object, **Explore** with What's Inside a Lightbulb?, and **Explain** with *How Things Works: Lightbulbs* Cloze and Read-Aloud

Day 2: **Explore** and **Explain** with Comparing Lightbulbs

Day 3: **Elaborate** with *Orion and the Dark* Read-Aloud, Nightlight Design Challenge, and Paper Circuits

Day 4 (or more): Continue Nightlight Design Challenge

Day 5: **Evaluate** with Nightlight Instruction Manual

Materials

For Mystery Object

- Paper bag with a question mark on it
- Large, clear incandescent lightbulb

For What's Inside a Lightbulb? (for teacher use only)

- Wire strippers

For What's Inside a Lightbulb? (per pair)

- Sticky notes
- Small clear flashlight-sized incandescent lightbulbs
- Hand lens
- D-cell battery
- 2 insulated copper wires (at least 20 cm long with the ends stripped)
- Masking tape

For Comparing Lightbulbs (for teacher use only)

- 3 lamp bases (with shades removed)
- 40-watt incandescent lightbulb with packaging
- 40-watt equivalent CFL lightbulb with packaging
- 40-watt equivalent LED lightbulb with packaging

For Nightlight Design Challenge and Paper Circuits (per student)

- Mini LED bulb
- Index card
- Conductive copper foil circuit tape (2 strips approximately 20 cm, 1 for paper circuits practice and 1 for the Nightlight Design Challenge)
- 3V lithium coin cell battery
- Scissors

SAFETY

- Use caution when working with glass bulbs. Bulbs can easily break and puncture the skin or eyes.
- Remember to follow the Environmental Protection Agency's recommendations for cleaning up broken CFLs *(ww.epa.gov/cfl/cleaning-broken-cfl)* and recycling or disposing of CFLs *(www.epa.gov/cfl/recycling-and-disposal-cfls)*.
- Use caution when working with batteries and bulbs in an active circuit. They can heat up and burn skin if touched.
- Use caution when working with wires, scissors, and other sharp objects to avoid cutting or puncturing your skin or eyes.
- Use caution when working with wire cutters. They can pinch and cut skin.

- Transparent tape
- Binder clip
- Markers
- A variety of nonconductive materials for building a nightlight base, such as cardstock, small boxes, wooden or plastic blocks, plastic bottles, and so on.

Student Pages

- How a Lightbulb Works
- Comparing Lightbulbs
- Paper Circuits
- Nightlight Design Challenge
- Nightlight Instruction Manual
- STEM at Home

Background for Teachers

The lightbulb is perhaps one of the most significant inventions in history. Before lightbulbs, people used candles, oil lamps, and gaslights to light buildings at night. These early sources of illumination could be smoky, smelly, and dangerous. The first electric light was the arc lamp, which had a spark of electricity that jumped across two rods called *electrodes*. Then came the *incandescent* bulb. In 1879, Thomas Edison invented an incandescent bulb that worked for a long time. An incandescent bulb works by *incandescence*, which is the emitting of light by heating a thin wire called *filament* (usually made of the metal tungsten). Inside the glass bulb is either a vacuum or an inert gas, such as argon, that keeps the filament from burning out. Early incandescent bulbs did not work well because the filament broke easily.

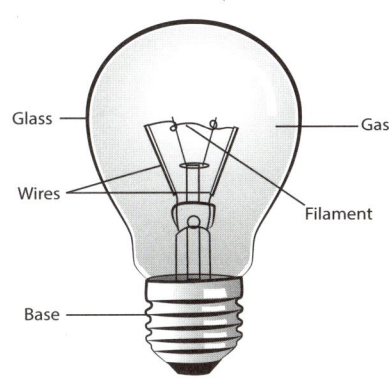

INCANDESCENT LIGHT BULB

Incandescent lightbulbs have been used widely for household and commercial lighting for more than 100 years, but other lightbulbs have been developed recently that are much more efficient. The halogen incandescent lightbulb works like a regular incandescent lightbulb, except it has a small capsule filled with halogen gas to increase bulb efficiency. The *CFL (compact fluorescent lamp)* is becoming more widely used today. A CFL uses about 70% less energy than an incandescent lightbulb and typically lasts for 6,000–15,000 hours, whereas an incandescent lightbulb usually lasts for only 750 to 1,000 hours. CFLs cost 3–10 times more than incandescent lightbulbs, but they lead to greater savings on the electric bill over time. However, the life of a CFL is significantly shorter if it is turned on and off frequently. The U.S. Energy Star program suggests leaving a CFL on when leaving the room for 15 minutes or less.

CFLs are essentially a curly version of long-tube fluorescent lights. In a CFL, an electric current is driven through a tube containing argon gas and a small amount of mercury vapor. This generates invisible ultraviolet light

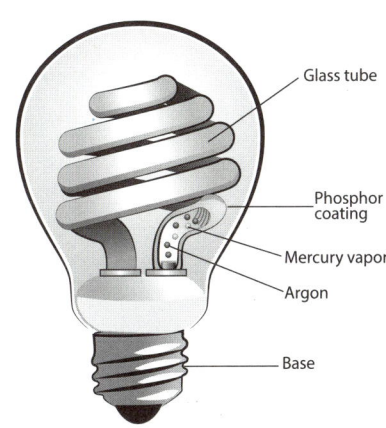

CFL (COMPACT FLUORESCENT LAMP)

that excites a fluorescent coating (called *phosphor*) on the tube, which then emits visible light. These energy-saving bulbs are sometimes a target of controversy because of their mercury content, but because most of the mercury is bound to the bulb, the mercury amount is considered harmless while in use. If a CFL breaks, however, people can consult the Environmental Protection Agency's special guidelines for safe clean up. (See the safety box on p. 116.) If you are concerned about using CFLs in this activity because of their mercury content, consider purchasing a shatter-resistant CFL, which has a coating on the outside that greatly reduces the risk of breakage.

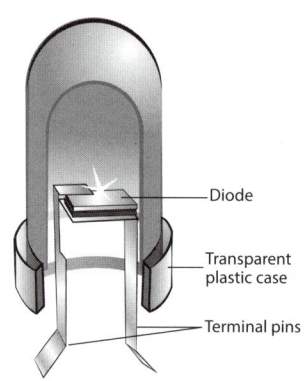

LED (LIGHT EMITTING DIODE)

Another type of bulb that is becoming more popular is the *LED (light-emitting diode)*. LEDs produce light by passing an electric current through a diode, which is a solid semiconductor material (crystal) that allows electricity to flow in only one direction. When electricity flows through the semiconductor material, light energy is emitted. LEDs consist of three main pieces, an anode (positive wire) a cathode (negative wire), and a diode in between. The diode is inside a transparent plastic case to protect it. LEDs are directional light sources, unlike the incandescent bulbs and CFLs that shine light in all directions. Common LED colors are amber, red, green, and blue. There is no "white" light produced by an LED. In an attempt to create whiter light, LEDs are mixed or coated with a phosphor that makes the light appear more white. LEDs are remarkably efficient; an LED can last from 25,000 to 100,000 hours and uses only about 10% of the energy of an incandescent lightbulb uses. LED bulbs do not get as hot as other kinds of lightbulbs because, although they produce heat within their semiconductor junctions, they contain metal "heatsinks" that dissipate the heat away from the LED diode and into the metal. LEDs are commonly used as indicator lights on electrical devices, street signs, light-up novelty items, and many other devices. LEDs are small, but many can be put together to make a full-sized lightbulb.

The U.S. Department of Energy is encouraging the United States to transition to more energy efficient lightbulbs by making consumers aware of the energy use of each bulb. The agency's Lighting Facts website states the following (see "Websites" section):

As of January 1, 2012, all packaging for medium screw base bulbs must include the Federal Trade Commission (FTC) Lighting Facts label with prominent information on the lumen output. The FTC Lighting Facts label emphasizes lumen output, and will help consumers choose the right bulb for their lighting needs, while moving them away from reliance on wattage when selecting a bulb. This effort is especially important with the Energy Independence and Security Act phase-out of inefficient bulbs beginning in 2012.

The lighting facts label requires the following on the front of the package:

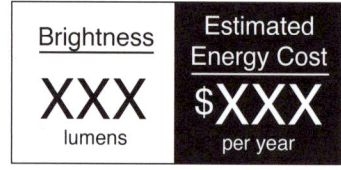

It requires the following on the back of the package:

> **Lighting Facts** Per Bulb, **Brightness** 870 lumens, **Estimated Yearly Energy Cost** $X.XX (Based on 3 hrs/day, 11¢/kWh. Cost depends on rates and use), **Life** X.X years (Based on 3 hrs/day), **Energy Used** XX watts, **Light Appearance** XXXX K
> **Contains Mercury:** For more on clean up and safe disposal, visit epa.gov/cfl.

You will need this information from the packaging and the retail prices of the lightbulbs for this lesson. Students will learn that the more efficient lightbulbs are more expensive to buy but can save money over time because they last longer and use less energy.

All three of these types of lightbulbs *convert,* or *transform,* electrical energy into light energy, so lightbulbs are used in this lesson to give students a context for energy transformations. Students first learn about these three types of bulbs and how they convert energy from one form (electrical energy) to another (light energy). Then, they use what they have learned to design a nightlight using LEDs and conductive copper foil tape. Students are challenged to design the nightlight given several criteria and constraints.

Note: We suggest that your students have some previous experience with electrical circuits before you teach this lesson. We recommend teaching Chapter 22, "Batteries Included," in *Picture-Perfect Science Lessons* before beginning this lesson (Ansberry and Morgan 2015). In that lesson, students investigate simple circuits using batteries, lightbulbs, and wires and learn that electricity needs a complete loop to flow. They also learn how a switch can be used to open and close a circuit.

engage

Mystery Object

Hide a lightbulb in a mystery bag (a paper bag with a question mark on it). Hold up the bag and ask students to guess what is inside. Give them the following clues:

- You use it every single day.
- It is something you have in your home.
- It is something we also have in our school.
- It was invented in 1879.
- The inventor, Thomas Edison, spent thousands of hours experimenting with it before he got it to work.
- It (literally) lights up our world!

Then, reveal the lightbulb inside the bag. *Ask*

? Have you ever thought about how a lightbulb works?

? How do you think it works?

Have students sketch a labeled diagram of an incandescent lightbulb on a sticky note as a pre-assessment, then share their ideas and sketches with a partner. Have them put the sticky note aside to revisit later. Next, show students the cover of *How Things Work: Lightbulbs* and introduce the author, Joanne Mattern. Read aloud only pages 14–19 of the book, which describe what it was like before the lightbulb was invented and discuss Thomas Edison's invention of the lightbulb.

 Turn and Talk

Have students discuss the following question with a partner:

? How would your life be different today if you had no electric lights?

Chapter 9

explore

What's Inside a Lightbulb?

Give each pair of students a small, flashlight-sized clear incandescent lightbulb to observe. Have students look carefully at the inside of the bulb, sketch it on another sticky note, identify any parts they know, and discuss with their partner how they think the lightbulb works. *Ask*

- **?** How do you think this kind of lightbulb produces light? (Answers will vary.)
- **?** What different parts do you see inside the bulb? (Answers will vary.)
- **?** What part of the lightbulb do you think is the part that glows? (Answers will vary.)
- **?** What would we need to do to get the bulb to light up? (provide a source of electricity)

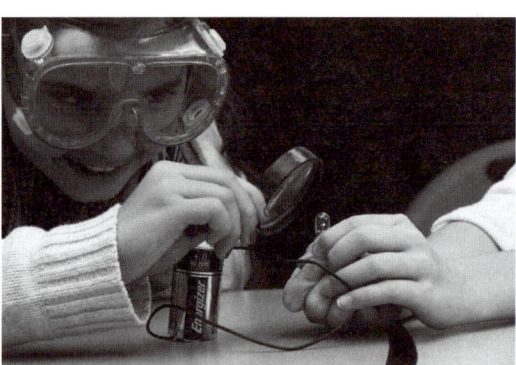

OBSERVING A LIGHTBULB

Figure 9.1. Lighting a Bulb

Give students two copper wires, a D-cell battery, and some masking tape (to hold the wires in place), and ask them to light up the bulb. If students have not had prior experience lighting a bulb with wires and a battery, you can read them the instructions on pages 28–29 of *How Things Work: Lightbulbs*. The key is to attach one wire to the positive (+) end of the battery and the other wire to the negative (−) end of the battery. Then, touch one of the wires to the bottom of the bulb and the other to the threaded side of the bulb (see Figure 9.1). Be sure to read the explanation on the bottom of page 29, which states, "You have created a complete path, or a circuit. This lets the electricity flow from the battery through the lightbulb."

Have students observe the glowing lightbulb with a hand lens. *Ask*

- **?** What part of the lightbulb glows? (small wire or coil in the center)
- **?** Why do you think there is glass surrounding it? (Answers will vary.)

explain

How Things Work: Lightbulbs Cloze and Read-Aloud

Connecting to the Common Core
Reading: Informational Text
KEY IDEAS AND DETAILS: 3.1, 4.1, 5.1

 Cloze Paragraph

Tell students that the book *How Things Work: Lightbulbs* can help them learn more about how the lightbulb they observed works. Give each student a copy of the How a Lightbulb Works cloze. Tell students that before reading, you would like them to cut out the lightbulb vocabulary cards at the bottom of the page and place each word in a space. After reading, they will have the opportunity to move the words.

After students finish placing their words, read aloud pages 1–13 and 20–27 of *How Things Work:*

120 National Science Teachers Association

Lightbulbs. After reading, students can move their words if necessary. The paragraph should read as follows:

> The curly piece of metal inside a(n) <u>incandescent</u> lightbulb is called the filament. It is held up by support wires. When you turn on the light, <u>electricity</u> travels from a wire to the metal base of the bulb to the filament. The filament gets so hot, it glows. The <u>glass bulb</u> keeps the filament safe. There cannot be any air inside or the filament will burn out! A <u>CFL</u> does not use a filament. Another type of lightbulb uses small <u>LED</u> lights.

Explain that, in the case of the small incandescent lightbulb they lit up earlier, the energy to light it came from the battery. When the battery is connected in a circuit (or loop), the electrical energy it produces is transferred through the wires, the metal base, the support wires, and the filament, thus making the filament glow. This glow produces light and heat.

As an exit ticket, have students revisit the lightbulb they drew on a sticky note, add any details they learned, and discuss how their ideas have changed.

explore

Comparing Lightbulbs

Ask

? What were the three kinds of lightbulbs described in the book? (incandescent, CFL, and LED)

Tell students that there are many more kinds of lightbulbs, such as neon, halogen, mercury vapor bulbs, and so on, but the three discussed in the book are the most common lightbulbs used in homes today.

Give each student the Comparing Lightbulbs chart and tell them that you would like them to compare the three different kinds of lightbulbs using four different criteria:

1. Light appearance (does it look "warm" or "cool," yellowish or bluish, etc.)
2. Price
3. Life (see package)
4. Energy cost per year (see package)

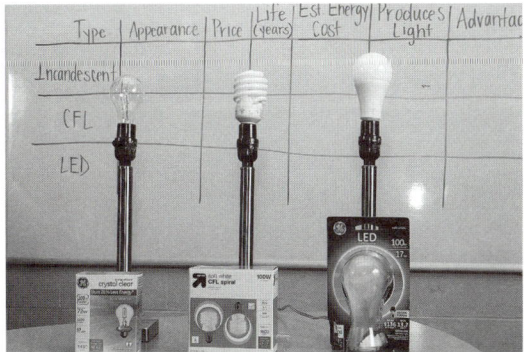

THREE TECHNOLOGIES THAT SOLVE THE SAME PROBLEM

COMPARING LIGHTBULBS

Show students a 40-watt incandescent lightbulb, a 40-watt equivalent CFL, and a 40-watt equivalent LED. Set up three lamp bases where all students can see them and plug them in. Use the following procedure to collect comparison data for each lightbulb:

Chapter 9

1. Place the lightbulb in a lamp with the shade removed and turn it on.
2. Have students observe the light and record their observations.
3. Tell students the price of the lightbulb.
4. Model how to use the lighting facts label to find out the life and energy cost per year.
5. Have students fill in the first four columns of the chart.

Repeat with each type of lightbulb.

explain

Comparing Lightbulbs

> **Connecting to the Common Core**
> **Mathematics**
> NUMBER AND OPERATIONS IN BASE TEN: 4. NBT.B.5, 4.NBT.B.6

Explain that the U.S. Department of Energy requires a lighting facts label to be on every lightbulb so that consumers can be aware of how much energy each one uses. Tell students that the amount of energy used by a lightbulb is measured in a unit of power called a *watt*. The incandescent bulb used in the demonstration uses 40 watts. In the past, looking at wattage was a useful way to buy a lightbulb. If your light fixture said, "100-watt max," you bought a 100-watt lightbulb. But today's energy-saving bulbs use much, much less wattage, and this wattage varies from brand to brand and bulb to bulb. So because comparisons based on wattage are no longer as meaningful, the strength of new energy-saving bulbs is expressed in lumens. (A *lumen* is a unit that measures the amount of light produced.) The higher the number of lumens, the brighter the light. All three bulbs used in the demonstration produce a similar amount of light (measured in lumens), as shown in Table 9.1.

Table 9.1. Comparison of Lightbulbs With Similar Light Output (Lumens)

Bulb Type	Lumens (approx.)	Watts (approx.)
Incandescent	450	40
CFL	450	9
LED	450	6

You can explain how energy cost per year is calculated using the following steps and formulas from Energy.gov (see "Websites" section):

1. **Find the daily energy consumption using the following formula:**

(Wattage × Hours Used Per Day) ÷ 1000 = Daily Kilowatt-hour (kWh) consumption

2. **Find the annual energy consumption using the following formula:**

Daily kWh consumption × number of days used per year = annual energy consumption

3. **Find the annual cost to run the appliance using the following formula:**

Annual energy consumption × utility rate per kWh = annual cost to run appliance

Explain that the reason the lighting facts label can only approximate the energy cost per year is that the actual cost depends on how much the lightbulb is used and how much the local electric company is charging the customer.

> **Connecting to the Common Core**
> **Reading: Informational Text**
> KEY IDEAS AND DETAILS: 3.1, 4.1, 5.1

 Rereading

Tell students that all three of these kinds of lightbulbs convert electrical energy into light and some heat. Different types of lightbulbs give off differing amounts of heat. Incandescent lightbulbs and CFLs

Table 9.2. Sample Comparing Lightbulbs Chart

Type of Lightbulb	Light Appearance	Life	Price	Estimated Energy Cost Per Year	How It Produces Light	Advantages	Disadvantages
Incandescent	Yellowish	[See package]	[See receipt]	[See package]	Electricity passes through the filament making it glow	• Less expensive to buy • Pleasant light	• Does not last as long as others • Uses a lot more energy than the others
CFL (compact fluorescent lamp)	Whitish	[See package]	[See receipt]	[See package]	Electricity passes into a gas inside the bulb making it glow	• Lasts longer • Energy efficient	• More expensive to buy
LED (light-emmitting diode)	Bluish	[See package]	[See receipt]	[See package]	Electricity flows through a very tiny chip gives off light	• Lasts longer • Energy efficient	• More expensive to buy • Can only shine in one direction

give off so much heat that they could burn someone's fingers. So it is important to let them cool before touching them. LEDs, however, do not give off as much heat. Reread the section in the book that explains how each lightbulb works and have students fill in the fifth column of the chart "How It Produces Light." The answers are as follows:

- (Page 9) Incandescent bulb—electricity passes through the filament making it glow
- (Page 21) CFL—electricity passes through a gas inside the bulb making it glow
- (Page 22) LED—electricity flowing through a very tiny chip gives off light

Explain that since the lightbulb was first invented by Thomas Edison in 1879, people have been trying to improve it. Each improved version has its advantages and disadvantages. Share the following examples with students:

- Incandescent lights have a more pleasant appearance than CFLs and LEDs (although that is a matter of opinion).
- Incandescent lights and CFLs generate much more heat than LEDs.
- CFLs and LEDs have much longer lifetimes than incandescent lightbulbs (although the lifetime of CFLs is significantly reduced if turned on and off frequently).
- CFLs and LEDs use less energy than incandescent lightbulbs.
- CFLs and LEDs are more expensive than incandescent lightbulbs.
- CFLs sometimes take a while to warm up and reach their maximum brightness.
- CFLs contain tiny amounts of mercury, which can be unsafe for humans and the environment.

Have students consider these and other advantages and disadvantages of each kind of lightbulb and fill in the last two columns of the chart (advantages and disadvantages). You may want to show students the video "Energy 101: Lighting Choices" from the U.S. Department of Energy, which explains the differences in efficiency in these three kinds of bulbs (see "Websites" section). Table 9.2 (p. 123) shows a sample comparing lightbulbs chart.

Finally, have students fill in the similarities and differences beneath the chart:

1. Similarities: Each of these bulbs converts <u>electricity</u> into <u>light</u> and some <u>heat</u>.
2. Differences: These bulbs use different <u>materials</u> to convert energy.

elaborate

Orion and the Dark Read-Aloud

 Questioning

Show students the cover of *Orion and the Dark* and introduce the author and illustrator, Emma Yarlett. *Ask*

? What do you think this book might be about? (Answers will vary.)

> **Connecting to the Common Core**
> **Reading: Literature**
> KEY IDEAS AND DETAILS: 3.1, 4.1, 5.1

Read the book aloud. *Ask*

? Have you ever been afraid of the dark? (Answers will vary.)

? Do you know anyone who is afraid of the dark? (Answers will vary.)

? How did Orion overcome his fear? (He became friends with the dark.)

? How could you help someone who is afraid of the dark? (talk to them, give them a special stuffed animal, make them a nightlight)

Nightlight Design Challenge

Tell students that you want to challenge them to design a nightlight for someone, like Orion, who is afraid of the dark. Review the criteria and constraints for the challenge. Note that you will need to define the time limit—for example, 60 minutes, one class period, or before a given due date (if they will be working for multiple class periods).

Criteria:

1. The nightlight must light up.
2. The nightlight must be able to rest on a nightstand.
3. The nightlight must have a switch to turn it on and off.

Constraints:

1. The nightlight must be built using only the materials provided or approved by your teacher.
2. The nightlight must be built within the given time limit. Time limit: _____

Paper Circuits

Tell students that to make their nightlights, they will be using small LED lights, small coin cell (or "button") batteries, and a special kind of tape that conducts electricity. This type of design is called a *paper circuit*. Explain that before they design the nightlight, they will have some time to practice making a paper circuit. Show students the supplies for making a paper circuit:

- Mini LED bulb
- Index card
- Conductive copper foil circuit tape (about 20 cm)
- 3V lithium coin cell battery
- Scissors
- Transparent tape
- Binder clip
- Markers

Explain that these supplies can be used just like the wires, battery, and flashlight bulb they used

> **SAFETY**
> - Have students wear safety glasses or goggles during this activity.
> - Use caution when working with glass bulbs. Bulbs can easily break and puncture the skin or eyes.
> - Use caution when working with batteries and bulbs in an active circuit. They can heat up and burn skin if touched.
> - Use caution when working with wires, scissors, and other sharp objects to avoid cutting or puncturing your skin or eyes.
> - Use caution when working with wire cutters. They can pinch and cut skin.

in the lesson before. The tape acts as the wire, the 3V battery supplies the energy, and the bulb has both an "in" wire and an "out" wire that can be bent. Point out the positive side and negative side of the battery. Give each student a copy of the the Paper Circuits student page, a small LED bulb, a 3V coin battery, a 20 cm strip of circuit tape, an index card, and a binder clip. Then, guide them through the following steps to practice making a complete circuit:

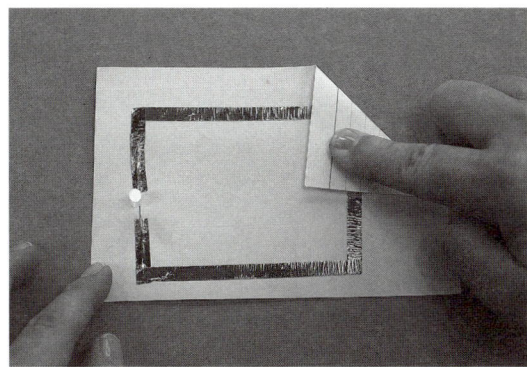

Making a paper circuit

1. Fold over one corner of the index card and trace the battery on each side of the fold.
2. Tape down one piece of circuit tape coming out of the center of one circle.
3. Tape down one piece of circuit tape coming out of the center of the other circle.
4. Place the bulb on the index card, and tape it down using two pieces of circuit tape, one on each side.
5. Connect all the pieces into a loop using the circuit tape. Fold the battery in the corner you created earlier. Does the bulb light? If not, flip the battery over. Be sure you are not touching the copper tape with your fingers or the circuit will not work.
6. Clip a binder clip on the corner to hold the battery in place.

After all students have successfully built a paper circuit, read and discuss the "How It Works" explanation at the bottom of the page together.

How It Works: Energy Is Converted and Transferred

When the paper circuit is complete, the chemical energy inside the battery is **converted** into electrical energy. The electrical energy is **transferred** through the circuit tape. When it reaches the bulb, the LED **converts** electrical energy to light energy (and some heat energy).

Next, give each student a copy of the Nightlight Design Challenge student page. Have them review the criteria and constraints, brainstorm ideas, and sketch a design in the box. Once they have completed the student page and you have signed off at the Teacher Checkpoint, students can use the supplies to create their nightlights. They will be able to use the same battery, LED bulb, and binder clip as they did when building their practice paper circuit, but they will need a new strip of circuit tape. They will also need scissors, markers, and a variety of nonconductive materials (such as cardstock, small boxes, wooden or plastic blocks, plastic bottles, etc.) to select from to build a base for their nightlights. To optimize their designs after their initial attempts, students may need additional circuit tape.

evaluate

Nightlight Instruction Manual

> Connecting to the Common Core
> **Writing**
> PRODUCTION AND DISTRIBUTION OF WRITING: 3.4, 4.4, 5.4

Writing

Once students are satisfied with their nightlight designs, give each student a copy of the Nightlight Instruction Manual student page. Have them decorate the cover and write their name on it. Inside, they will list the parts, draw a labeled diagram, give instructions for use, explain how it works, and give troubleshooting tips.

> Connecting to the Common Core
> **Speaking and Listening**
> PRESENTATION OF KNOWLEDGE AND IDEAS: 3.4, 4.4, 5.4

Finally, have students share their nightlight prototypes and instruction manuals with others.

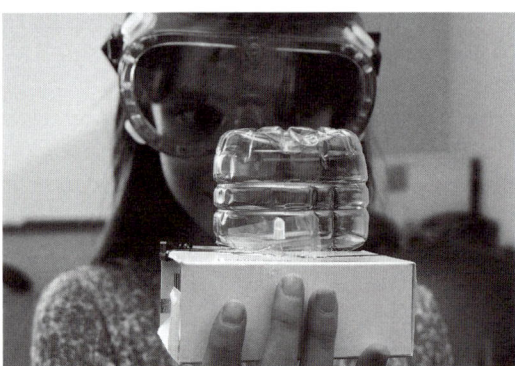

NIGHTLIGHT DESIGN

STEM at Home

Have students complete the "I learned that …" and "My favorite part of the lesson was …" portions of the STEM at Home student page as a reflection on their learning. They may choose to do the following at-home activity with an adult helper and share their results with the class. If students do not have access to the internet or these materials at home, you may choose to have them complete this activity at school.

"At home, we can watch a short video called "How It's Made—Incandescent Lightbulbs," which shows how lightbulbs are manufactured.

Search "How It's Made Incandescent Lightbulbs" to find the Science Kids video at www.sciencekids.co.nz/videos/engineering/lightbulbs.html.

"As we watch the video together, we can listen for the following information:

1. What are some of the materials used to make a lightbulb? (glass, wires, metal)
2. Why are some lightbulbs white? (They are coated on the inside with a white powder.)
3. What kind of gas was put in these lightbulbs? (argon)
4. How are the lightbulbs tested? (They are lit up multiple times throughout the process.)"

National Science Teachers Association

For Further Exploration

This section is provided to help you encourage your students to use the science and engineering practices in a more student-directed format. This box lists questions and challenges related to the lesson that students may select to research, investigate, or innovate. Students may also use the questions as examples to help them generate their own questions. After selecting one of the questions in the box or formulating their own questions, students can individually or collaboratively make predictions, design investigations or surveys to test their predictions, collect evidence, devise explanations, design solutions, or examine related resources. They can communicate their findings through a science notebook, at a poster session or gallery walk, or by producing a media project.

Research
Have students brainstorm researchable questions:

? What is the world record for the longest lasting lightbulb? (See the "Bulb-cam" at *www.centennialbulb.org*!)

? When and why was the CFL invented, and who invented it?

? What are some common uses for LEDs?

Investigate
Have students brainstorm testable questions to be solved through science or math:

? What happens when you use two batteries to light up a lightbulb?

? Can you use more than one LED in a paper circuit?

? What kinds of objects can be used as a switch in a paper circuit?

Innovate
Have students brainstorm problems to be solved through engineering:

? Can you design a piece of artwork using an LED, circuit tape, and a battery?

? Can you light every room in a dollhouse using batteries, circuit tape or wires, and LEDs or flashlight bulbs?

? What are some other practical uses of a paper circuit other than a nightlight?

Reference

Ansberry, K., and E. Morgan. 2010. *Picture-perfect science lessons: Using children's books to guide inquiry, 3–6.* 2nd ed. Arlington, VA: NSTA Press.

Websites

"Energy 101: Lighting Choices" (video)
http://energy.gov/energysaver/lighting-choices-save-you-money

Energy.gov: Estimating Appliance and Home Electronic Energy Use
www.energy.gov/energysaver/estimating-appliance-and-home-electronic-energy-use

Lighting Facts
www.lightingfacts.com/Library/Content/FTCLabel

U.S. Department of Energy
www.energy.gov

U.S. Department of Energy Lighting Facts
www.lightingfacts.com

More Books to Read

Brown, D. 2010. *A wizard from the start: The incredible boyhood and amazing inventions of Thomas Edison.* New York: Houghton Mifflin.
Summary: This "storyography" depicts Thomas Edison's life from his humble beginnings as a farmer's son selling newspapers on trains and reading through public libraries shelf by shelf, to his inventing career, to eventually his becoming a world-renowned legend.

Editors at Time for Kids. 2005. *Thomas Edison: A brilliant inventor.* Time for Kids Biographies. New York: HarperCollins.
Summary: This book gives readers an up-close look at the life and work of Thomas Edison and is illustrated with historical and contemporary photographs.

Oxlade, C. 2012. *Making a circuit.* Chicago: Heinemann Library.
Summary: From the *It's Electric* series, this simple book full of helpful photographs provides an introduction to circuits. It includes information on conductors, insulators, switches, series and parallel circuits, and much more.

Name: _____

How a Lightbulb Works

Lightbulb

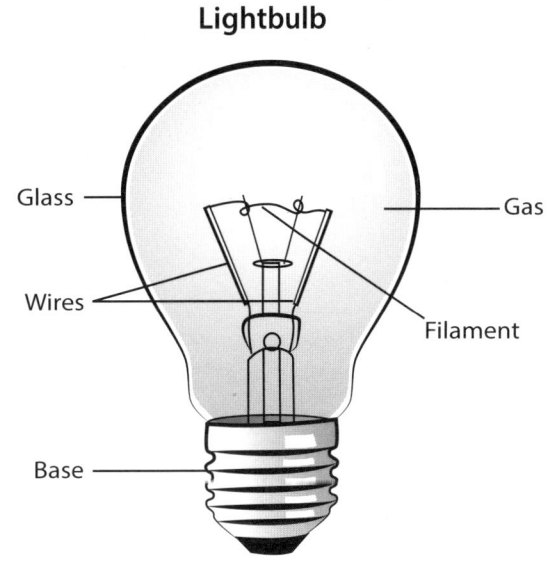

Directions: Cut out the lightbulb vocabulary words. Before reading *How Things Work: Lightbulbs*, place each word in a space. After reading, you will have the opportunity to move the words.

The curly piece of metal

inside a(n) _____

lightbulb is called the filament. It is held

up by support wires. When you turn on the light, _____

travels from a wire to the metal base of the bulb to the filament. The filament

gets so hot, it glows. The _____ keeps the filament safe.

There cannot be any _____ inside or the filament will burn out! A

_____ does not use a filament. Another type of lightbulb

uses small _____ lights.

Lightbulb Vocabulary Words

incandescent	CFL	glass bulb
LED	air	electricity

Picture-Perfect STEM Lessons, 3–5

Name : _____

Comparing Lightbulbs

Type of Lightbulb	Light Appearance	Life	Price	Estimated Energy Cost Per Year	How It Produces Light	Advantages	Disadvantages
Incandescent							
CFL (compact fluorescent lamp)							
LED (light-emitting diode)							

1. Similarities: Each of these bulbs converts _____ into _____ and some _____.

2. Differences: These bulbs use different _____ to convert energy.

Name : _____

Paper Circuits

Directions: Follow the steps below to make a paper circuit.

Materials

- Mini LED bulb
- Coin battery
- Circuit tape
- Index card
- Scissors
- Binder clip

1. Fold over one corner of the index card and trace the battery on each side of the fold.

2. Tape down one piece of circuit tape coming out of the center of one circle.

3. Tape down one piece of circuit tape coming out of the center of the other circle.

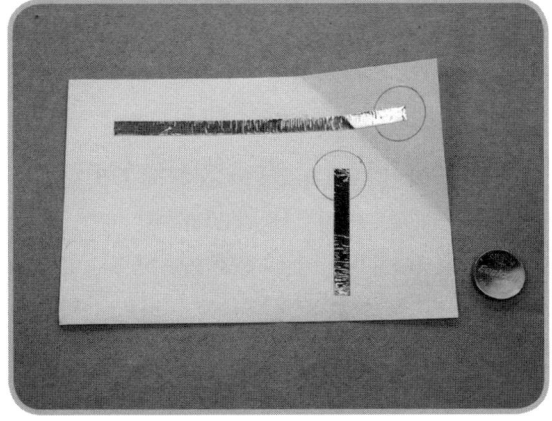

4. Place the bulb on the index card, and tape it down using two pieces of circuit tape, one on each side.

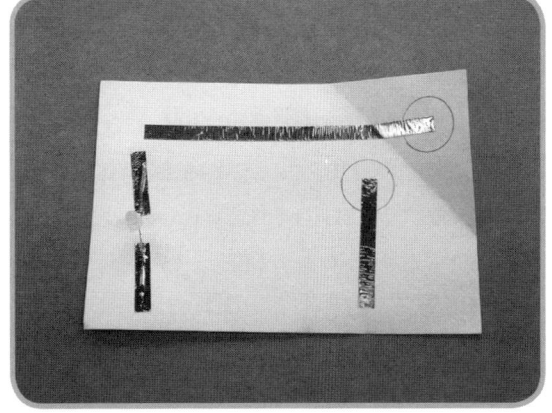

Picture-Perfect STEM Lessons, 3–5

5. Connect all the pieces into a loop using the circuit tape.

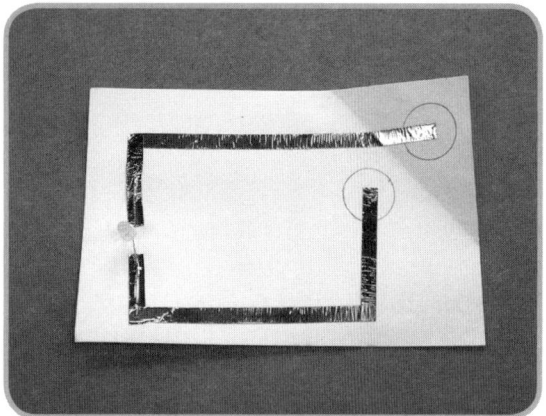

6. Fold the battery in the corner you created earlier. Did the bulb light? If not, flip the battery over. Be sure you are not touching the copper tape with your fingers or the circuit will not work.

7. Clip a binder clip on the corner to hold the battery in place.

How It Works: Energy Is Converted and Transferred
When the paper circuit is complete, the chemical energy inside the battery is **converted** into electrical energy. The electrical energy is **transferred** through the circuit tape. When it reaches the bulb, the LED **converts** electrical energy to light energy (and some heat energy).

Name: _____

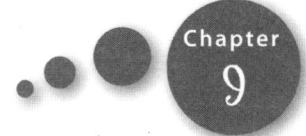

Nightlight Design Challenge

Problem: Design a nightlight for someone who is afraid of the dark.

Criteria (desired features or outcomes):
1. The nightlight must light up.
2. The nightlight must be able to rest on a nightstand.
3. The nightlight must have a switch to turn it on and off.

Constraints (limits on materials and time):
1. The nightlight must be built using only the materials provided or approved by your teacher.
2. The nightlight must be built within the given time limit.
   Time limit: _____

Brainstorm ideas, then sketch your design below:

[sketch box]

How will you know whether your nightlight design is successful?

_____

_____

Teacher Checkpoint ☐

Picture-Perfect STEM Lessons, 3–5

Nightlight Instruction Manual

By: _____

Troubleshooting Tips

If the nightlight does not work, try:

National Science Teachers Association

Using the Nightlight
How to use the nightlight:

_____

_____

_____

How It Works
In this nightlight, _____ from

the battery flows through a very tiny chip in the

_____ LED and gives off _____ so

you can see!

Parts

_____

_____

_____

_____

Labeled Diagram

Name: _____

STEM at Home

Dear _____,

At school, we have been learning about **how lightbulbs work.**
I learned that:

_____

_____

My favorite part of the lesson was:

_____

_____

At home, we can watch a short video called "How It's Made—Incandescent Lightbulbs," which shows how lightbulbs are manufactured.

🔍 Search "How It's Made Incandescent Lightbulbs" to find the Science Kids video: *www.sciencekids.co.nz/videos/engineering/lightbulbs.html.*

As we watch the video together, we can listen for the following information:

1. What are some of the materials used to make a lightbulb?

_____

2. Why are some lightbulbs white?

_____

3. What kind of gas was put in these lightbulbs?

_____

4. How are the lightbulbs tested?

_____

Burn

Description

After reading about Michael Faraday's famous lecture for children called "The Chemical History of a Candle," students explore the extraordinary chemical and physical changes that occur as an ordinary candle is burned. They elaborate on their understandings about combustion and the conservation of matter by learning about wildfires and firefighting technologies. Then, they compare a variety of solutions used to combat fires.

Suggested Grade Levels: 3–5

LESSON OBJECTIVES Connecting to the *Framework*		
Science and Engineering Practices	**Disciplinary Core Ideas**	**Crosscutting Concept**
Obtaining, Evaluating, and Communicating Information	**PS1.B:** Chemical Reactions **ETS2.B:** Influence of Engineering, Technology, and Science on Society and the Natural World	Cause and Effect

Featured Picture Books

TITLE: ***Burn: Michael Faraday's Candle***
AUTHOR: **Darcy Pattison**
ILLUSTRATOR: **Peter Willis**
PUBLISHER: **Mims House**
YEAR: **2016**
GENRE: **Narrative Information**
SUMMARY: *This book takes readers back in time to December 28, 1848, in London, England, as British scientist Michael Faraday gives one of his famous juvenile science Christmas lectures at the Royal Institution. The whimsical artwork helps clarify Faraday's explanations of the science behind a burning candle.*

TITLE: ***National Geographic Kids: Wildfires***
AUTHOR: **Kathy Furgang**
PUBLISHER: **National Geographic Children's Books**
YEAR: **2015**
GENRE: **Non-Narrative Information**
SUMMARY: *Full of fascinating facts about how wildfires happen, why they are ecologically important, when they are dangerous, and how they are controlled, this book is sure to ignite the interest of young readers.*

Time Needed

This lesson will take several class periods. Suggested scheduling is as follows:

Day 1: Engage with *Burn: Michael Faraday's Candle* Read-Aloud, **Explore** with Candle Observations and **Explain** with Capillary Action Demonstration

Day 2: Explain with Birthday Candles Probe: Part 1 and "The Chemistry of a Candle" Article

Day 3: Elaborate with *National Geographic Kids: Wildfires* Read-Aloud

Day 4: Evaluate with *National Geographic Kids: Wildfires* Questions, "A Green Way to Fight Fires" Video, and Birthday Candles Probe: Part 2

Materials

For Burn: Michael Faraday's Candle *Read-Aloud*

- Photo or projection of Royal Institution of Great Britain Faraday Theatre from a Google images search
- Strip of white paper towel
- Small container of water darkly colored with food coloring
- Safety goggles for each student

For Candle Observations (per group of 3–4 students)

- Aluminum pie pan
- Small votive candle in a metal base (with an unburnt wick)
- Colored pencils

For teacher use only

- Child-resistant candle lighter
- Metal candle snuffer

Student Pages

- Candle Observations
- Birthday Candles: Part 1
- The Chemistry of a Candle
- Wildfires
- Birthday Candles: Part 2
- STEM at Home

SAFETY

- Roll up your sleeves, secure any loose clothing, and tie back long hair.
- Wear safety goggles over your eyes.
- Never reach over or touch the flame.
- Keep your work area clean and clear of flammable materials.
- Remind students that lighted candles and melting wax are hot and can burn skin.
- Have a fire extinguisher available, and be trained on how to use it.

Background for Teachers

What Is Matter?

Matter is all around us. It is defined as anything that has mass and takes up space. The paper this book is written on, the water in your bottle, the air you are breathing—it's all made of matter! Atoms are the smallest particles of matter. They are so small that you cannot see them with just your eyes or even with a standard microscope. Atoms combine to form *molecules,* and these molecules make up a variety of substances. Matter can be described by its properties. Some properties of matter include color, texture, hardness, solubility (ability to dissolve in other substances), reactivity (ability to chemically react with other substances), and state.

Matter Can Change

Most matter on Earth is found in one of three states: solid, liquid, or gas. Matter can change from one state to another. A common example of matter changing state is liquid water *freezing* to form solid ice, or *boiling* or *evaporating* to form a gas called *water vapor.* Water vapor can even be produced directly from ice; this is called *sublimation.* Water vapor can change back into liquid water through *condensation.* The driving force behind the changes in the states of matter is heat energy. Heating a substance adds energy to the molecules, causing them to move more. For example, when solid wax is heated, the molecules begin to flow under and over each other. At the right temperature, the wax melts and becomes liquid. These are all examples of *physical changes*, which create different states, or forms, of the same matter. Water is still water if it changes from a liquid state into ice, which is a solid state. Wax is still wax if it changes from a solid state into melted wax, which is a liquid state.

This lesson addresses disciplinary core idea PS1.B. Chemical Reactions from the *Framework*, which states that by the end of grade 5, students understand that "When two or more substances are mixed, a new substance with different properties may be formed" (NRC 2012, p. 110) This core idea refers to a change in matter known as a *chemical change.* Chemical changes create entirely new substances. After a chemical change occurs, physical methods, such as drying, filtering, or changing temperature, cannot undo the change. In a chemical change, or *reaction*, the molecules of different materials rearrange to form entirely new *compounds.* The new compounds have different properties. Observing any of the following when you combine two or more substances can give you clues that a chemical change has occurred:

- Gas produced
- Heat given off or absorbed
- Solid formed or disappeared
- Odor changed
- Color changed
- Light produced

Any of those phenomena may be evidence of a chemical change after combining substances, but a physical change can sometimes have similar results. For example, boiling water causes gas bubbles to appear. The bubbles contain water vapor—liquid water that has physically changed into a gas—so no *new* substance is produced. Another example is mixing paint. Although the resulting color may be different from the original colors, the chemical properties of the paint are the same. No new substance has been produced—it's still paint! A common misconception about distinguishing between physical and chemical changes is that after a physical change you can "change it back." That is not always true.

For example, after you tear a piece of paper into a thousand pieces, you can't return it to its original form. But tearing paper is a physical change because the small pieces are still the same substance as the whole piece of paper was, whereas burning paper is a chemical change because the products of burning are completely new substances. In fact, some chemical reactions *can* be reversed. Perhaps the most useful defining characteristic of a chemical change is the presence of new substances with chemical and physical properties that are entirely different from the starting substances. Sometimes, distinguishing between a chemical change and a physical change can be difficult, but the important idea is that matter can be changed in different ways.

Burning Matter

One way to chemically change a substance is to burn it. Burning something forms new substances—invisible gases and tiny airborne particles that you can see as smoke. Burning, or *combustion*, requires three things: oxygen, heat, and *fuel* (a combustible substance). When a *hydrocarbon* (chemically bonded carbon and hydrogen atoms) such as wood is heated to the point that it burns, it is chemically changed. The molecules of the original substance are rearranged during combustion and are released into the air as new compounds—mainly carbon dioxide gas (CO_2) and water vapor (H_2O). Heat and light energy are also produced during combustion. Some *ash*, a powdery substance composed mainly of calcium carbonate and minerals, may be left behind. *Soot*, the impure carbon particles remaining from the incomplete combustion of hydrocarbons, is also left behind.

In the case of a burning candle, a physical change occurs in the wax as the heat of the flame melts the solid wax at the base of the wick. The liquid wax, essentially a cup of hydrocarbon fuel for the flame, is then drawn up the wick by *capillary action*. The flame at the top of the wick *vaporizes* the liquid wax (turns it into a hot gas) and draws the vaporized hydrocarbon molecules up into the flame, where the chemical change occurs. The wax vapor reacts with oxygen from the air in a chemical reaction to produce carbon dioxide gas, water vapor, and energy in the form of heat and light. So when you see a candle flame, it's not solid wax or even liquid wax that is burning—it's wax vapor!

In this lesson, students are introduced to the crosscutting concept of cause and effect by observing a burning candle and wondering how wax moves up the wick to the flame. The great scientist Michael Faraday, in the book *Burn: Michael Faraday's Candle* (which is based on one of his famous Christmas lectures for children) remarks, "I hope you will always remember that whenever a result happens, especially if it be new, you should say, 'What is the cause? Why does it occur?' And you will, in the course of time, find out the reason." And through observations and reading, students do find the reason. In Michael Faraday's words, "It is by what is called *capillary action* that the fuel is carried to the part where the combustion or burning goes on." Likewise, the book poses the question of why a flame is an oblong shape, and students learn the cause: Currents of hot air draw out its shape.

Conservation of Matter

This lesson also addresses another part of the the disciplinary core idea PS1.B. Chemical Reactions, which states, "No matter what reaction or change in properties occurs, the total weight of the substances does not change" (NRC 2012, p. 111). When any physical or chemical change occurs, including the chemical reaction of burning, matter is conserved. This conservation is easy to demonstrate with most physical changes. For example, if you measure the mass of a stick, then break it in half and measure it again, the mass remains the same. The conservation of matter during a combustion reaction is nearly impossible to demonstrate in the elementary classroom. Although you could find the mass of a stick before burning it, and the mass of the remaining ashes, you could not measure the mass of the oxygen used in the reaction, and you could not measure the mass of the gases released into the air during burning. In theory, when wood is burned the mass of the *reactants* (wood and oxygen) is equal to the mass of the *products* (carbon dioxide gas, water vapor, and leftover ashes) (see Figure 10.1). Matter cannot be created or destroyed, only changed, even when it is burned. This is known as the *law of conservation of matter*. In a chemical reaction, the amount of matter present in the reactants is always equal to the amount of matter present in the products.

Figure 10.1. Conservation of Matter When Wood Burns

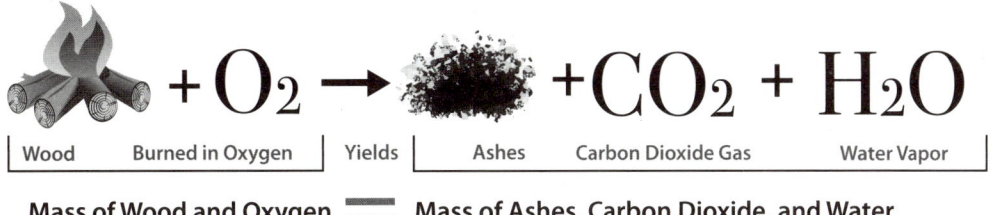

Wildfires

The crosscutting concept of cause and effect is reinforced as students explore the effects of fire on an ecosystem. They learn that wildfires can happen almost anywhere in the world, but they happen most often in places with hot, dry weather. They learn that although wildfires can cause a lot of damage, they are also an important part of some ecosystems. Wildfires can remove sick plants and harmful insects that kill trees. Some evergreen trees cannot reproduce unless the extreme heat of a wildfire releases their seeds. Wildfires can thin out crowded forests so that other plants can grow, and the ashes that result from combustion add nutrients to the soil.

In the elaborate phase of this lesson, students apply what they have learned about the chemical change of burning by comparing the reactants and products of a burning candle to the reactants and products of a burning wildfire, as well as evaluating different wildfire-fighting methods and technologies. Only three things are needed for a wildfire (or any other fire for that matter): fuel, heat, and oxygen. In a wildfire, sources of fuel could be grasses, trees, sticks, and other plant life. Fire spreads very quickly when plant life is dry, such as during a drought. Sources of heat could be natural (lightning strikes or hot lava) or manmade (an unattended campfire or a lit match or cigarette). The source of oxygen is the air all around. To burn, fuel must react with oxygen in the air. The combination of fuel, heat, and oxygen is known as the *fire triangle*. Removing any part of the triangle can slow or stop a wildfire. Firefighters can do a *prescribed burn* of vegetation or remove brush and dig ditches to create a *fire line* to take away the source of fuel for a wildfire. They can reduce heat by spraying water from hoses or airplanes on a wildfire. They can remove the oxygen supply by dropping fire retardant material on a wildfire. Those types of fire control methods are studied in fire science labs, where scientists and engineers work together to create new technologies to fight fires. The more we understand about wildfires, the better we will be at keeping people safe and ecosystems healthy.

Chapter 10

engage

Burn: Michael Faraday's Candle Read-Aloud

Show students the cover of *Burn: Michael Faraday's Candle* and introduce the author, Darcy Pattison, and illustrator, Peter Willis. Tell students that this book is based on a famous lecture given more than 160 years ago to children in London, England, at the Royal Institution of Great Britain. The speaker was a brilliant British scientist named Michael Faraday. He was known as one of the greatest science experimenters in history. Faraday had a passion for finding the answers to the most basic questions of science: "What is the cause?" and "Why does it occur?"

 Visualizing

Before reading, tell students that the Royal Institution still stands in London. In fact, the theater inside is named after Michael Faraday. Show students a photograph of the Royal Institution's theater (Google images search: Royal Institution of Great Britain Faraday Theatre). Tell them to imagine it is a cold day in December 1848, and they are headed inside. Then, read pages 4–9. Have students close their eyes and imagine 4,000 children and adults crowded onto hard wooden benches inside the three-story theater. *Ask*

- ? What would it look like?
- ? What would it sound like?
- ? Have you ever been inside a crowded theater just before a show or lecture of some sort? What was it like?
- ? How would you feel if you were one of the lucky children there that day who got to see a famous scientist?

Have them visualize Michael Faraday standing behind the experiment desk and holding a burning candle. Then, read pages 10–13, stopping after "This is a wonderful thing about a candle." *Ask*

- ? What *is* a wonderful thing about a candle?
- ? Have you ever *closely* watched a candle burning?

ENGAGING WITH BURN: MICHAEL FARADAY'S CANDLE

- ? What did you notice?
- ? Did you ever wonder how the wax moves up the wick where it can be burned?

explore

Candle Observations

 Stop and Try It

Tell students that, rather than reading Michael Faraday's explanation, you would like them to first observe a burning candle closely (and safely!). At this point, the students will most likely be very excited, so before passing out the materials make sure you have discussed the following fire safety precautions:

- Roll up your sleeves, secure any loose clothing, and tie back long hair.

142   National Science Teachers Association

- Wear safety goggles over your eyes.
- Never reach over or touch the flame.
- Keep your work area clean and clear of flammable materials.
- Do not touch lighted candles and melting wax; they are hot and can burn skin.
- Be sure to follow this guideline from the third edition of *Safety in the Elementary Science Classroom* (American Chemical Society 2011), "Teachers should never leave the room while any flame is lighted or other heat source is in use." Also, know the location of the nearest fire extinguisher, and know how to use it.

Tell students that you have a challenge for them. Their challenge is to **make as many observations of a candle as they can in 10 minutes!** Before they begin, review the difference between an observation and an inference.

- Making an *observation* involves using one or more of the senses to find out about objects or events. (Today, you will be using only your senses of sight and smell!)

OBSERVING A CANDLE

- Making an *inference* involves logical reasoning—drawing a conclusion using prior knowledge to explain your observations.

Next, hand out the Candle Observations student page to each student. Tell students that they will be drawing a close-up, detailed picture of the flame in the oval provided and writing their candle observations in the first column. Colored pencils will come in handy for drawing different colors of the flame. Remind them to write only observations; inferences should be phrased as questions in the Wonderings column. For example, an observation written in the first column might be, "The flame has an oblong shape." A corresponding inference written in the second column might be "I wonder why the flame has an oblong shape?" or "I wonder if air currents affect the shape of a flame?"

Then, give each group of three to four students an aluminum pie pan containing a votive candle in a metal base and reiterate the task: **"You will be making as many observations of a candle as you can in 10 minutes!"** Encourage students to get ready to make observations right away, because some very interesting things happen the second a candle is lit! Light each candle using a child-resistant candle lighter and start a timer. Students will likely be surprised by all of the different observations of an ordinary votive candle they can make. Encourage them to write as many observations as they can in the first column and as many wonderings (inferences) as they can in the second column.

After 10 min. have passed, discuss the observations and wonderings they recorded. Then, tell students that they will be able to make even more observations after you put out the candles! Go around to each group and extinguish the candles using a metal candle snuffer. Give students another minute to write any additional observations, then discuss what they observed. *Ask*

? Were you able to observe any liquid wax moving up the wick where it could be burned?

? What is the cause? Why does it occur?

Picture-Perfect STEM Lessons, 3–5

CAPILLARY ACTION DEMONSTRATION

explain

Go back to the book *Burn: Michael Faraday's Candle* and read pages 14–18, stopping after "There is a beautiful point about that—capillary action." *Ask*

? Have you ever heard the term *capillary action*?

? What do you think it means?

Capillary Action Demonstration

 Stop and Try It

Invite students to closely observe as you demonstrate the phenomenon of capillary action. Place the bottom of a long strip of white paper towel into a clear container of darkly colored water. Students should notice the colored water being very slowly pulled up the paper towel by capillary action.

> **Connecting to the Common Core**
> **Reading: Informational Text**
> Key Ideas and Details: 3.1, 4.1, 5.1

 Questioning

Then, read pages 19–21, and *ask*

? How does Michael Faraday explain capillary action? (when two substances won't dissolve in each other, but instead hold together)

? How did the water move up the paper towel by capillary action? (The water moved to the top of the paper towel; other water particles followed because the water particles are attracted to each other.)

? How does melted wax move to the top of a wick? (The melted wax climbs to the top of the cotton wick; other wax particles follow because the wax particles are attracted to each other.)

? Why do you think the candle does not burn the wick all the way down to the melted wax? (Answers will vary.)

Next, read pages 22–23, which explain, "The only reason why the candle does not burn all down the side of the wick is that the melted wax extinguishes, or puts out the flame."

Then, read pages 24–25, stopping after "Now as to the shape of the flame …" without reading the rest of the sentence. *Ask*

? What *is* the shape of a candle flame? What did you observe? (oblong, pointy at the top, etc.)

? What is the cause? Why does it occur? (Answers will vary.)

Read pages 25–27, then *ask*

? So why does the flame appear the way it does? (A current of hot air streaming upward draws the flame out.)

Continue reading to the end of the book, including the end matter about Michael Faraday and the Royal Institution's Christmas Lecture Series. Then, read the paragraph under "Observe the Changes: Solid, Liquid, Gas" on the last page. *Ask*

? Does burning a candle demonstrate a physical change or a chemical change? (Answers will vary.)

Tell students that during the next class period, they will learn much more about the chemistry of a candle!

Birthday Candles Probe: Part 1

(*Note:* This activity is adapted from Keeley and Tugel 2009.)

Give each student a copy of Birthday Candles Probe: Part 1. Have them circle the response they agree with and explain their thinking. Collect the papers, and use them to assess students' preconceptions about conservation of matter in chemical changes. Write this question on the board:

Where does candle wax go when it burns?

Tell students that you would like them to think about this question as they read an article called "The Chemistry of a Candle." Tell students that they will have a chance to respond to the probe again later.

"The Chemistry of a Candle" Article

 *Pairs Read*

> Connecting to the Common Core
> **Reading: Informational Text**
> Range of Reading and Level of Text Complexity 3.10, 4.10. 5.10

Give each student a copy of "The Chemistry of a Candle" article and have students do a pairs read. In a pairs read, one student reads a paragraph while the other listens and then makes comments ("I think …"), asks questions ("I wonder …"), or shares new learning ("I didn't know …").

> Connecting to the Common Core
> **Reading: Informational Text**
> Key Ideas and Details: 3.1, 4.1, 5.1

 Questioning

After the reading, have students answer the questions on the third page. Then, discuss students' responses to the questions. Answers are as follows:

1. What is a physical change that occurs when a candle burns? (The wax melts.)
2. What are the signs that a chemical change is occurring when a candle burns? (color change, odor change, heat and light produced)
3. What is the difference between a physical change and a chemical change? (During a chemical change, a new substance with new properties is produced.)
4. What is combustion? (burning)
5. What are the new substances that form when wax combines with oxygen from the air in a combustion reaction? (carbon dioxide gas and water vapor)
6. What is meant by the conservation of matter? (Matter cannot be created or destroyed, only changed.)
7. When you burn a candle, is matter created or destroyed? Explain. (Neither, but new substances are formed from the original substances.)
8. If you weighed a birthday candle before and after burning it on your birthday cake, would it weigh the same? Explain. (No, it would weigh less. The solid matter changes into gases and goes into the air.)

elaborate

National Geographic Kids: Wildfires Read-Aloud

Turn and Talk

Pass out the Wildfires student page. Show students the cover of the book *National Geographic Kids: Wildfires* and introduce the author, Kathy Furgang. Tell students that they can apply what they have learned about combustion reactions and the law of conservation of matter to better understand the causes and prevention of wildfires, including the various technologies that are used to fight wildfires. Ask students to discuss what they know or have heard about wildfires with a partner.

Then, have students take the short "before reading" quiz on the student page (adapted from pages 44–45 of *Wildfires*) by underlining their answers on the first page and answering the bonus question at the bottom. Tell them that they will have a chance to revisit the quiz after you read the book aloud.

Next, read aloud *National Geographic Kids: Wildfires*, stopping to have students check their answers as you read the information that pertains to each item on the quiz. They can then circle the correct answers. Be sure to show the pictures and share the captions as you read. After reading, you may want to show the 2:43 min. video from PBS Learning Media called "Wildfire," which illustrates many of the concepts presented in the book (see "Websites" section).

> Connecting to the Common Core
> **Reading: Informational Text**
> Key Ideas and Details: 3.1, 4.1, 5.1

 Questioning

After reading the book and watching the video about wildfires, *ask*

? What are some of the wildfire management methods and technologies you learned about? (prescribed burns or back burns, fire lines, smoke jumping, sky Jell-O, Pulaski tool, "ping pong balls," water sprayed from hoses and dropped from airplanes, fire retardant foam, removal of ground cover, etc.)

? What are some of the advantages and disadvantages of these different methods and technologies? (Answers will vary.)

? What do they all have in common? (They all remove part(s) of the fire triangle.)

? What kind of engineer designs the chemicals to put out fires, such as the sky Jell-O? (chemical engineers)

The before-reading quiz answers are bold and underlined below:

1. Which of these are ways that wildfires can start?

   a. lightning strike

   b. hot lava

   c. matches

   d. All of the above

2. Which of these is **not** part of the "fire triangle?"

   a. heat

   b. fuel

   c. water

   d. oxygen

3. Which of these is a beneficial effect of a wildfire?

   a. They thin out crowded forests.

   b. Ashes add nutrients to the soil.

   c. They remove sick plants.

   d. They remove insects that kill trees.

   e. All of the above

   f. Wildfires are never beneficial.

4. What fraction of wildfires are caused by humans?

   a. 1 out of 100

   b. 1 out of 10

   c. 4 out of 5

   d. None of the above

5. Which **does not** describe a method of fighting wildfires?

   a. Digging a ditch to create a fire line

   b. Setting part of a forest on fire

   c. Parachuting into a fire zone

   d. Dropping water from airplanes

   e. Covering a fire zone with plastic

   f. Dropping fire retardant "sky Jell-O" from airplanes

6. How can you help keep wildfires from starting?

   a. Never leave a campfire alone.

   b. Put out a campfire before leaving.

   c. Never start a campfire during a dry spell.

d. All of the above

Bonus: What is Smokey Bear's fire safety message?

"Only you can prevent wildfires."

evaluate

National Geographic Kids: Wildfires Questions

 Synthesizing

After reading *National Geographic Kids: Wildfires*, have students complete the first three "after reading" questions on the second page as an assessment of their understandings about combustion reactions and the law of conservation of matter as they apply to wildfires. Answers are as follows:

1. What three things are needed for a fire to burn? (fuel, heat, and oxygen)
2. Think back to what you have learned about *combustion*, or burning, from observing a lighted candle. What new substances do you think are formed when a wildfire burns? (carbon dioxide gas, water vapor, and ashes)
3. When a tree burns in a wildfire, it can be reduced to a pile of ashes. What happens to the rest of the matter that makes up the tree? (It changes into invisible gases—CO_2 and H_2O—and spreads out into the air.)

"A Green Way to Fight Fires" Video

After students have finished questions 1–3, tell students that a new firefighting technology is being tested. Show the 5:31 min. NOVA video from PBS Learning Media called "A Green Way to Fight Fires." The video is about TetraKO, an environmentally friendly product. Have them think about how this technology compares with the ones they heard about in the book *National Geographic Kids: Wildfires*, and then have them answer question 4.

4. Now, watch as your teacher shows a video about a new technology for fighting fires called TetraKO. After watching, think about the firefighting technologies you read about in the book *Wildfires*. What are the advantages of this new technology? (It puts out fires faster than anything else; it is all natural; it is a gel that can be sprayed out of a hose as a liquid, but turns into a solid when it hits a surface; it can be applied up to a day ahead of an approaching wildfire; and it is safe for plants, wildlife, and humans.)

Birthday Candles Probe: Part 2

Finally, administer the Birthday Candles: Part 2 assessment probe to see if students' conceptions about conservation of matter in chemical reactions have changed. The best answer is Lily's: "I think the candles are smaller because when they burned, some of the wax changed into invisible gases that went into the air." Students should be able to explain why her answer is correct and the other answers are incorrect.

Mom: I think the candles are smaller because when they burned, some of the wax was destroyed.

Mom is incorrect because matter can never be destroyed.

Grandpa: I think the candles are smaller because when they burned, all of the wax near the tops melted and ran down the sides of the candles.

Grandpa is incorrect because wax doesn't just melt when it burns. It combines with oxygen from the air to form invisible gases (carbon dioxide and water vapor) that go into the air.

Sam: I think the candles are smaller because when they burned, they became heavier and sank into the cake.

Sam is incorrect because matter can never be created and/or when wax burns it combines with oxygen from the air to form invisible gases (carbon dioxide and water vapor) that go into the air.

Lily: I think the candles are smaller because when they burned, some of the wax changed into invisible gases that went into the air.

Lily is correct because wax goes through a chemical change when it burns, combining with oxygen from the air to form invisible gases (carbon dioxide and water vapor) that go into the air.

STEM at Home

Have students complete the "I learned that …" and "My favorite part of the lesson was …" portions of the STEM at Home student page as a reflection on their learning. They may choose to do the following at-home activity with an adult helper and share their results with the class. If students do not have access to the internet or these materials at home, you may choose to have them complete this activity at school.

"At home, we can do a fire safety inspection. First, we will need to print out the fire safety checklist."

The checklist is available at www.sparky.org/pdf/sparkychecklist.pdf.

"Then, we can go from room to room together, answering the questions on the checklist. For each question we answer 'yes' to, we get a point! When we're finished, we can add up the points to find out our score.

Note: For all questions to which you answered "no," make sure your family takes the steps needed to make those answers a "yes" so you can all score a fire safety home run!"

For Further Exploration

This section is provided to help you encourage your students to use the science and engineering practices in a more student-directed format. This box lists questions and challenges related to the lesson that students may select to research, investigate, or innovate. Students may also use the questions as examples to help them generate their own questions. After selecting one of the questions in the box or formulating their own questions, students can individually or collaboratively make predictions, design investigations or surveys to test their predictions, collect evidence, devise explanations, design solutions, or examine related resources. They can communicate their findings through a science notebook, at a poster session or gallery walk, or by producing a media project.

Research

Have students brainstorm researchable questions:

- What were some of Michael Faraday's greatest discoveries?
- What kind of chemical change causes steel wool to rust? What are the products of the reaction?
- What were the biggest wildfires in U.S. history? What caused them? How were they fought? Have there been any recent wildfires near your area?

Investigate

Have students brainstorm testable questions to be solved through science or math:

- What ratio of cornstarch and water makes the best Oobleck? Does this mixture result in a physical or chemical change?

- **?** Which rusts in the shortest amount of time: steel wool in tap water, salt water, or vinegar? Graph the results, then analyze your graph. What can you conclude?

- **?** Survey your friends: Would you want to be a firefighter? Graph the results, then analyze your graph. What can you conclude?

Innovate

Have students brainstorm problems to be solved through engineering:

- **?** Can you design a way to prove that matter is conserved when steel wool rusts?

- **?** Can you design a fire safety plan for your home?

- **?** How do you think a candle would burn in microgravity? How could astronauts study combustion reactions safely on the International Space Station? Find out how by searching the internet for information on NASA Saffire (Spacecraft Fire Experiment).

References

American Chemical Society. 2011. *Safety in the elementary science classroom.* 3rd ed. Washington, DC: American Chemical Society.

Keeley, P., and J. Tugel. 2009. "Burning Paper." In *Uncovering student ideas in science volume 4: 25 new formative assessment probes*, 23–29, P. Keeley and J. Tugel. Arlington, VA: NSTA Press.

National Research Council (NRC). 2012. *A framework for K–12 science education: Practices, crosscutting concepts, and core ideas.* Washington, DC: National Academies Press.

Websites

"A Green Way to Fight Fires" (video)
www.pbslearningmedia.org/resource/nvmms.sci.eng.fire/a-green-way-to-fight-fires

"Wildfire" (video)
www.pbslearningmedia.org/resource/idptv11.sci.ess.earthsys.d4kwfir/wildfire

More Books to Read

Collard, S. B. 2014. *Fire birds: Valuing natural wildfires and burned forests.* Missoula, MT: Bucking Horse Books.
Summary: In this intriguing book for readers in grades 4–8, award-winning science author Sneed B. Collar III challenges society's negative views toward natural forest fires. Large print, glossy pages, and numerous full-page, up-close color photos of the dozens of bird species that depend on natural forest fires give readers a keener sense of the complex relationships between fire and thriving plant and animal communities.

Maurer, T. 2013. *Changing matter: Understanding physical and chemical changes.* North Mankato, MN: Rourke Educational Media.
Summary: Informative text and diagrams, real-world examples, and full-color photographs bring physical and chemical changes to life for upper-elementary readers.

Ochiltree, D. 2012. *Molly, by golly! The legend of Molly Williams, America's first female firefighter.* Honesdale, PA: Calkins Creek.
Summary: This true story chronicles how Molly Williams, an African American cook for New York City's volunteer Fire Company 11, jumped in to help a skeleton crew of firefighters put out a house fire during the 1818 blizzard. Working tirelessly alongside the men to battle the raging blaze, Williams secured both a job as "Volunteer No. 11" and a place in history.

Simon, S. 2016. *Wildfires.* 2nd ed. New York: HarperCollins Children's Books.
Summary: Clear, informative text and full-page photographs give readers a comprehensive picture of the destruction that wildfires cause, as well as information on wildfires' surprising benefits to ecosystems.

Candle Observations

Flame Drawing

Fire Safety Precautions
- Roll up your sleeves, secure any loose clothing, and tie back long hair.
- Wear safety goggles over your eyes.
- Never reach over or touch the flame.
- Keep your work area clean and clear of flammable materials.
- Do not touch lighted candles and melting wax; they are hot and can burn skin.

What I OBSERVE	What I WONDER

Name: _____

Birthday Candles

Part 1

Sam's family and friends were celebrating his birthday. Sam's mom lit all 10 candles on his cake and everyone sang "Happy Birthday to You." After Sam made a wish and blew out the candles, he noticed that they seemed smaller than they were before being lit. Sam asked, "Why are the candles smaller?"

The group all had different ideas about why the candles were smaller after being burned. This is what they said:

Mom: I think the candles are smaller because when they burned, some of the wax was destroyed.

Grandpa: I think the candles are smaller because when they burned, all of the wax near the tops melted and ran down the sides of the candles.

Sam: I think the candles are smaller because when they burned, the candles became heavier and sank into the cake.

Lily: I think the candles are smaller because when they burned, some of the wax changed into invisible gases that went into the air.

Which person do you agree with and why? Explain your thinking.

_____

_____

_____

_____

Chapter 10

The Chemistry of a Candle

Study a flickering candle flame, and you will find that there is much more to it than meets the eye! In fact, people have been fascinated by the mystery of flames for hundreds of years. In 1848, the great British scientist Michael Faraday gave a series of famous talks to schoolchildren on the physics and chemistry of a candle. These talks were part of a series of lectures for young people given each year at the Royal Institution in London—a tradition that continues to this day.

MICHAEL FARADAY

Let us explore the mystery of a candle! You know that a candle gets smaller as it burns. You might think this happens because most of the wax melts and drips down the sides or because the matter that makes up the candle is completely destroyed. But neither of these things is true. To discover the secret, you will need to observe a burning candle closely. Use only your senses of sight and smell, and do not touch the flame. This observation should be done with the help of your teacher or another adult. Observe all safety precautions, and NEVER use a lighter or matches yourself!

The first thing you will likely notice when a fresh candle has been lit is some of the wax melting and dripping down the wick. Wax changes from a solid to a liquid when it gets hot. This is a **physical change**—a change in matter that might change the form or appearance of a substance but does not produce any new substances. The liquid wax is still wax.

Very quickly, the melting wax exposes the cotton wick and the wick begins to burn. Burning, or **combustion**, causes a **chemical change**. A chemical change, or **chemical reaction**, is a change in matter that produces new substances. One sign that a chemical change has occurred in a **combustion reaction** is a change in color, and you will see that the top part of the wick has changed from white to black. Other signs include the production of heat, light, and a change in odor. All of these signs are present as the wick begins to burn.

The next thing you will notice is a cup of wax forming at the base of the candle. The air surrounding the candle is drawn upward by the heat of the flame and cools the sides of the candle. The cup that is formed contains melted wax, which becomes the **fuel** that keeps the candle burning. This is where something surprising happens. The liquid wax climbs up the wick to meet the flame! This happens because tiny liquid wax particles, called **molecules**, are attracted to one another and follow each other up the absorbent wick. This process is known as **capillary action.**

You might think that wax in liquid form is what burns when it meets the flame. Actually, another physical change happens first—the liquid wax becomes so hot that it **vaporizes**, or turns into a gas. Then, a chemical change occurs. The wax vapor **reacts** with oxygen from the air because of the heat of the flame. This chemical reaction produces new substances with new properties. The new substances are invisible gases—carbon dioxide (CO_2) and water vapor (H_2O). Energy in the form of heat and light is also released by the burning candle.

So where does the wax go when a candle burns? It melts into a liquid, vaporizes into a gas, and finally burns to form carbon dioxide gas and water vapor. You can't see these new substances, but if you could capture them and measure their mass, you would find that they weigh the same as the wax that burned and the oxygen used in the

reaction. This is a wonderful thing about matter. It cannot be created or destroyed, only changed, even when burned. This fact is known as the **law of conservation of matter.**

For fans of magic tricks, you may be disappointed to read that matter cannot simply appear out of nowhere, and likewise matter cannot disappear. Matter may change forms, however, giving the illusion of nothing out of something or vice versa, but the mass of the matter is always the same before and after the change. If 100 g of wax and oxygen is burned, then 100 g of carbon dioxide and water vapor must be produced.

$$\text{Wax} + O_2 \longrightarrow CO_2 + \text{Water}$$
(Burned in Oxygen)   *(Invisible Gases)*
100 g   =   100 g

Look again at a candle that has burned for a bit, and think about where the wax has gone. Though much of the wax is no longer in its original form, the matter that made up the missing wax, along with the oxygen burned, is still in the room in the form of invisible gases. We have solved the mystery!

The Chemistry of a Candle

Questions

1. What is a physical change that occurs when a candle burns?

2. What are the signs that a chemical change is occurring when a candle burns?

3. What is the difference between a physical change and a chemical change?

4. What is combustion?

5. What are the new substances that form when wax combines with oxygen from the air in a combustion reaction?

6. What is meant by the conservation of matter?

7. When you burn a candle, is matter created or destroyed? Explain.

8. If you weighed a birthday candle before and after burning it on your birthday cake, would it weigh the same? Explain.

Wildfires

Directions: Before reading *National Geographic Kids: Wildfires,* make your best guess by underlining the answer for each question. After reading, circle the correct answers.

1. Which of these are ways that wildfires can start?
   a. lightning strike
   b. hot lava
   c. matches
   d. All of the above

2. Which of these is **not** part of the "Fire Triangle?"
   a. heat
   b. fuel
   c. water
   d. oxygen

3. Which of these is a beneficial effect of a wildfire?
   a. They thin out crowded forests.
   b. Ashes add nutrients to the soil.
   c. They remove sick plants.
   d. They remove insects that kill trees.
   e. All of the above
   f. Wildfires are never beneficial.

4. What fraction of wildfires are caused by humans?
   a. 1 out of 100
   b. 1 out of 10
   c. 4 out of 5
   d. None of the above

5. Which **does not** describe a method of fighting wildfires?
   a. Digging a ditch to create a fire line
   b. Setting part of a forest on fire
   c. Parachuting into a fire zone
   d. Dropping water from airplanes
   e. Covering a fire zone with plastic
   f. Dropping fire retardant (sky Jell-O) from airplanes

6. How can you help keep wildfires from starting?
   a. Never leave a campfire alone.
   b. Put out a campfire before leaving.
   c. Never start a campfire during a dry spell.
   d. All of the above.

Bonus: What is Smokey Bear's fire safety message?

_____

_____

_____

Name: _____

After Reading *National Geographic Kids: Wildfires*

Learning about wildfires helps us protect nature, ourselves, and our homes, pets, and property. At fire science labs, scientists and engineers are studying new ways to prevent and fight wildfires. They are developing technologies that remove one or more components of the **fire triangle**. The fire triangle is made up of the three things that any fire needs to burn, whether an immense wildfire or the tiny flame of a candle.

1. What three things are needed for a fire to burn?

   _____, _____, and _____

2. Think back to what you have learned about **combustion**, or burning, from observing a lighted candle. What new substances do you think are formed when a wildfire burns?

   _____

3. When a tree burns in a wildfire, it can be reduced to a pile of ashes. What happens to the rest of the matter that makes up the tree?

   _____
   _____
   _____

4. Now, watch as your teacher shows a video about a new technology for fighting fires called TetraKO. After watching, think about the firefighting technologies you read about in the book, *Wildfire*. What are the advantages of this new technology?

   _____
   _____
   _____
   _____

Picture-Perfect STEM Lessons, 3–5

Name : _____

Birthday Candles

Part 2

Sam's family and friends were celebrating his birthday. Sam's mom lit all ten candles on his cake and everyone sang "Happy Birthday to You." After Sam made a wish and blew out the candles, he noticed that they seemed smaller than they were before being lit. Sam asked, "Why are the candles smaller?"

The group all had different ideas about why the candles were smaller after being burned. This is what they said:

Mom: I think the candles are smaller because when they burned, some of the wax was destroyed.

Grandpa: I think the candles are smaller because when they burned, all of the wax near the tops melted and ran down the sides of the candles.

Sam: I think the candles are smaller because when they burned, the candles became heavier and sank into the cake.

Lily: I think the candles are smaller because when they burned, some of the wax changed into invisible gases that went into the air.

First, check ☑ *correct* or *incorrect* for each answer. Then, explain your thinking for each one.

Mom: I think the candles are smaller because when they burned, some of the wax was destroyed.

Mom is   ☐ *correct*   ☐ *incorrect*   *because* _____

_____

_____

_____

_____

National Science Teachers Association

Name: _____

Grandpa: I think the candles are smaller because when they burned, all of the wax near the tops melted and ran down the sides of the candles.

Grandpa is ☐ *correct*   ☐ *incorrect*   *because* _____

_____

_____

_____

_____

Sam: I think the candles are smaller because when they burned, they became heavier and sank into the cake.

Sam is ☐ *correct*   ☐ *incorrect*   *because* _____

_____

_____

_____

_____

Lily: I think the candles are smaller because when they burned, some of the wax changed into invisible gases that went into the air.

Lily is ☐ *correct*   ☐ *incorrect*   *because* _____

_____

_____

_____

_____

STEM at Home

Dear _____,

At school, we have been learning about the **chemistry of candles** and the science behind **preventing and fighting wildfires**.

I learned that: _____
_____

My favorite part of the lesson was:
_____
_____
_____

At home, we can do a fire safety inspection. First, we will need to print out the fire safety checklist.

The checklist is available at www.sparky.org/pdf/sparkychecklist.pdf.

Then, we can go from room to room together, answering the questions on the checklist. For each question we answer "yes" to, we get a point! When we're finished, we can add up the points to find out our score.

We can record our fire safety "batting average" below:

_____ YES to all 22 questions … we've hit a fire safety home run!

_____ YES to 15–21 questions … we've made it to third base.

_____ YES to 10–14 questions … we've hit a double.

_____ YES to 0–9 questions … we need to make many changes around our home to be fire safe.

Note: For all questions to which you answered "no," make sure your family takes the steps needed to make those answers a "yes" so you can all score a fire safety home run!

Chapter 11

From Edison to the iPod

Description

Students read about famous inventors of the past (Thomas Edison) and the present (Tony Fadell), and explore the evolution of sound-recording devices, beginning with the invention of Edison's phonograph and ending with Fadell's iPod. They learn how the first analog recordings worked by exploring with Talkie Tapes and then design and build a three-dimensional prototype for a music player of the future.

Suggested Grade Levels: 3–5

LESSON OBJECTIVES Connecting to the *Framework*		
Science and Engineering Practices	**Disciplinary Core Ideas**	**Crosscutting Concept**
Constructing Explanations and Designing Solutions	**PS4.C:** Information Technologies and Instrumentation **ETS1.C:** Optimizing the Design Solution **ETS2.B:** Influence of Engineering, Technology, and Science on Society and the Natural World	Stability and Change

Featured Picture Books

- TITLE: **Timeless Thomas: How Thomas Edison Changed Our Lives**
- AUTHOR: **Gene Barretta**
- ILLUSTRATOR: **Gene Barretta**
- PUBLISHER: **Henry Holt & Company**
- YEAR: **2012**
- GENRE: **Narrative Information**
- SUMMARY: *This clever book shows modern-day devices that had their beginnings in Edison's lab. Whimsical illustrations depict Edison and his team of employees working in the lab, while the opposite side of each page shows present-day versions of his inventions. End matter includes a timeline of Edison's most famous inventions as well as short bios of some of his employees.*

- TITLE: *iPod and Electronics Visionary Tony Fadell*
- AUTHOR: **Anastasia Suen**
- PUBLISHER: **Lerner**
- YEAR: **2014**
- GENRE: **Non-Narrative Information**
- SUMMARY: *Part of the STEM Trailerblazer Bios series, this informative book shares how Tony Fadell went from a curious kid to the designer of the iPod.*

Chapter 11

Time Needed

This lesson will take several class periods. Suggested scheduling is as follows:

Day 1: **Engage** with *Timeless Thomas* Read-Aloud, **Explore** with Edison's First Recording and Talkie Tapes, and **Explain** with How Records Work

Day 2: **Explore** with Comparing Music Players and **Explain** with Music Player Time Line

Day 3: **Elaborate** with *iPod and Electronics Visionary Tony Fadell* Read-Aloud and Fadell's TED Talk

Day 4 and beyond: **Evaluate** with Music Player Design Challenge

Materials

Per class

(*Tip:* Contact parents and other teachers to borrow any music players you do not already own.)

- Record player
- Record albums
- Cassette tape player
- Cassette tapes
- Walkman
- Compact discs (CDs)
- CD player
- MP3 player
- iPod

Per student

- Talkie tape or talking strip (available at *www.stevespanglerscience.com*)
- Small piece of scotch tape
- Plastic cup
- Music Player Info Cards
- Hand lens
- Foamboard or cardboard
- Construction paper
- Markers

Student Pages

- Music Player Design Challenge
- The Pitch
- The Pitch Scoring Rubric
- STEM at Home

Background for Teachers

Since Thomas Edison's groundbreaking invention of the phonograph in 1877, engineers and designers have been innovating and improving the way we record and listen to music. (For more about Thomas Edison, see Chapter 6, "The Inventor's Secret"). Early music players (which played records first and cassette tapes later) recorded and played with *analog* technology. For those players, the sound was recorded as grooves in a record or as magnetic pulses on a magnetized plastic strip.

In this lesson, students explore how analog technology works by observing the grooves embossed on the surface of a "talkie tape." When they drag a thumbnail along the grooves, they can hear the "talk" recorded on one side of the strip ("Science is fun!"). This experience is similar to how a record player works. But instead of a thumbnail, a record player uses a stylus to ride along the grooves that have been pressed into the spinning record, turning mechanical energy into sound energy. The sound vibrations are then turned into electrical signals that are amplified by a loudspeaker. Students explore how using a plastic cup can amplify the sound their talkie tapes make.

Analog technology was used for many years, but limited the amount of music that could be held and the quality of the sound. However, things changed dramatically when music could be recorded and played with *digital* technology. Digitization translates sound into a numerical code that can be processed by a computer. Compact discs (CDs) and MP3 players are digital technologies that record and play music. In a CD, the digits are represented by microscopic bumps arranged in a spiral. A motor spins the disc and a laser reads the bumps. In MP3 technology, the digitized code is stored on a hard drive. These digital technologies greatly improved the amount of music that could be stored as well as the quality of the sound produced. With analog storage, such as records and cassette tapes, the quality would diminish as the device was repeatedly used. With digital storage, however, the songs can be played repeatedly without degradation.

Another revolutionary development in music players was made in 2001 when Apple introduced the iPod. The iPod uses MP3 technology, which had been around for several years, but the big changes came with the amount of music that could be stored, the ease of use, and the way the music was purchased. The first-generation iPod had a 5 GB, super-thin hard drive and could store up to 1,000 songs. The wheel on the front made finding the song you wanted to play easy, and the integration with iTunes made downloading music easier than ever. The iPod and iTunes together transformed not only the way people played music but also the music industry itself. Since the invention of the first-generation iPod, Apple has made many improvements in size, memory, sound quality, and other features. Once iPod technology was integrated into the iPhone, however, sales began declining steadily as consumers began using their mobile phones to access music.

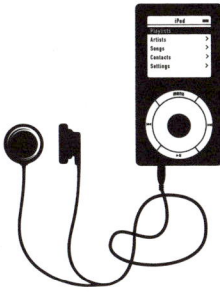 

A Framework for K–12 Science Education suggests that by the end of 5th grade, students understand that "Digitized information can be stored for future recovery or transmitted over long distances without significant degradation. High-tech devices, such as computers or cell phones, can receive and decode information—convert it from digitized form to voice—and vice versa" (NRC 2012, p. 137). Students explore these concepts in the context of music players, where music is the information being digitally received and decoded. This lesson begins with the first sound-recording device, the phonograph, and ends with the most current devices, with students comparing some of the innovations in between. Looking at the development of these devices over the years gives students the opportunity to explore the crosscutting concept of stability and change in the context of technology.

engage

Timeless Thomas Read-Aloud

 Making Connections: Text to Self

Start a conversation about music with your students. *Ask*

? What is your favorite song? (Answers will vary.)

? Who is your favorite singer, or what is your favorite band? (Answers will vary.)

? What kinds of devices do you use to listen to music? (iPod, smartphone, CDs, etc.)

Then, show students a vinyl record (or a cassette tape), and ask them if they know what it is. (Some may know, and others may have never seen one.) Tell students that before iPods, MP3 players, and CDs, people (maybe even you?) listened to music by using a record player. Tell students they are going to learn more about the person who invented the very first music player. Show them the cover of the book, *Timeless Thomas: How Thomas Edison Changed Our Lives*.

> **Connecting to the Common Core**
> **Reading: Informational Text**
> KEY IDEAS AND DETAILS: 3.1, 4.1, 5.1

 Determining Importance

Tell them that as you read, you would like them to listen for the other inventions that Thomas Edison designed (most of them more than 100 years ago) and how these inventions are still part of our lives today. After reading, call on students to share one of Edison's inventions that they enjoy a version of today, and discuss how the invention has been improved over the years. See Table 11.1 for examples.

Table 11.1. Sample Modern-Day Versions of Edison's Inventions

Edison	Now
Phonograph	iPod
Telephone transmitter	Cell phone
Electric pen	Photocopier
Nickel-iron battery	Batteries of all shapes and sizes

explore

Edison's First Recording

Tell students that although Edison's most famous invention was the lightbulb, the phonograph was quite a revolutionary invention as well. Explain that before the phonograph, if people wanted to hear music, they had to go hear a live band or orchestra.

 Making Connections: Text to Self

Ask

? How often do you listen to music each day? (Answers will vary.)

? Can you imagine what it would be like to have to go to a live performance *every* time you wanted to listen to music? (Answers will vary.)

Play Edison's first recording, which he made with his phonograph in 1877 (see "Websites" section). *Ask*

? What do you think about the quality of the recording compared with the quality of recordings you hear nowadays? (The phonograph recording has a lot of background noise and is not as clear as the recordings we hear today.)

Reread the bottom of page 11 of *Timeless Thomas,* which explains how the phonograph recorded sound and played it back. *Ask*

? How can a needle riding over grooves produce sound? (Answers will vary.)

Talkie Tapes

Tell students that you have a device that can help them understand how the original phonograph worked. Give each pair of students a "talkie tape," a thin plastic strip with grooves embossed along the recorded edge. Have them use hand lenses to observe the grooves on the tape. Then, demonstrate how you can make sound with the talkie tape by pulling it through your index finger and thumbnail so that your thumbnail moves along the grooves. Have them try this with their talkie tapes. Ask if they can feel their thumbnail vibrating when the sound is produced. Tell students that when they think they hear the secret message, don't repeat it! Instead, write down what they think the tape is saying. That way, everyone can have a chance to figure it out for themselves. For fun, go around the room and have each student call out what they wrote. You will hear a lot of interesting variations of the phrase "Science is fun!"

Tell students that the phonograph worked on the same principle as a talkie tape. Instead of a plastic strip, the grooves were made on a cylinder covered with tin; instead of a fingernail, a needle rode over the grooves, causing the needle to vibrate. Show students a photo of the original phonograph, pointing out the cone-shaped part. *Ask*

? What was the purpose of the cone-shaped part of the phonograph? (Answers will vary.)

Exploring Talkie Tapes

Tell students that they can discover the answer to this question by attaching a cup to their talkie tape. Have students tape a cup to the pointed end of their talkie tape. Then, have them play the message again the same way they did before, but now with the cup on their ear. *Ask*

? How did the cup change the sound? (It made the sound louder.)

Tell students that the cup-shaped part of the phonograph was called the *horn,* and its purpose was to *amplify* the sound vibrations, or make them louder. Students can try using different types and sizes of cups as amplifiers. They can even experiment with surfaces such as paper, a whiteboard, or a blown-up balloon to amplify the sound.

explain

How Records Work

Explain that a record player works in a similar way as the talkie tape. Have students closely observe the grooves on a record with a hand lens. Explain that when the record spins and the needle (stylus) runs over the grooves, different vibrations are created that produce different sounds. This phenomenon is very similar to how the talkie tape makes sound when a thumbnail runs over the grooves, but the grooves on a record are a lot more detailed. Play a song for students on the record player so they can see and hear the player in action. Demonstrate how to change songs by reading the label, counting the grooves, and placing the needle on the right groove. Point out the speaker that is amplifying the sound and explain that Thomas Edison's phonograph had a horn attached to amplify the sound, but the record player uses electronic speakers.

COMPARING MUSIC PLAYERS

explore

Comparing Music Players

Tell students that since Thomas Edison's revolutionary phonograph, scientists and engineers have been improving sound-recording devices. Provide some examples of the following audio players from over the years: a record player, cassette tape player, Walkman, CD player, MP3 player, and iPod. Be sure to tell students that many more music players have been invented over the years, but these represent some of the most commonly used ones. Line the devices up in random order for students to see. Explain that each of these devices was designed by engineers to take information, in this case music or sound, and store it to share with others. Also explain that each device stores information in a different way. Then, demonstrate how each device works, including how to start and stop the music, go forward and backward, and so on. Give students a chance to try the devices out and compare the sound quality of each one.

explain

Music Player Time Line

 *Card Sequencing*

Give each pair of students a set of Music Player Info Cards. Explain that these cards contain information about how each music player stores information, how the music is played, how many minutes of music it can hold, and how to switch from one song to another. Have them read each card and sequence them in the order in which they think the music players were invented. Allow students to share their sequences with each other and explain why they think the devices were invented in that order. Then, share the correct year each one was invented and have them re-sequence their cards if necessary. The correct order is shown in Table 11.2.

Table 11.2. Order for Time Line of Music Players

Music Player	Year Invented
Record player	1895
Cassette tape player	1963
Walkman	1979
Compact disc player	1982
MP3 Player	1993
iPod	2001

Next, have students figure out how many years passed in between each invention. For example, 68 years passed between the invention of the record player and the invention of the compact cassette tape player, and only 11 years passed between the compact disc player and the MP3 player. Ask students to use the info cards to find out how each device is better than the one that preceded it. For example, a compact cassette tape player is better than a record player because it is easier to transport and holds more music. The CD is better than the cassette tape because it has better sound quality, holds more songs, and doesn't require rewinding to listen to a song again. *Ask*

? What criteria do you think people considered when purchasing a music player when record players were first sold? (music could be heard, ability to switch between songs, affordability)

? How do you think the criteria have changed over the years. In other words, what are people looking for now when they purchase a music player? (sound quality, portability, capability for holding a lot of music, ease of use, affordability, etc.)

Explain that the latest invention is not always better than the previous one. A music player called the 8-track player was invented just two years after the cassette tape player and is a good example of this. The 8-track player was on the market for only about 5 years (1965–1970). The cassette tape player was in common use for many years beyond that. In fact, the cassette tape player was the most common device used until the CD player was invented nearly 20 years later. Another important fact to note is that each of these devices underwent many design improvements over the years. For example, the Walkman uses the same technology as cassette players, but it was revolutionary because its size allowed people to take their music wherever they wanted to go. *Ask*

? How has the size of the music players changed over the years? (They have gotten smaller over time.)

? How can something smaller (such as the iPod) hold more music than something much larger (such as a vinyl record)? (Answers will vary.)

Explain that the answer is *digitization*. On a vinyl record, a cassette tape, and an 8-track tape, music is stored as grooves or magnetic pulses. However, the CD, MP3, and iPod store music *digitally*. This means that the music is actually converted into computer code (digits). With a CD, the digital code is burned into the CD as tiny bumps. (See "Websites" section for a short video explaining how a CD player works.) In the MP3 players and iPods, the code is stored on a hard drive. As hard drives become smaller, our music players can also become smaller.

Another advantage of digitization is that digitized information can be sent over long distances without significant *degradation*, or reduction in sound quality. For example, when you talk into a cell phone, your voice is converted to digital code, which can be sent over a long distance to someone else's phone, decoded by their phone, and translated back so they hear the sound of your voice. Even if you are talking to someone halfway around the world, the sound quality remains relatively good. In the olden days, when calls were made over traditional analog telephones that sent information across wires, significant degradation could occur. Sometimes, making out what a caller was saying was difficult during a long-distance phone call—the sound was faint and crackly.

Explain that the opposite of digital is *analog*. Have students sort their cards into two piles: analog and digital. Table 11.3 shows how the cards should be sorted.

Table 11.3. Answers for Digital vs. Analog Card Sort

Analog	Digital
Record player	Compact disc player
Cassette tape player	MP3 player
Walkman	iPod

Students may be aware that that there has been a resurgence of interest in analog music, specifically vinyl records. For some people, excellent sound quality is not the most important criterion. A growing number of people prefer the warmer, more natural, and sometimes scratchy "retro" sound of analog recordings over digital recordings.. Because of this "vinyl revival," record players are being sold and used again.

elaborate

iPod and Electronics Visionary Tony Fadell Read-Aloud

Connecting to the Common Core
Reading: Informational Text
KEY IDEAS AND DETAILS: 3.1, 4.1, 5.1

Tell students that each of the sound devices on the cards transformed how we share music. The most recent of those devices was the iPod, which was invented in 2001. *Ask*

? Do you know what the first-generation iPod looked like?

? Do you know who designed the iPod?

Show students the cover of the biography titled *iPod and Electronics Visionary Tony Fadell. Ask*

? What is a visionary?

Explain that a *visionary* is a person with original ideas about what the future could or will be like. Tell students that as you read the book, you would like them to think about Fadell and the characteristics that made him an electronics visionary.

 **Visualizing**

Read the book aloud, stopping when you get to page 19, which describes Fadell's meeting with Steve Jobs to present his prototypes. For these two pages (pp. 19–20) have students close their eyes and visualize the scene. After reading, *ask*

? How do you think Fadell felt during the meeting? (excited, nervous, etc.)

? Why do you think Fadell placed his last prototype under the wooden bowl? (to add a dramatic effect or the feeling of surprise)

? Do you think Fadell rehearsed his presentation before the meeting? (Answers will vary.)

Explain that Fadell is a designer. He did not invent the MP3 technology used in the iPod; rather, he designed a new kind of MP3 device. Fadell's device was smaller and lighter than other MP3 players, could hold thousands of songs, and was easier to use.

Read the rest of the book aloud. After reading, *ask*

? What in Fadell's childhood inspired his future career? (He built, took apart, and fixed things with his grandfather.)

? What characteristics of Fadell led to him being considered an electronics visionary? (He was creative and determined, he took risks, and he was not afraid of failure.)

On page 28, students will learn that Fadell was the first recipient of the Alva Award. This award named after Thomas Alva Edison recognizes a remarkable inventor who influences the world. *Ask*

? What does Fadell have in common with Thomas Edison? (He made a new kind of music player, kept trying different ideas until one was successful, worked with a team of people, was creative, took risks, and made mistakes.)

Convey to students that people are always going to be improving designs and inventing new things. Tell students that since the iPod was released in 2001, engineers have improved its design, and many other companies have made similar music players. *Ask*

? What do you think music players might be like in the future?

? How could they be designed to be better than the ones we use today?

? What problems do you have with your current music player?

? What would you do to make it better?

Tell students that they are going to have an opportunity to design a three-dimensional (3-D) model of a new music player out of cardboard or foam board like Fadell did when he designed the iPod.

> **Connecting to the Common Core**
> **Reading: Informational Text**
> Key Ideas and Details: 3.2, 4.2, 5.2

Fadell's TED Talk

 Synthesizing

Tell students that for inspiration, you would like them to watch Fadell's TED Talk "The First Secret of Design … Noticing" (see "Websites" section). Before watching the video, *ask*

? What do you think that title means—that the secret of design is noticing? (Answers will vary.)

Have students listen for the answer to that question. After watching the video, *ask*

? Now what do you think the title of Fadell's TED Talk means? (Noticing the small problems that no one else notices is a great way to come up with product design ideas.)

In the TED Talk, Fadell also advises designers to think broader, look closer, and think younger.

? What does "think broader" mean? (Look at the whole product or process. Think about the steps leading up to or after the device is used.)

? What does "look closer" mean? (Look at the smallest details of the design and think about how they can be improved.)

? What does "think younger" mean? (Be creative and imaginative. See all the possibilities—even the outrageous ones!)

evaluate

Music Player Design Challenge

Tell students you would like them to "think broader, look closer, and think younger" to design "The Music Player of the Future!" Together, brainstorm ideas about how music might be listened to in the future. Encourage out-of-the-box thinking by asking,

? How could music players be more convenient?
? How could they hold more music?
? How could the sound quality be better?

Next, challenge students to design a 3-D model for a music player of the future. But before they get started, go back to page 17 of *iPod and Electronics Visionary Tony Fadell* and read the paragraph about how he built a new MP3 player model with fishing weights inside, so the model would feel more substantial. Explain that they can use any materials they want to build their 3-D model. Have scissors, cardboard, foam board, construction paper, and markers available. Students are not limited to creating a handheld device like the one in the book. They might consider headphones, an in-home streaming device, a wearable device, or another design.

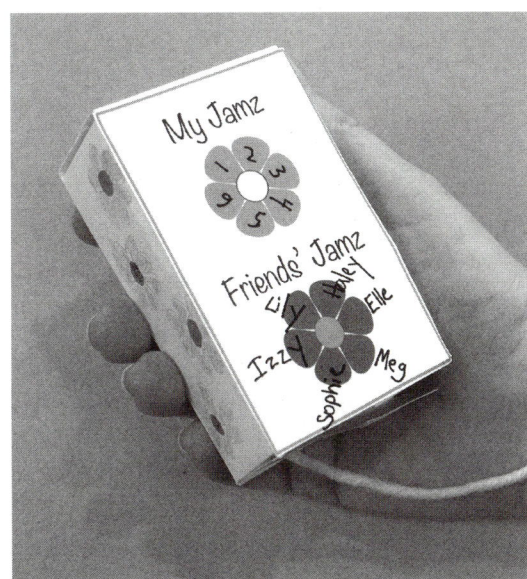

Music player 3-D model

Pass out a copy of the Music Player Design Challenge student page to each student. In pairs or individually, students should design a 3-D model that illustrates their ideas of how music players can be improved or totally reimagined. Have them use the design challenge student page to record ideas and sketch a diagram of their models. After you sign off on the Teacher Checkpoint, give students sufficient time to build their models (and redesign or rebuild if necessary). To optimize their 3-D models after their initial attempts, students may need additional materials.

> Connecting to the Common Core
> **Speaking and Listening**
> PRESENTATION OF KNOWLEDGE AND IDEAS: 3.4, 4.4, 5.4

Tell students that once they have finished their models, they are going to have a chance to "pitch" their idea to a company or investor (as Fadell did when he presented his model to Steve Jobs at Apple!). Their presentations can be done live or pre-recorded. Student can also create visual aids, such as posters, props, or slideshows to use in their pitch. Pass out both The Pitch student page and The Pitch Scoring Rubric to help pairs plan their pitch and evaluate student performance using the rubric.

STEM at Home

Have students complete the "I learned that …" and "My favorite part of the lesson was …" portions of the STEM at Home student page as a reflection on their learning. They may choose to do the following at-home activity with an adult helper and share their results with the class. If students do not have access to the internet or these materials at home, you may choose to have them complete this activity at school.

"At home, we can use the website Kids Think Design to explore various areas of design."

🔍 *Find the Kids Think Design website at* www.kidsthinkdesign.org.

"I can choose the category of design featured on the website that interests me most and 'meet' a designer in that field. (The categories are fashion, graphics, interiors, books, film and theater, architecture, animation, and environment.)"

For Further Exploration

This section is provided to help you encourage your students to use the science and engineering practices in a more student-directed format. This box lists questions and challenges related to the lesson that students may select to research, investigate, or innovate. Students may also use the questions as examples to help them generate their own questions. After selecting one of the questions in the box or formulating their own questions, students can individually or collaboratively make predictions, design investigations or surveys to test their predictions, collect evidence, devise explanations, design solutions, or examine related resources. They can communicate their findings through a science notebook, at a poster session or gallery walk, or by producing a media project.

Research

Have students brainstorm researchable questions:

- ? How has the way we watch movies changed over the years? What is a VCR? What is a DVD player?
- ? What kind of education and training do you need to become a designer?
- ? How are video games designed and produced?

Investigate

Have students brainstorm testable questions to be solved through science or math:

- ? Survey your friends and family: What types of music players do they use? Graph the results, then analyze your graph. What can you conclude?

- ? What is inside of a speaker or headphones that converts electricity into sound? Take the device apart to find out.

- ? What happens to the sound of the music played on a smartphone or iPod (with an internal speaker) when you place it inside a plastic cup? A glass cup? A cereal bowl? A cardboard box? A paper towel roll?

Innovate

Have students brainstorm problems to be solved through engineering:

- ? How could your classroom computers or tablets be improved?

- ? How could you improve a set of headphones?

- ? Can you design a new hands-free holder for a smartphone or iPod?

Reference

National Research Council (NRC). 2012. *A framework for K–12 science education: Practices, crosscutting concepts, and core ideas.* Washington, DC: National Academies Press.

Websites

Edison's First Sound Recording: "Mary Had a Little Lamb" (audio)
http://archive.org/details/EDIS-SCD-02

"How CDs Work" (video)
http://electronics.howstuffworks.com/cd.htm

TED Talk: "The First Secret of Design … Noticing" (video)
www.ted.com/speakers/tony_fadell

More Books to Read

Allen, K. 2010. *The science of a rock concert: Sound in action.* Mankato, MN: Capstone Press.
Summary: This book describes the science behind rock concerts, including sound waves, instruments, sound systems, and acoustics.

Barretta, G. 2008. *Now and Ben: The modern inventions of Benjamin Franklin.* New York: Henry Holt and Company.
Summary: From the same author as *Timeless Thomas*, this colorfully illustrated picture book details the inventions and discoveries of Benjamin Franklin and how they are integrated into today's society.

Barretta, G. 2009. *Neo Leo: The ageless ideas of Leonardo da Vinci.* New York: Henry Holt & Company.
Summary: From the same author as *Timeless Thomas*, this book cleverly shows how da Vinci's ideas, many inspired by nature, foreshadowed modern inventions.

Various authors. 2014. *STEM trailblazer bios series.* Minneapolis, MN: Lerner Publications.
Summary: These high-interest biographies include alternate reality game designer Jane McGonigal, astrophysicist and space advocate Neil deGrasse Tyson, Flickr cofounder and web community creator Caterina Fake, Google Glass and robotics innovator Sebastian Thrun, and YouTube founders Steve Chen, Chad Hurley, and Jawed Karim.

Music Player Info Cards

Name : _____

Directions: Cut out the cards. Read the information on each card and sequence them in order from the first invention to the most recent invention. Later, you will learn which year each music player was invented and have a chance to re-sequence the cards.

iPod

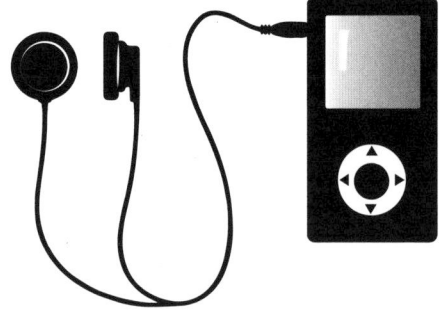

Music Stored As: Digital computer files

Music Played By: Computer processor that deciphers the computer code

Storage Capacity: First generation holds over 10 hours of music, now holds more depending on the size of the memory

To Find a Song: Scroll through titles with wheel

Portability: Pocket-sized

Power: Rechargeable with charging cord

mp3 Player

Music Stored As: Digital computer files

Music Played By: Computer processor that deciphers the computer code

Storage Capacity: Varies depending on size of the memory

To Find a Song: Click through the titles with up and down buttons

Portability: Pocket-sized

Power: Batteries needed

Name: _____

Record Player

Music Stored As: Grooves on a vinyl disc

Music Played By: Needle running over the grooves as the record spins

Storage Capacity: 40–60 min. (about 20–30 min. on each side)

To Find a Song: Move the needle to the right groove

Portability: Difficult to transport

Power: Needs electrical outlet

Compact Disc (CD) Player

Music Stored As: Digital code represented by microscopic bumps on a thin disc of metal and plastic

Music Played By: Laser beam bounces off the spinning disc

Storage Capacity: About 80 min.

To Find a Song: Click to the song number

Portability: Early models not portable; later models could be held in-hand

Power: Early models needed electrical outlets; later models ran on batteries

Picture-Perfect STEM Lessons, 3–5

Name: _____

Cassette Tape Player

Music Stored As: Magnetic pulses on a thin plastic tape coated with a magnetic material

Music Played By: Running the tape through an magnetic field created by an electromagnet inside the player

Storage Capacity: 90 min. (45 min. on each side)

To Find a Song: Read the label to find the order then rewind or fast forward and listen

Portability: Early models not portable, but became more portable over time

Power: Electrical outlet or batteries required

Walkman

Music Stored As: Magnetic pulses on a plastic tape coated with magnetic material

Music Played By: Running the tape through an magnetic field created by an electromagnet inside the player

Storage Capacity: 90 min. (45 min. on each side)

To Find a Song: Read the label to find the order then rewind or fast forward and listen

Portability: Easily portable, pocket-sized

Power: Batteries required

Name: _____

Music Player Design Challenge

Challenge: Design a device for listening to music.

Name of Device: _____

Designer(s): _____

Ideas:

_____

_____

_____

_____

Sketch:

Teacher Checkpoint ☐

Picture-Perfect STEM Lessons, 3–5

175

Name: _____

The Pitch

Imagine that you are meeting with a company or investor who can help you manufacture and sell your new music player. Create a "pitch" to present during the meeting. Your pitch can be presented live or be pre-recorded. During your pitch, you can use visual aids, such as a poster or a slideshow, and your 3-D model. You must include these four elements:

Name — *What do you call your device? Does the name clearly communicate what your device does? How will you display the name in your pitch?*

Problem — *What problem or problems with current music players does this device solve? What is the target market for the device? In other words, what group(s) of people would most likely want to buy the device?*

Features — *What special features does your device have? For example, How many songs does it hold? How do you play it? How do you listen to it? What powers it? How do you charge it? Use the 3-D prototype to show its features.*

Comparison — *What are some similar products on the market? How is your music player better than the products currently being sold?*

Name : _____

The Pitch Scoring Rubric

Imagine that you are meeting with a company or investor who can help you manufacture and sell your new music player. Create a "pitch" to present during the meeting. Your pitch can be presented live or be pre-recorded. During your pitch, you can use visual aids, such as a poster or a slideshow, and your 3-D model. Your presentation will be scored using the rubric below.

Score	Criteria
_____ 4 _____ 3 _____ 2 _____ 1	**Name** • The name clearly communicates what the device does. • The name is clearly displayed or announced.
_____ 4 _____ 3 _____ 2 _____ 1	**Problem** • The problem(s) the device solves is (are) clearly described. • The target market (or group[s] of people who would most likely want to buy the device) is identified.
_____ 4 _____ 3 _____ 2 _____ 1	**Features** • The features of the device are clearly communicated. • The 3-D model effectively shows the features of the device.
_____ 4 _____ 3 _____ 2 _____ 1	**Comparisons** • Similar products on the market are identified. • The device is shown to be better than similar products on the market.

4—Excellent   3—Above Average   2—Average   1—Below Average

_____ Total/16 Points

Picture-Perfect STEM Lessons, 3–5

Name : _____

STEM at Home

Dear _____,

At school, we have been learning about **designing new technologies.**

I learned that: _____
_____

My favorite part of the lesson was: _____
_____

At home, we can use the website Kids Think Design to explore various areas of design.

Find the Kids Think Design website at *www.kidsthinkdesign.org.*

I can choose the category of design featured on the website that interests me most and "meet" a designer in that field. (The categories are fashion, graphics, interiors, books, film and theater, architecture, animation, and environment.)

Design Category: _____

Featured Designer (bottom right corner):
_____

What did this person design?
_____

Chapter 12

Better Together

Description

After learning about some of the many collective nouns used to describe groups of animals, students observe a species that is known for having one of the most complex social structures of all mammals—the elephant. Students explore how scientists compile animal behaviors into inventories called *ethograms*, and then they use simulated field data to graph the frequency of certain elephant behaviors. That activity provides evidence for an argument that some animals form groups that help members survive. Analyzing their graphs also helps students understand that individual elephants have different roles within groups. Through a variety of text and digital media, students also learn about the benefits that other species get from living in groups and the way scientists use technology to track individual animals within larger groups.

Suggested Grade Levels: 3–5

LESSON OBJECTIVES Connecting to the *Framework*		
Science and Engineering Practices	**Disciplinary Core Ideas**	**Crosscutting Concept**
Analyzing and Interpreting Data Using Mathematics and Computational ThinkingEngaging in Argument From Evidence	**LS2.D:** Social Interactions and Group Behavior **ETS2.A:** Interdependence of Science, Engineering, and Technology	Cause and Effect

Featured Picture Books

TITLE: **An Ambush of Tigers: A Wild Gathering of Collective Nouns**
AUTHOR: **Betsy R. Rosenthal**
ILLUSTRATOR: **Jago**
PUBLISHER: **Millbrook Press**
YEAR: **2015**
GENRE: **Narrative Information**
SUMMARY: *Clever illustrations accompany rhyming couplets of collective nouns that describe groups of animals. A glossary offers further explanations.*

TITLE: **Animals That Live in Groups**
AUTHOR: **Kelsi Turner Tjernagel**
PUBLISHER: **Capstone Press**
YEAR: **2013**
GENRE: **Non-Narrative Information**
SUMMARY: *Clear text, full-color illustrations, and fascinating insets describe the advantages some kinds of animals gain from living in groups.*

Picture-Perfect STEM Lessons, 3–5

Chapter 12

Time Needed

This lesson will take several class periods. Suggested scheduling is as follows:

Day 1: Engage with *An Ambush of Tigers* Read-Aloud and **Explore** with "Elephant Behavior" Video and Elephant Ethogram

Day 2: Explore with Field Journal Notes and **Explain** with Graphing Elephant Behaviors

Day 3: Explain with *Animals That Live in Groups* Read-Aloud, *Kratt's Creatures* Videos, and The Social Lives of Elephants

Day 4: Elaborate with Wildlife Tracking Technology and **Evaluate** with Better Together Booklets or Posters

Materials

For Graphing Elephant Behaviors

- Two different colors of markers, crayons, or colored pencils

Student Pages

- Elephant Ethogram
- Field Journal Notes
- Elephant Behavior Graph
- The Social Lives of Elephants
- Better Together Booklet or Pamphlet
- Better Together Scoring Rubric
- STEM at Home

Background for Teachers

From social butterflies to solitary tigers, virtually all animals gather in groups at some point in their lives. The benefits include being protected from predators, obtaining food, raising young, mating, and even huddling together for warmth. But regardless of what draws animals together, one thing is true of many animal groups—there is a descriptive, clever, or sometimes silly collective noun for them! This lesson engages students in the topic of animal group behavior with a clever rhyming book that describes everything from an ambush of tigers to a rumba of rattlesnakes. Although many of these names are rarely used, even by scientists, students enjoy learning about them, and the discussion leads to the bigger question, "Why do some animals live in groups?"

Many animals are social only to the extent that males and females interact to mate or parents and offspring bond. The term *social animal* is usually applied when there is a high level of social organization and long-term relationships among members. Groups of elephants are particularly well-known for their complex social structure, so these majestic animals are used as a focus of study in this lesson. A group of elephants is called a *herd*, but within the herd can be one or more family groups. A family is usually made up of a mother, her sisters, her daughters, and their babies, or *calves*. Males, or *bulls*, live with the herd until they are about 13 years old, and then they join other males to form "bachelor herds" or roam on their own. Sometimes, herds of elephants combine with other herds to form even larger *clans*. The *matriarch* is the oldest and most respected female in the herd, and typically the largest as well. She decides when to move, how fast to travel, and when to stop. The herd stays close to the matriarch, and she defends it from predators. The matriarch teaches her daughters how to care for their babies, and all of the adult females in a herd teach the younger elephants how to eat and behave. Females will even nurse calves that are not their own. Elephants are very affectionate with their friends and family members, and they form strong bonds that can last a lifetime.

Because of their highly social nature, elephants have a sophisticated system of communication, involving calls, gestures, and postures. Researchers have catalogued hundreds of these behaviors and their descriptions into databases called *ethograms*. In this lesson, students are introduced to the exciting work of one of the world's premier elephant *ethologists*, Dr. Joyce Poole. Students "ride along" as Dr. Poole and her brother, cinematographer Bob Poole, observe and photograph elephant behaviors in Mozambique's Gorongosa National Park. Dr. Poole and her team have created an organization called ElephantVoices, whose mission is "to inspire wonder in the intelligence, complexity and voices of elephants, and to secure a kinder future for them."

Because elephants have one of the most complex social structures of all mammals, they are good to study in support of elementary education standards. *A Framework for K–12 Science Education* states that, by the end of grade 5, students should be able to understand that many animals live in groups and that "groups can be collections of equal individuals, hierarchies with dominant members, small families, groups of single or mixed gender, or groups composed of individuals similar in age. Some groups are stable over long periods of time; others are fluid, with members moving in and out. Some groups assign specialized tasks to each member; in others, all members perform the same or a similar range of functions." (NRC 2012, p. 156). Students learn how elephant researchers use ethograms to help them observe and identify patterns in elephant behavior. They familiarize themselves with some common elephant behaviors through an "elephant charades" game, then they analyze a page of simulated field journal notes on elephant behaviors.

Students learn that data on animal behavior can be collected in many different ways. A common method is *interval sampling*. In interval sampling, an observer notes what an animal is doing at preset, evenly spaced time intervals. In our highly-simplified fictitious example, a researcher has recorded which behavior (as defined by the ethogram) he or she observed each elephant doing every 30 min. for an entire day. These observations allow the researcher to get a clear "snapshot" of an elephant's behavior at a precise moment in time. When multiple observations are combined and analyzed, researchers can begin to draw conclusions about which types of behaviors might be most common or when certain types of behaviors are most likely to occur. Likewise, students use the field journal notes to graph the frequency of each behavior, then they look for patterns and make inferences about the causes of certain behaviors in elephants. In this way, the crosscutting concept of cause and effect is introduced as well as the science and engineering practice of using mathematics and computational thinking.

After exploring the elephant ethogram, reading a nonfiction book about a variety of animals that live in groups, and watching several videos about elephants, other social animals, and even solitary animals, students are able to construct an argument that being part of a group benefits elephants in at least three ways. One benefit for elephants is that members of a herd take care of each other. When a calf is born, the herd raises it. When an elephant is sick or injured, the herd helps it recover. Another benefit of elephants living in a group is that they help each other find enough food to fuel their massive bodies. Because elephants need more than 300 lb. of food a day and spend 16 hours a day eating it, cooperative foraging is a tremendous advantage for elephant herds. A third benefit is safety in numbers. When danger is near, elephants often form a *defensive circle,* with the youngest elephants in the middle and the adult females facing outward. The matriarch then investigates the threat.

Chapter 12

For the technology component of this lesson, students explore how satellite technology is used to track wildlife. Sometimes, researchers tape or glue a larger *satellite transmitter* onto an animal's back, head, or other body part. Other times, they outfit the animal with a custom-made harness or collar that carries a satellite transmitter. Researchers are very careful not to harm animals when placing the devices on them, and they are also careful not to use a device that will interfere with the animal's motion or hinder their social interactions with other animals in their group. This kind of technology is making it much easier to estimate animal populations, track migration patterns, map the entire *range* of a species (the geographical area within which that species can be found), and study an individual animal's day-to-day movements. By watching footage of conservationists in Kenya outfitting elephants with GPS collars and watching video of elephant movements being tracked on Google Earth, students explore the disciplinary core idea of the interdependence of science, engineering, and technology. From these videos, they also learn how the data these technologies provide assist wildlife managers in making decisions that reduce human–elephant conflicts, thus helping to preserve elephants and their habitats as human populations grow ever bigger.

engage

An Ambush of Tigers Read-Aloud

Connecting to the Common Core
Reading: Informational Text
KEY IDEAS AND DETAILS: 3.1, 4.1, 5.1

Before reading, introduce the author, Betsy R. Rosenthal, and the illustrator, a man who goes by one name: Jago. *Ask*

? Have you ever heard of an *ambush* of tigers? (Answers will vary.)

 **Inferring**

Ask

? What do you think this book is about? (Answers will vary, but some students will have an idea that the book is about names for animal groups.)

Read aloud the inside flap, which explains that the book is about the different collective nouns we use for animal groups. Tell students that a *collective noun* is a name for a collection, or number of people, animals, or things. Then, read aloud this quote from the author about how collective nouns inspired her to write the book *An Ambush of Tigers*:

> *When I was a little girl, I collected glass and ceramic animals, including groups of particular species. I ended up with a large* pack of dogs *of various breeds, and even a* scurry of ceramic squirrels. *Between my cherished collection of groups of different kinds of animals and much later in life learning some of the more unusual names for those groups, particularly the bizarre* a murder of crows, *I just had to write about collective animal nouns. And I used to have a* flock of children *at home, but they've all flown the coop!*

Then, read the book aloud, pointing out the clever illustrations as you read the poem. After reading, *ask*

? What were some of the collective nouns for the animal groups in the book? (tower of giraffes, raft of otters, leap of leopards, etc.)

? Which of the collective nouns might be used to describe a collection of things besides animals? (a bouquet of pheasants or flowers, a band of gorillas or musicians, an army of herring or soldiers, etc.)

? Which illustrations did you think were the most clever? (Answers will vary.)

Turn and Talk

Then, reread page 4, which says, "Do you ever wonder what animals do when they gather in groups of more than two?" Ask students to turn to a partner and discuss the following questions:

? What do animals do in groups?

? Why do some animals live in groups?

Tell students that you will be exploring these two questions by observing one of the most social mammals on earth—the elephant! *Ask*

? Do you know a collective noun for a group of elephants? (a herd)

explore

"Elephant Behavior" Video

Tell students that we can learn a lot about why some animals live in groups by observing elephants in herds. One of the world's leading elephant experts, Dr. Joyce Poole, has been studying the behavior of elephants for more than 40 years. Joyce grew up in Africa and spent her school holidays on safari, sleeping in a tent in the national parks. From an early age, she loved spending time with and observing the behavior of animals. When she was 11, she attended a lecture by Jane Goodall, who described her early research on chimpanzees. After the lecture, Joyce told her mother that she wanted to study animal behavior when she grew up! Now, Dr. Poole has a PhD in elephant behavior and gets to live out her dream as she studies the elephants of Gorongosa National Park in Mozambique.

Show Mozambique on a world map or globe. *Ask*

? What continent is Mozambique on? (Africa)

? What kind of elephants live there? (African elephants)

Tell students that there are two kinds of elephants: African and Asian. African elephants are larger and have bigger ears. Asian elephants are the ones more likely to be seen in zoos. Both kinds of elephants have a very complex system of communication. They use different kinds of *calls*, or sounds, to express their ideas and feelings. Many of these sounds are so low that humans can't hear them! They also use many different *gestures*, or movements, to express their ideas and feelings. Dr. Poole has spent a lot of time studying these calls and gestures and building large databases of what she and others have observed.

Determining Importance

Connecting to the Common Core
Reading: Informational Text
KEY IDEAS AND DETAILS: 3.1, 4.1, 5.1

Have students watch for different elephant behaviors as they "ride along" on safari through Gorongosa National Park with Dr. Poole and her brother, wildlife cameraman Bob Poole. Show the 6:15 min. *Elephant Behavior* video, and then *ask*

? What elephant behaviors did you observe or hear Dr. Poole describe? (periscope-sniff, trunk-twisting, head-shaking, foot-back, etc.)

? Why do you think she said that some of these elephants have a "bad attitude" toward people? (Answers will vary, but explain that more than 20 years ago Mozambique was in a civil war, and many elephants and other animals were killed. The elephants that are old enough to remember the war think of people as a threat. Elephants can live to be 70 years old, so many of the older elephants in the herd remember the violence and are fearful of people.)

Tell students that Dr. Poole and many other people in Mozambique have been working very hard to help the elephant population rebound and to help the elephants trust people again.

Elephant Ethogram

Explain that in the video, Dr. Poole is in the midst of a multiyear project to catalog, or list, all of the elephants' calls and behaviors. Because there are literally hundreds of different calls and behaviors, she has quite a job! Making a list like this can be very useful for scientists. Pass out the Elephant Ethogram student page to each student, and read the description at the top together:

> An **ethogram** *is a list of possible behaviors for a particular kind of animal, including descriptions of each behavior. Codes help scientists record data easily when they are observing animals in the wild. One way of using an ethogram is to observe a single animal's behavior at regular intervals, for example, every minute for 10 min. or every 30 min. for 10 hours. Below is a small sample of the hundreds of common African elephant behaviors observed by scientists in the Gorongosa National Park in Mozambique.*

Tell students that ethograms are like dictionaries of animal behavior. Scientists use ethograms to help them collect data and look for patterns. The behaviors can be divided into different categories, depending on what a scientist wants to study. Tell students that because we are interested in answering the questions, "What do animals do in groups?" and "Why do some animals live in groups?," we are going to analyze some feeding, social, and protection-related behaviors to see if we can find any patterns or make any *inferences* (conclusions based on our prior knowledge and our observations).

Next, read each behavior out loud, and then have students act out the behavior. This is a lot of fun, but it also a good way for students to familiarize themselves with the different behaviors and categories on the ethogram because they will be using the ethogram later to analyze some simulated field data. You can lead the whole class through the behaviors or have students work with a partner to play "elephant charades." As one student acts out a behavior, the other uses the ethogram to interpret what his or her partner is doing or trying to communicate. After they finish, remind them that elephants have *hundreds* of different calls and gestures that scientists have observed!

ELEPHANT BEHAVIOR CHARADES

Field Journal Notes

Next, pass out the Field Journal Notes student page. Explain that this page from a simulated (fictitious) field journal is similar to what a scientist in Mozambique might create during a day at work observing elephants out on the hot savanna or in the shady fever tree forest. *Ask*

? How often did this scientist record what the elephants were doing? (every 30 min.)

? How long did he or she spend making observations? (from 8:00 a.m. to 6:00 p.m., or 10 hours!)

? How many individual elephants was this scientist observing? (two)

Then, have students spend a few minutes working with a partner to make some observations about the field journal data. They will need to use the Elephant Ethogram to interpret the codes and read the descriptions. Some observations might include the following:

- The two elephants were sometimes doing the same behavior and sometimes doing different behaviors at the moment they were observed.

- They seemed to eat and drink more in the early morning and late afternoon.

explain

Graphing Elephant Behaviors

Connecting to the Common Core
Mathematics
MEASUREMENT AND DATA: 3.MD.3

ELEPHANT BEHAVIOR GRAPH

Tell students that it is often helpful to graph data to make meaning from it. Pass out the Elephant Behavior Graph student page. Using the Field Journal Notes page, students should tally how many times each behavior was observed for Elephant A. Then, they should repeat this step for Elephant B. They will use these totals to make a double bar graph, using two different colors, of the frequency of behaviors for each elephant. (Depending on the age and skills of your students, you may want to make the graph as a class.) The completed graphs will look like the "Elephant Behavior Graph" photo.

Questioning

When students are finished, have them use the ethogram, Field Journal Notes, and completed graph to look for patterns and make inferences by answering the questions on the second page of the Elephant Behavior Graph student page. The questions and answers are as follows:

1. What were the two most frequently observed behaviors for Elephant A? For Elephant B? (Elephant A: eating and periscope-sniff, Elephant B: eating and playing)
2. What time of day were the elephants most likely to be observed eating? (early morning and late afternoon)
3. What category of behaviors was observed around noon? What do you think may have occurred then? (protection behaviors; possibly an interaction with a predator)
4. The leader of the elephant herd is called the **matriarch**, the oldest and most respected female. Which elephant do you think might be the matriarch? What is your evidence? (Elephant A is the matriarch. She was observed leading the group and engaged in defensive behaviors more frequently.)
5. Which elephant do you think might be a young **bull** (male)? What is your evidence? (Elephant B was never observed leading the group but was observed playing and sparring)
6. What advantages do you think elephants gain from living in groups? What is your evidence? (They keep each other company and engage in social behaviors such as sparring, playing, and greeting each other. They help protect each other, for example by forming a defensive circle. From the data, you might also infer that they help each other find food and water—one elephant was perhaps leading the others to food and water.)

Finally, ask students if there are any other inferences they can draw from the data (answers will vary). Then, explain that there are many different ways to collect data on animal behavior. The data they saw in the Field Journal Notes were collected using a method called *interval sampling*. In interval sampling, scientists note what an animal is doing at pre-set, evenly spaced intervals (in this case, every 30 min.). These observations allow the scientist to get a "snapshot" of an animal's behavior at a precise moment in time. The observations don't tell them what the animal is doing in between those moments, but they give them enough data, over time, to begin to draw conclusions about an animal's behavior. Then, tell students that they were given only two pages of a field journal. The fictitious scientist most likely recorded other information. *Ask*

? What information is missing that could be useful in helping us make inferences about the elephants' behavior? (what the two elephants looked like and their sizes, the exact location in the park, the number of elephants in the herd, what other kinds of animals were seen in the area, the weather conditions and temperature that day, etc.)

Then, *ask*

? After analyzing the ethogram and the field journal notes, are you beginning to draw some conclusions about how elephants benefit from being in groups? What benefits do you think elephants get from living in groups? (Answers will vary.)

? What are you still wondering? (Answers will vary.)

Animals That Live in Groups Read-Aloud

Connecting to the Common Core
Reading: Informational Text
CRAFT AND STRUCTURE: 3.4, 4.4, 5.4

 **Inferring**

Show students the cover of the book, *Animals That Live in Groups*, and introduce the author, Kelsi Turner Tjernagel. Tell students that this book will give them much more information about why some animals live in groups, including how elephants benefit from living in a herd. Explain that the book has some bold-print words. They can use clues from the text and the illustrations to figure out what those words mean. Read the book aloud, including the Animal Fact! insets. Stop at each bold-print word to have students discuss what they think it means. Then, read each definition at the bottom of the page.

 Questioning

Connecting to the Common Core
Reading: Informational Text
KEY IDEAS AND DETAILS: 3.1, 4.1, 5.1

 Turn and Talk

After reading, ask students to turn to a partner and recall specific examples of how some other animals, besides elephants, benefit from being in a group. For example, they can discuss the following:

- Herring live in schools. Light reflects from their scales, making them blend together. Predators can't pick one fish out of the school.
- Budgerigars preen, or clean, each other's feathers.
- Emperor penguins huddle together for warmth.
- Young chimpanzees learn from adult chimps how to swing through trees, build nests, and catch termites.

Kratt's Creatures Videos

Next, tell students that you have some short (~1:30 min.) video segments to share from the *Kratt's Creatures* television show. They may have seen reruns of *Kratt's Creatures* (it was produced more than 20 years ago), but they will likely be more familiar with the Martin Kratt and Chris Kratt from their latest series, *Wild Kratts*. The Kratt brothers are not just television hosts; they are scientists, too, with degrees in zoology (Martin) and biology (Chris).

 Determining Importance

Have students listen for more evidence of why animals live in groups as you show the first segment, "Kratt's Creatures: Family on the African Savannah" (see "Websites" section).

> **Connecting to the Common Core**
> **Reading: Informational Text**
> INTEGRATION OF KNOWLEDGE AND IDEAS: 3.9, 4.9, 5.9

 Turn and Talk

After watching, have students discuss what new evidence they have from the video that wasn't in the book. (Meerkats will sometimes cover their young with their own bodies when startled by predators; all of the members of a wild dog pack help raise the pups; baboon troops can be as large as 200, and friendly baboon males will sometimes act as "godfathers"; elephants will suckle, or feed, each other's calves; etc.)

 Determining Importance

Then, have students listen for why some animals *don't* live in groups as you show the second segment, "Kratt's Creatures: Solitary on the African Savannah" (see "Websites" section).

 Turn and Talk

After watching, have students discuss how a solitary animal, such as the cougar, benefits from living alone for most of its adult life. (Living alone may help an animal be stealthier as it hunts, it only needs to feed itself, it can hide from predators more easily, and it has the flexibility to move around whenever and wherever it wants.)

The Social Lives of Elephants

 Pairs Read

> **Connecting to the Common Core**
> **Reading: Informational Text**
> RANGE OF READING AND LEVEL OF TEXT COMPLEXITY 3.10, 4.10, 5.10

Distribute The Social Lives of Elephants student page, and have students read the article independently or as a pairs read. In a pairs read, one student reads a paragraph while the other listens and then makes comments (I think …), asks questions ("I wonder …"), or shares new learning ("I didn't know …"). Alternately, you may want to use this article as a homework assignment.

 Questioning

> **Connecting to the Common Core**
> **Reading: Informational Text**
> KEY IDEAS AND DETAILS: 3.1, 4.1, 5.1

After reading, have students annotate the Elephant Facts section and respond to the questions at the bottom. Answers are as follows:

1. What are three ways elephants benefit from living in a group? (They care for each other, they help each other get food, and they protect each other from predators.)

2. What disadvantage might there be for an animal living in a group? (it might have to compete against other members of the group for food, it might have to battle other members of the group to move up in the social order, the group might be so large that it attracts predators, etc.)

3. A cougar lives most of its adult life alone. Can you think of another solitary animal? What benefit does it get from living alone? (Some examples include the tiger, koala, leopard, tortoise, tapir, sloth, scorpion, giant anteater, wolverine, Tasmanian devil, arctic fox, etc. Living alone may help an animal be stealthier as it hunts, it only needs to feed itself, it can hide from predators more easily, and it has the flexibility to move around whenever and wherever it wants.)

elaborate

Wildlife Tracking Technology

(*Note:* In advance, you may want to select some photographs from Google image searches of

Chapter 12

"penguin satellite transmitter" and "wildlife satellite tracking" to use during this discussion.)

Reread the main text on page 20 of *Animals That Live in Groups*, which mentions how radio transmitters are used to track penguins. Show some pictures of penguins outfitted with radio transmitters. *Ask*

? Why would scientists use such a device on a penguin? (It would be difficult to keep track of a single penguin in a group of hundreds or thousands by just watching it.)

Explain that there is no substitute for some of the "low-tech" work that wildlife scientists do in the field, such as developing ethograms of animal behaviors. But more and more, researchers are using "high-tech" tools and instruments to help them answer scientific questions. For example, GPS technology is helping scientists study the *range* of elephant herds (the geographical area where they can be found) and even keep track of the location of individual elephants. Large animals, such as elephants, bears, and some primates, are big enough to wear collars that hold a high-tech but hefty GPS tracker. These devices are similar to the computers in newer cars, which can give you your exact location at all times. The trackers, which continually "talk" back and forth with a satellite, require a very large battery but can tell researchers where an individual elephant is at any given moment.

Turn and Talk

Have students turn to a partner and discuss the following question:

? How would you attach a GPS satellite collar to a wild, 10,000 lb. bull elephant?

Determining Importance

Tell students that you have a video that will show them exactly how this is done. Explain that in the Tsavo National Park in Kenya, researchers are using GPS satellite collars to map elephants' movements along corridors, or traditional routes, elephants take as they travel between national parks, reserves, and other protected areas. Sometimes, the corridors go through areas where people live, and this can cause problems. Have students listen for the steps that researchers and conservationists go through to collar elephants as they watch the first 4 min. of the 5 min. "Mission: Tracking Elephants" video from the International Fund for Animal Welfare.

Connecting to the Common Core
Reading: Informational Text
KEY IDEAS AND DETAILS: 3.1, 4.1, 5.1

Ask

? What are the steps the people in the video went through to collar the elephant? (Answers should include a variation of the following steps: [1] Locate the elephant using a spotter plane. [2] Immobilize the elephant with a dart. [3] Keep the elephant safe by pushing it onto its side to protect its lungs. [4] Place the collar on it, adjust the length of the collar, and bolt the two ends of the collar together. [5] Take measurements, draw blood, etc. [6] Revive the elephant, and get out of the way!)

The elephants are revived very quickly using an antidote to the tranquilizer. The GPS collars they use weigh about 40 lb., but they don't bother the elephants. The information provided from the GPS satellite collars assists wildlife managers in reducing human–elephant conflicts and will help to preserve these majestic creatures and their habitats as human populations grow ever bigger. For more on this, students can watch a 1:15 min. video called "The Daily Movements of Samburu's Elephants" (see "Websites" section), which shows how an organization called Save the Elephants has paired with Google Earth to create a useful technology for visualizing elephant movements. In this time-lapse video, the elephants can be seen gathering at a river during the heat of the day and dispersing to grazing areas on higher ground at nighttime. As they watch the video, have students make observations and inferences about the movements of the elephants in this group, including the differences between the behaviors of the male elephants (blue) and the female elephants (pink).

evaluate

Better Together Booklets or Posters

Connecting to the Common Core
Writing
Text Types and Purposes: 3.2, 4.2, 5.2

Writing

Ask students to think back to some of the different collective nouns for animal groups they learned about in the book *An Ambush of Tigers*. Some examples are as follows:

- Tower of giraffes
- Raft of otters
- Pod of whales
- Leap of leopards
- Pack of wolves
- Cast of hawks
- Crash of rhinos
- Shiver of sharks
- Prickle of porcupines

Then, tell students that they may choose their favorite animal from the book or select a different animal (as long as there is a collective noun used to describe a group of them). They will be writing a booklet or creating a poster focusing on the animal's group behaviors and how the animal benefits from living in a group. (A student page for the booklet is provided on pp. 199–200.) Next, pass out the Better Together Scoring Rubric and go over the criteria:

1. **Better Together:**
   - Write the collective noun used for your animals and the name of the animal on the title page.
   - Describe three ways your animal benefits from being part of a group.

Better Together poster

2. **Habitat and Facts:**
   - Draw or attach a map of your animal's range in the wild.
   - List three fascinating facts about your animal.

3. **Sketch:**
   - Draw a detailed sketch of your animal.

4. **Ethogram:**
   - By reading text and/or observing videos, research or observe at least four common behaviors of your animal. Include a detailed description for each behavior.

Use the rubric to evaluate student work on the booklet or poster.

STEM at Home

Have students complete the "I learned that …" and "My favorite part of the lesson was …" portions of the STEM at Home student page as a reflection on their learning. They may choose to do the following

Chapter 12

at-home activity with an adult helper and share their results with the class. If students do not have access to the internet or these materials at home, you may choose to have them complete this activity at school.

"At home, we can choose an animal video to watch from the San Diego Zoo Kids website. While we watch the video, we can start an **ethogram**, or inventory of all the behaviors we observe. We can also look for a 'live cam' of our animal!"

🔍 *Go to* www.kids.sandiegozoo.org, *and click on the "Animal Cams and Videos" button.*

For Further Exploration

This section is provided to help you encourage your students to use the science and engineering practices in a more student-directed format. This box lists questions and challenges related to the lesson that students may select to research, investigate, or innovate. Students may also use the questions as examples to help them generate their own questions. After selecting one of the questions in the box or formulating their own questions, students can individually or collaboratively make predictions, design investigations or surveys to test their predictions, collect evidence, devise explanations, design solutions, or examine related resources. They can communicate their findings through a science notebook, at a poster session or gallery walk, or by producing a media project.

Research

Have students brainstorm researchable questions:

- ? What are the differences between African and Asian elephants? Do they use the same gestures and calls to communicate?

- ? How do scientists track the movements of very small migrating animals such as birds, bats, or butterflies?

- ? What does GPS stand for? How many GPS satellites orbit Earth? Who owns and operates GPS?

Investigate

Have students brainstorm testable questions to be solved through science or math:

- ? Make an ethogram for a pet or backyard animal by observing it frequently, at different times of day, over the course of several days. What behaviors did you observe? How can you describe them? Then, make observations at intervals (e.g., every minute for 10 min.). What patterns do you see?

- ? Survey your friends and family: What is your favorite social animal? Graph the results, then analyze your graph. What can you conclude?

- ? Using the Field Journal Notes of elephant observations, make a "time budget" pie chart. What percentage of the observations involved feeding behaviors? Social behaviors? Protection behaviors?

Innovate

Have students brainstorm problems to be solved through engineering:

? Can you invent your own system of communication using only calls and gestures? Teach the system to a friend!

? Can you design a zoo habitat for an elephant? What would it need to survive? What would it need for enrichment? Do you think a zoo elephant should be kept by itself? Why or why not?

? Can you design a campaign to support conservation efforts for an endangered animal?

Reference

National Research Council (NRC). 2012. *A framework for K–12 science education: Practices, crosscutting concepts, and core ideas*. Washington, DC: National Academies Press.

Websites

ElephantVoices
www.elephantvoices.org

"Elephant Behavior—Gorongosa Park: Rebirth of Paradise" (video)
www.pbslearningmedia.org/resource/gorongosa-clip-03/elephant-behavior-gorongosa-park-rebirth-of-paradise

"Kratt's Creatures: Family on the African Savannah" (video)
www.pbslearningmedia.org/resource/1e5eff7a-6606-4889-9ffd-40626ade7c9c/1e5eff7a-6606-4889-9ffd-40626ade7c9c

"Kratt's Creatures: Solitary on the African Savannah" (video)
www.pbslearningmedia.org/resource/48130444-f7c3-4766-820e-66297b2d64bc/48130444-f7c3-4766-820e-66297b2d64bc

"Mission: Tracking Elephants" (video)
https://vimeo.com/23661067

"The Daily Movements of Samburu's Elephants" (video)
http://savetheelephants.org/project/ste-and-google-earth

More Books to Read

Blewett, A. 2014. *Elephant rescue: All about elephants and how to save them*. National Geographic Kids Mission. New York: National Geographic Children's Books.
Summary: This book for older readers (grades 5–7) begins with the harrowing rescue of an orphaned elephant and follows her journey to sanctuary. Full of facts, figures, and photographs, it also explains why elephants are endangered and how children can help elephant conservation efforts. The book includes an interview with elephant scientist Dr. Joyce Poole.

Esbaum, J. 2015. *Animal groups*. New York: National Geographic Kids.
Summary: Gorgeous, full-color photographs accompany clear, simple text that describes a variety of animal groups. The book includes information on these groups' habitats and behaviors.

Jenkins, S., and R. Page. 2008. *Sisters and brothers: Sibling relationships in the animal world*. Boston: Houghton Mifflin Company.
Summary: This book provides a fascinating look at animal family groups—from falcons that learn to hunt with their sisters and brothers, to elephant sisters that babysit their younger siblings, to wild turkey brothers that stay together for life.

Kalman, B. 2016. *Animals that live in social groups*. New York: Crabtree Publishing Company.

Summary: This book is an interesting read about the families, social behaviors, and communication within different animal species, including humans! The text includes colorful photographs, insets, and a glossary.

McDonnell, P. 2011. *Me ... Jane*. New York: Little, Brown, & Company.

Summary: This picture book tells the inspirational story of the young girl who would grow up to be Jane Goodall—primatologist, environmentalist, humanitarian, and United Nations Messenger of Peace. The book includes some of Goodall's childhood drawings.

O'Connell, C. 2014. *A baby elephant in the wild*. New York: HMH Books for Young Readers.

Summary: In this true story for younger readers, a newborn elephant and her family journey through the scrub desert of Namibia. The text is accompanied by beautiful photographs of elephants bathing, rolling in the mud, and moving from place to place. The book includes a "Did you know?" section in the back, which provides more facts about elephants.

O'Connell, C., and M. Jackson. 2011. *The elephant scientist*. New York: Houghton Mifflin.

Summary: Part of the excellent *Scientists in the Field* series, this book for older readers (grades 5–7) describes the work of elephant scientist Caitlin O'Connell as she studies the elephants of the Namibian desert, including her groundbreaking discovery about elephant communication: Elephants actually listen with their limbs.

Smith, L. 2016. *There is a tribe of kids*. New York: Roaring Brook Press.

Summary: This enchanting story follows a little boy through a series of fanciful encounters with groups of animals and natural objects. Each animal and object has a "tribe" of its own, but it is up to the reader to decide if the boy is lost and searching for his own tribe or if he is an explorer bringing news of his adventures home.

Winter, J. 2011. *The watcher: Jane Goodall's life with the chimps*. New York: Schwartz & Wade Books.

Summary: A moving biography of Jane Goodall, the woman who lived with the chimps and who still speaks out to protect animals and their homes.

Wright, A. *A tower of giraffes: Animals in groups*. Watertown, MA: Charlesbridge.

Summary: This beautifully illustrated picture book combines art and science in an exploration of the collective nouns we use to describe animal groups.

Elephant Ethogram

Name: _____

An **ethogram** is a list of possible behaviors for a particular kind of animal, including descriptions of each behavior. Codes help scientists record data easily when they are observing animals in the wild. One way of using an ethogram is to observe a single animal's behavior at regular intervals, for example, every minute for 10 minutes or every 30 minutes for 10 hours. Below is a small sample of the hundreds of African elephant behaviors observed by scientists in the Gorongosa National Park in Mozambique.

Category	Behavior	Code	Description
Feeding	Eating	E	Eating food by picking it up with trunk
	Drinking	D	Sucking up water with trunk, then pouring it into mouth
Social	Sparring	S	Testing strength playfully by pushing against each other's heads
	Playing	P	Pushing on each other gently or kicking objects playfully
	Let's-go	LG	A female "rumbling" to herd, facing in direction she wishes to travel
	Head-raising	HR	Raising the head while greeting other elephants excitedly
	Head-shaking	HS	Shaking head and flapping ears to show annoyance
	Periscope-sniff	PS	Detecting predators or gathering information by raising trunk and sniffing the wind
Protection	Trunk-twisting	TT	Twisting the tip of the trunk back and forth when nervous or unsure
	Defensive-circle	DC	Encircling young with heads out
	Charging	C	Running at a predator or rival aggressively (can be real or mock)

Source: Adapted from Poole, J. H., and P. K. Granli. 2009. Gestures database. ElephantVoices. www.elephantvoices.org.

Picture-Perfect STEM Lessons, 3–5   193

Field Journal Notes

Name: _____

Elephant Behaviors Observed at 30-Minute Intervals, July 1, 2016, Gorongosa National Park

Elephant A Behaviors		Elephant B Behaviors	
8:00 am	LG	8:00 am	E
8:30 am	E	8:30 am	E
9:00 am	D	9:00 am	D
9:30 am	E	9:30 am	P
10:00 am	TT	10:00 am	E
10:30 am	PS	10:30	S
11:00 am	E	11:00	E
11:30 am	PS	11:30	P
12:00 pm	DC	12:00	P
12:30 pm	C	12:30	DC
1:00 pm	H/S	1:00	TT
1:30 pm	LG	1:30	H/S
2:30 pm	PS	2:30	E
3:00 pm	D	3:00	D
3:30 pm	H/R	3:30	H/R
4:00 pm	E	4:30	S
5:00 pm	LG	5:00	P
5:30	E	5:30	E
6:00	PS	6:00	PS

Name: _____

Elephant Behavior Graph

KEY
Elephant A = ☐
Elephant B = ☐

Frequency of Observed Behaviors

7	
6	
5	
4	
3	
2	
1	
0	

Number of Observations

Behaviors: Eating, Drinking, Sparring, Playing, Let's-Go, Head-Raising, Head-Shaking, Periscope-Sniff, Trunk-Twisting, Defensive-Circle, Charging

Picture-Perfect STEM Lessons, 3–5

Chapter 12

Questions

Name : _____

Using the elephant ethogram, the field journal, and your elephant behaviors graph, answer the following questions:

1. What were the two most frequently observed behaviors for Elephant A? For Elephant B?

   Elephant A: _____ and _____   Elephant B: _____ and _____

2. What time of day were the elephants most likely to be observed eating?

3. What category of behaviors was observed around noon? What do you think may have occurred then?

4. The leader of the elephant herd is called the **matriarch**, the oldest and most respected female. Which elephant do you think might be the matriarch? What is your evidence?

5. Which elephant do you think might be a young **bull** (male)? What is your evidence?

6. What advantages do you think elephants gain from living in groups? What is your evidence?

The Social Lives of Elephants

A Herd of Elephants

Elephant **herds** are well-known for their complex social structure. Within the herd can be one or more **family** groups. A family is usually made up of a mother, her sisters, her daughters, and their babies, or **calves**. Males, or **bulls**, live with the herd until they are about 13 years old, and then they join other males or

roam on their own. Sometimes, herds of elephants combine with other herds to form even larger **clans**. Elephants are very affectionate with their friends and family members and form strong bonds that can last a lifetime.

Mama Know Best

The **matriarch** is the oldest and most respected female in the herd and usually the largest too. She decides when to move, how fast to travel, and when to stop. The herd stays close to the matriarch, and she defends it from **predators**. The matriarch teaches her daughters how to care for their babies, and all of the adult females in a herd teach the younger elephants how to eat and behave.

Better Together

Animals live in a group for many reasons. One benefit for elephants is that members of a herd take care of each other. When a calf is born, the herd raises it. When an elephant is sick or injured, the herd cares for it. Another advantage of living in a group is that elephants help each

other find enough food to fuel their massive bodies. Adult elephants can eat more than 300 pounds of food a day, and they spend about 16 hours a day eating! Living in a group also provides safety in numbers. When danger is near, elephants often form a **defensive circle**, with the youngest elephants in the middle and the adult females facing outward. No predator wants to mess with a circle of elephants!

Chapter 12

Name : _____

Elephant Facts: Put a check (✓) next to facts you knew, an exclamation point (!) next to facts that are new to you, and a star (★) next to the most interesting fact!

- African elephants are the largest land animals on Earth.
- A **trunk** is actually an elephant's nose. It is used for smelling, breathing, drinking, trumpeting, caressing, greeting, and also for grabbing things—especially food. When elephants swim, they use their trunks as snorkels.
- After drinking at a watering hole, an elephant will sometimes take a mud bath or spray itself with water to cool off and remove biting insects.
- An elephant's ears are filled with blood vessels. By holding their ears out in the wind or flapping them, the ears **radiate** excess heat, which keeps the elephant's whole body cooler.
- Elephants use their tusks to dig for food and water and strip bark from trees. Males use their tusks to battle one another.
- Elephants have a sophisticated system of communication, involving both sounds and **gestures**—movements that express meanings.
- Because people in some places value ivory, elephants are being **poached**, or illegally killed, for their tusks.

1. What are ways elephants benefit from living in a group?

_____

_____

_____

2. What disadvantage might there be for an animal living in a group?

_____

3. A cougar spends most of its adult life alone. Can you think of another solitary animal? What benefit does it get from living alone?

_____

A _____

of _____

By: _____

Sketch

Ethogram

Behavior	Description

Name: _____

Range

Facts
_____
_____
_____
_____
_____
_____
_____
_____

Better Together

A _____ of _____ is better together because
_____
_____
_____
_____
_____
_____
_____
_____
_____
_____

National Science Teachers Association

Name : _____

Chapter 12

Better Together Scoring Rubric

Collective Noun: A _____ **of** _____

Score	Criteria
____ 4 ____ 3 ____ 2 ____ 1	**Better Together:** • Write the collective noun used for your animals. • Describe three ways your animal benefits from being part of a group.
____ 4 ____ 3 ____ 2 ____ 1	**Habitat and Facts:** • Draw or attach a map of your animal's range in the wild. • List three fascinating facts about your animal.
____ 4 ____ 3 ____ 2 ____ 1	**Sketch:** • Draw a detailed sketch of your animal.
____ 4 ____ 3 ____ 2 ____ 1	**Ethogram:** • By reading text and/or observing videos, research or observe at least four common behaviors of your animal. Include a detailed description for each behavior.

4—Excellent   3—Above Average   2—Average   1—Below Average

_____ Total/16 Points

STEM at Home

Dear _____,

At school, we have been learning about **why some animals live in groups** and **how scientists study animal behavior**.

I learned that: _____

_____

_____

My favorite part of the lesson was:

_____

_____

_____

At home, we can choose an animal video to watch from the San Diego Zoo Kids website. As we watch the video together, we can start building an **ethogram**, or list of all the behaviors of our animal. We can also look for a "live cam" of our animal!

*Go to **www.kids.sandiegozoo.org,** and click on the "Animal Cams and Videos" button.*

Ethogram for a _____

Behavior	Description

National Science Teachers Association

Chapter 13

Spider Science

Description

This lesson explores the secret lives of spiders, focusing on how they spin silk, build webs, and capture their prey. By undertaking a design challenge, students discover that spiderwebs are carefully engineered, intricate structures that spiders know how to build without ever being taught. Students also learn how genetic information is passed from parent to offspring, how inherited and acquired traits differ, and how scientists design experiments to study animal behavior in space.

Suggested Grade Levels: 3–5

LESSON OBJECTIVES Connecting to the *Framework*		
Science and Engineering Practices	**Disciplinary Core Ideas**	**Crosscutting Concept**
Constructing Explanations and Designing Solutions	**LS3.A:** Inheritance of Traits **LS3.B:** Variation of Traits **ETS2.A:** Interdependence of Science, Engineering, and Technology	Cause and Effect

Featured Picture Books

TITLE: *Next Time You See a Spiderweb*
AUTHOR: **Emily Morgan**
PUBLISHER: **NSTA Press**
YEAR: **2015**
GENRE: **Non-Narrative Information**
SUMMARY: *Stunning, up-close photography and clear text reveal the surprising secrets of spiders and their webs.*

TITLE: *Nefertiti, the Spidernaut: The Jumping Spider Who Learned to Hunt in Space*
AUTHOR: **Darcy Pattinson**
ILLUSTRATOR: **Valeria Tisnés**
PUBLISHER: **Mims House**
YEAR: **2016**
GENRE: **Narrative Information**
SUMMARY: *The extraordinary story of a tiny jumping spider that spent 100 days on the International Space Station learning how to hunt in microgravity.*

Time Needed

This lesson will take several class periods. Suggested scheduling is as follows:

Day 1: **Engage** with *Next Time You See a Spiderweb* Read-Aloud, Part 1, and **Explore** with Spiderweb Design Challenge

Day 2: **Explain** with Spiderweb Videos and *Next Time You See a Spiderweb* Read-Aloud, Part 2

Day 3: **Explore** with Inherited Versus Acquired Traits Card Sort and **Explain** with Inherited Versus Acquired Traits Discussion

Day 4: **Elaborate** with *Nefertiti, the Spidernaut* Read-Aloud and Videos

Day 5: **Evaluate** with Animals in Space Research Proposal and Animals in Space Research Proposal Presentations

Materials

For Next Time You See a Spiderweb *Read-Aloud*

- Spiderweb Cards (1 precut set per group of 4 students)

For Spiderweb Design Challenge

- White acrylic yarn (approximately 10 yards per student, approximately 1 skein per class)
- Scissors
- Tape
- Plastic spider (*Note*: 1 ½ in. plastic spiders [72-pack] are available at Amazon.com for about $10.00.)
- Paper plates (black or other dark color)
- Disposable cups
- Straws
- Cotton balls

SAFETY
Use caution when working with scissors. They are sharp and can puncture the skin or eyes.

Student Pages

- Spiderweb Design Challenge
- Inherited Versus Acquired Traits Sorting Cards
- Animals in Space Research Proposal Rubric
- Animals in Space Research Proposal
- STEM at Home

Background for Teachers

The characteristics that an organism inherits from its parents are called *traits*. In humans, traits include such things as hair color, eye color, skin color, blood type, and even the shape of the earlobes. These traits are determined by the information passed from parents to their children by *genes*, which are the fundamental units of inheritance in living organisms. Genes are regions of *DNA*, long double-helix structures that form a cell's chromosomes. Genes control how an organism looks, behaves, and reproduces. It is currently estimated that humans have around 19,000 different genes (Ezkurdia et al. 2014). The branch of science that studies how traits are passed from one generation to the next is called *genetics*. Genetics affects all living things, from miniscule bacteria to massive blue whales.

Inheritance is the reason that offspring resemble their parents—kittens look like cats, puppies look like dogs, snakelets look like snakes, and spiderlings look like spiders. Baby animals such as tadpoles and caterpillars don't look exactly like their parents at first; they must go through a metamorphosis. But once they grow into adults, such organisms do look like their parents. Because each parent (in most multicellular organisms) contributes only half of the genetic information that makes up its offspring, offspring are not exact replicas of their parents.

Inheritance isn't the whole story, however. Genes may determine the *inherited traits* that are passed from parent to offspring, but the environment can also shape the development, appearance, behavior, and survival of an individual organism. The chemicals they are exposed to, the foods they eat, and the experiences they have all help shape organisms' *acquired traits*. For example, a normally tall sunflower plant can be stunted by lack of sunlight. A tomato plant can bear more fruit if given the right fertilizer. A dog can become obese by eating too many dog treats. A dolphin that lost part of its tail can learn to swim again by moving in a different way. A garden spider exposed to caffeine spins irregular webs. A jumping spider can learn to hunt in a microgravity environment.

Not all of the traits that organisms inherit can be modified by environmental influences. In humans, for example, blood type is an inherited trait that will not change, whereas the tendency to be tall or short can be modified by how fast you grow, what you eat, childhood illnesses, and other factors.

Upper-elementary students should understand that many characteristics of organisms are inherited from their parents, other characteristics result from individuals' interactions with the environment, and some characteristics involve both inheritance and the environment. To engage students in these core ideas, this lesson begins with a read-aloud of a book about spiders and their webs. Students learn through a fun design challenge that spiderwebs are elaborate and difficult to build. Through reading, they learn that spiders can build these carefully engineered structures without having a single lesson. This skill is called an *inherited trait* (also known as an *instinctive behavior*). Each kind of spider builds the same type of web that its parents build. Other traits that spiders inherit from their parents are the physical characteristics that they are born with: eight legs, fangs, spinnerets, body shape and coloring, and so on. As opposed to inherited traits, *acquired traits* are shaped by an individual organism's environment. Students are introduced to acquired traits in animals through the true story of a jumping spider named Nefertiti. NASA sent her to the International Space Station (ISS) as part of a research project proposed by an 18-year-old science whiz kid named Amr Mohammed. Amr's hypothesis was that the spider would not be able to hunt in the microgravity environment of the ISS. Astonishingly, after many failed attempts, she developed a hunting behavior that allowed her to catch fruit flies in space! Nefertiti not only learned to hunt in a new way but also learned to re-adapt to her inherited, or instinctive, method of hunting once she returned to Earth.

The crosscutting concept of cause and effect is developed in this lesson as students learn that some traits in organisms are caused by their genetics (inherited traits) and others are caused by factors in an organism's environment (acquired traits). They also explore an unusual cause of acquiring new behaviors—being sent into orbit and living in microgravity! Students are engaged in the science and engineering practice of constructing explanations and designing solutions throughout this lesson. First, they are challenged to design and build a model of a spiderweb. Then, they learn how scientists designed a habitat that could support the basic needs of a spider at the ISS and how an experiment was set up to test Amr's hypothesis. Through this example, students learn how science and technology support each other. Finally, they submit an Animals in Space Research Proposal in which they design an experiment to study learned behaviors of animals in microgravity.

engage

Next Time You See a Spiderweb Read-Aloud, Part 1

> Connecting to the Common Core
> **Reading: Informational Text**
> CRAFT AND STRUCTURE: 3.6

Show students the cover of *Next Time You See a Spiderweb* and introduce the author, Emily Morgan. Tell students that, although many people are afraid of spiders, she is not! The author wanted to write a book about spiders and their webs so more people would realize that spiders are fascinating and beneficial animals with some cool adaptations. In other words, she wants people to say "Ooh!" when they spot a spider instead of "Ick!" She also wants her books to inspire adults and children to spend time together outdoors, experiencing the joy, excitement, and mystery of the natural world together.

Making Connections: Text to Self

Before reading, ask students to turn to a neighbor and discuss the following questions:

? Do you like to spend time outside observing nature?

? What kinds of spiders have you observed (either inside or outside)?

? What kinds of spiderwebs have you seen?

? How do *you* feel about spiders?

Then, pass out a set of Spiderweb Cards to each group of four students. (*Note:* Cut them out for each group beforehand.) Have students discuss the similarities and differences they notice about the webs. Tell students that you would like them to signal when they hear about each kind of web by holding up the picture of the web as you read.

Next, read pages 4–21 of *Next Time You See a Spiderweb*, pausing to discuss the characteristics of each type of web as you read about it. Stop after reading, "It is astonishing that these small creatures are able to make such elaborate and efficient traps without a single lesson! Spiders are born knowing how to build webs—they don't need anyone to teach them." *Ask*

? How do spiders know how to spin webs?

? Do you think it is easy or difficult for a spider to build a web?

? Do *you* think you could build a spiderweb?

explore

Spiderweb Design Challenge

Tell students that you have a challenge for them! Their challenge is to design and build a model of a spiderweb, using no more than 10 yards of yarn and some tape. They may also use paper plates, cups, straws, and cotton balls if desired. They may choose to build any of the webs from the Spiderweb Cards: orb web, funnel web, sheet web, or cobweb. Remind students that, for any engineering challenge, they should first brainstorm possible solutions and then design a solution by sketching it out on paper before building it. Pass out the Spiderweb Design Challenge student page. Tell students that after they design their web, they must get approval from you at the Teacher Checkpoint before building.

As students brainstorm and make sketches, give suggestions and feedback. One hint is to cut small slits around the edges of the paper plate to serve as places to hold the yarn as they loop it around. If they are attempting an orb web, they should form the "radial lines" first and secure them to the paper plate with tape. Once they do that, they can loop segments of yarn around the radial lines to make the spiraling "orb lines" that go around and around the web. To make a cob web, students can crisscross the yarn around the plate in a more random, disorganized way. To make a funnel web, students can try cutting out the center of the paper plate to create a circular frame at the opening of the funnel and then wrap a cup in yarn to create the bottom of the funnel. To make a sheet web, students can crisscross lines of yarn above the plate using straws to support it and then pull apart cotton balls to create a sheet on another plate below. As you approve each design, allow students to begin building their models. You may want to set a time limit for building.

A SPIDERWEB DESIGN

After students have completed their spiderweb models and secured the plastic spiders to them, have each student make a label with his or her name and the type of web built and attach it to the model. You may want to display the finished models on a bulletin board, hang them from the ceiling, or display them in the hallway. Halloween is a great time of year to do this activity!

explain

Spiderweb Videos

Ask

? How easy or difficult was it to build your spiderweb model?

? Did you try any designs that didn't work?
? What would you do differently?
? Do you think building webs is difficult for spiders?

Connecting to the Common Core
Reading: Informational Text
KEY IDEAS AND DETAILS: 3.1, 4.1, 5.1

Orb Web Video

Have students watch the 1 min. time lapse video of a garden spider spinning an orb web (see "Websites" section). After watching, *ask*

? What did you notice about the spider's web-building technique?
? How did it compare with your technique?

Remind students that the book said, "Spiders are born knowing how to build webs—they don't need anyone to teach them." Explain that most animals are born "knowing" how to do many things. These instincts are known as *inherited traits*. Offspring *inherit*, or get, these behaviors from their parents. The information is passed from parent to offspring through tiny structures in cells called *genes*. Baby spiders, or spiderlings, have the instinct to build webs "pre-programmed" into their genes. Their parents don't have to teach them!

Spiderlings Video

Show the 2 min. video called "Young Garden Spiders Emerging and Spinning Webs" (see "Websites" Section), and have students watch as dozens of garden spiderlings emerge from their eggs and begin spinning webs without a single lesson from their parents. After watching, *ask*

? Did you observe any adult spiders teaching the spiderlings how to spin webs? (no)
? What do you call animal behaviors that are instinctive (do not have to be learned)? (inherited traits)
? What physical characteristics of the baby spiders did you notice as you watched? (The

Chapter 13

spiderlings had eight legs, spinnerets, a yellowish color, round abdomens with a dark spot, etc.)

Show the picture of an adult garden spider on page 11 of *Next Time You See a Spiderweb*. Explain that baby garden spiders look pretty much like their parents (except they are smaller and their color and markings are a little different). Their physical characteristics, such as their body shape and having eight legs, fangs, spinnerets for making silk, and so on are inherited from their parents through genes. Because of genetic inheritance, all living things look a lot like their parents. Genes are passed from generation to generation. So garden spider spiderlings grow into adult garden spiders. Jumping spider spiderlings grow into adult jumping spiders. Trapdoor spider spiderlings grow into adult trapdoor spiders. (It is important to note that some baby animals such as tadpoles and caterpillars don't look exactly like their parents at first; they must go through a metamorphosis. But once they grow into adults, they do look like their parents.)

Explain that inherited traits are only one piece of the puzzle, however. It is true that many characteristics of organisms are inherited from their parents. For example, tiger cubs are born with stripes. Baby sea turtles are born with the instinct to scurry to the ocean after hatching. Sunflowers tend to grow tall. But other characteristics can result from an organism's interactions with its environment. These characteristics are known as *acquired traits*. The following are examples of acquired traits:

- A spider exposed to caffeine spins irregular webs.
- A tomato plant can bear more fruit if given the right fertilizer.
- A cat can become obese by eating too many cat treats.
- A sea turtle can lose a flipper in a collision with a boat.
- A normally tall sunflower plant can be stunted by being grown in the dark.
- A tiger cub learns to hunt prey by watching its parents.
- A chimpanzee can learn sign language.
- A dolphin that loses part of its tail can learn to swim again by moving in a different way.

The organisms in those examples of acquired traits did not inherit the traits from their parents. The traits were not passed through genes. Instead, they resulted from learning or from the individual organism's interactions with its environment. All of these things can help shape an organism's acquired traits: the chemicals they are exposed to (the spider exposed to caffeine, the tomato plant given fertilizer), the foods they eat (the cat eating too many cat treats), the things that happen to them (the dolphin losing part of its tail, the sea turtle losing a flipper, the sunflower being put in the dark), and the behaviors they learn (the tiger cub learning to hunt, the chimpanzee learning sign language, the dolphin learning to swim a different way). Acquired traits *cannot* be passed on through genes. For example, a sea turtle that loses a flipper in an accident does not have three-flippered babies. A chimpanzee that learns sign language does not give birth to a baby chimpanzee that knows sign language.

Some characteristics of organisms involve *both* genetic inheritance and environment. For example, a fair-skinned boy inherits his skin color from a combination of his parents' genes for skin color. If this child spends too much time in the sun, he can get a bad sunburn. The reddened skin that results is a combination of an inherited trait (being born with fair skin) and a trait acquired from his environment (a sunburn from spending too much time in the sun).

Next Time You See a Spiderweb Read-Aloud, Part 2

 Questioning

Next, tell students you are going to read the rest of the book *Next Time You See a Spiderweb*, and have them watch for examples of inherited traits. Before reading, *ask*

? Do all spiders inherit the ability to make webs?

208 National Science Teachers Association

? Do all spiders inherit the trait of producing silk?

? How strong is spider silk?

Connecting to the Common Core
Reading: Informational Text
KEY IDEAS AND DETAILS: 3.1, 4.1, 5.1

Then, read the rest of the book aloud (pages 22 to end.) After reading, *ask*

? Do all spiders inherit the ability to make webs? (no)

? What kinds of spiders do *not* spin webs? (jumping spiders, fishing spiders, trapdoor spiders—about half of spider species have other clever ways to catch food)

? How do jumping spiders catch food? (They use their excellent vision to detect prey, and then they pounce on it.)

? Do you think a jumping spider's way of catching food is inherited or learned? (inherited)

? Do all spiders inherit the trait of producing silk? (yes)

? How strong is spider silk? (Some spider silk is stronger than a thread of steel, and some can stretch up to three times its length without breaking.)

? Why do scientists and engineers study spider silk? (Scientists are trying to mimic spider silk to create ultra-strong materials; engineers are studying it to design structures that can withstand disasters such as earthquakes.)

 **Synthesizing**

Connecting to the Common Core
Reading: Informational Text
KEY IDEAS AND DETAILS: 3.2, 4.2, 5.2

Next, *ask*

? What is the main idea of *Next Time You See a Spiderweb*? (A spiderweb is a trap.)

? How is the main idea supported by details? (The book has many photographs of spiderwebs, describes different ways that spiders use webs to trap food, and explains the strength of spider silk.)

explore

Inherited Versus Acquired Traits Card Sort

The following activity will help students explore and understand the difference between inherited traits and acquired traits. Give each pair of students a sheet of Inherited Versus Acquired Traits Sorting Cards, and have students cut them out. Students should work with their partners to sort the cards into two groups: inherited traits and acquired traits. Remind students that inherited traits are those

SORTING THE CARDS

genetic characteristics an organism gets from its parents; acquired traits are those characteristics that an organism learns receives from its environment.

explain

Inherited Versus Acquired Traits Discussion

Visit each pair as they work, and have students explain their answers to the following questions:

? Why did you place that card where you did?

Picture-Perfect STEM Lessons, 3–5

Table 13.1. Inherited Traits Versus Acquired Traits: Card Sort Answers

Inherited Traits	Acquired Traits
A trout swimming in a stream	A wolf cub learning to hunt in a pack with other wolves
A dalmatian puppy being born with spots	A parrot saying "Hello!"
A sunflower plant in a window growing toward the sunlight	A girl speaking English and Spanish
A garden spider hatching from an egg and spinning a web	A beagle eating so many dog treats that it gets obese
A robin laying an egg in a nest	A bean plant in a dark closet turning yellow
A tiger cub being born with stripes	A jumping spider losing a leg to a centipede
A boy having brown eyes	A goldfish swimming to the top of a bowl when it sees your hand holding food

? Did you get stuck on any of the examples? Why?

? Can you think of other examples of inherited traits?

? Can you think of other examples of acquired traits?

? What are you wondering about inherited traits and acquired traits?

Answers to the card sort are shown in Table 13.1.

Next, have students think about their own traits. *Ask*

? What inherited traits do you have? (hair color, eye color, and so on)

? What acquired traits do you have? (language, learned skills, and so on)

You may want to have students make a chart of their own called Inherited Traits Versus Acquired Traits.

elaborate

Nefertiti, The Spidernaut Read-Aloud and Videos

Connecting to the Common Core
Reading: Informational Text
CRAFT AND STRUCTURE: 3.6

Show students the cover of *Nefertiti, the Spidernaut: The Jumping Spider Who Learned to Hunt in Space* by Darcy Pattison (illustrated by Valeria Tisnés). Explain that this book is a true story about a jumping spider that was sent to the International Space Station (ISS) as part of a research project. The idea for the experiment was proposed by an 18-year-old Egyptian student named Amr Mohammed through a YouTube Space Lab global competition. Amr's hypothesis was that a jumping spider would not be able to catch its prey in the microgravity environment of the ISS. Explain that *microgravity* is the condition in which people or objects appear to be weightless, as is the case on the ISS orbiting 200–250 miles above Earth. Show the 2:18 min. video called "Meet Amr From Egypt," which is about Amr and his proposal (see "Websites" section).

Next, read the following quote from the author of *Nefertiti*:

> When I interviewed Sunita Williams, the astronaut who cared for Nefertiti, she called the spider her 'scary friend.' I know that spiders are often scary to kids, but here's an interesting fact. Worldwide, there are around 50,000 spider species and only about a dozen are dangerous. In the U.S., only the black widow and brown recluse can be deadly to humans. The chances of your meeting a spider that is dangerous is unlikely. I love the book's cover, a close-up of the Johnson jumping spider. She's beautiful. Sometimes spiders are scary; but they can also be a friend. For Sunita, it was comforting to have another living creature with her on the space station, especially when the spider stopped and watched her moving around. (personal correspondence)

Then, *ask*

? How do you think the author, Darcy Pattison, feels about spiders? (Answers will vary, but students will likely say that the author likes spiders.)

? How do you think Emily Morgan, the author of *Next Time You See a Spiderweb*, feels about spiders? (Answers will vary, but students will likely say that the author thinks spiders are fascinating.)

? What do their feelings tell you about how authors of nonfiction books might choose their topics? (They are probably very interested in, curious about, or even fond of the subjects of their books.)

Tell students that it would be hard to write a book about something you didn't really care about!

Inferring

Before reading *Nefertiti*, ask

? What did you learn about how jumping spiders catch their prey from the book *Next Time You See a Spiderweb*? (They use their excellent vision to detect prey, and then they pounce on it.)

? Is this behavior inherited or acquired? (inherited)

? In microgravity, objects are almost weightless, but not quite. How do you think microgravity would affect a jumping spider's ability to hunt and catch prey? (Answers will vary.)

Then, read the book aloud. After reading, you may want to show the 57 sec. video called "Jumping Spider, Nefertiti, Onboard the International Space Station" so that students can observe for themselves the spider's attempts at hunting in microgravity and her re-adaptation to Earth's gravity (see "Websites" section).

Next, read the following excerpt from Amr's SpaceLab experiment description.

> Jumping spiders have very good vision that they use to track and stalk prey. Unlike orb weavers, the jumping spider does not spin a web to capture food. Jumping spiders are hunters. They move around during the day seeking prey. Once it visually identifies prey, [the jumping spider] may stalk it for some distance prior to catching it. Once the jumping spider is within close proximity of its prey, it will secure a drag line using its silk and then jump with great speed onto the prey securing it with a lethal bite. The drag line acts as a safety harness in case the spider should miss its target and fall. This experiment seeks to determine if the jumping spider alters its predation technique in a microgravity environment. (YouTube Space Lab 2016)

Questioning

> Connecting to the Common Core
> **Reading: Informational Text**
> KEY IDEAS AND DETAILS: 3.1, 4.1, 5.1

Scientists usually present a problem statement, or something they want to investigate, when they write a research proposal. In this proposal, scientists wanted to find out if a jumping spider could hunt in microgravity.

Next, have students think back to the book *Nefertiti*. Ask

- ? What did Amr think would happen if a jumping spider was sent to space? (He predicted it would not be able to hunt in microgravity.)
- ? How did scientists investigate this? What were their methods? (They chose a spider to use in the experiment. They designed a special habitat for her. They tested to see if she could survive in the habitat on Earth. Then, they sent her to space on an unmanned cargo carrier. After she arrived on the ISS, they videotaped her attempts to hunt.)
- ? What new technology had to be invented to do this investigation? (A special habitat had to be designed for the spider to live in.)
- ? What did scientists discover? (A jumping spider could adapt to microgravity and learn to hunt in a new way.)
- ? Do you think this new way of hunting would be considered an inherited trait or an acquired trait? (acquired trait)
- ? What surprised the scientists when she returned to Earth? (She re-learned how to hunt on Earth.)
- ? How do you think Amr felt about the results of the experiment? (Answers will vary.)

evaluate

Animals in Space Research Proposal

> Connecting to the Common Core
> **Writing**
> TEXT TYPES AND PURPOSES: 3.2, 4.2, 5.2

Writing

Ask students to think about Amr's ingenious idea for an experiment and what scientists learned from it. Tell them that they are going to have an opportunity to propose their own experiment to test animal behavior in space! They will first write their proposal and then record a video (2 min. or less) for others to see. Their proposal won't actually be part of a global competition, but it will be used to evaluate what they have learned about organisms' inherited versus acquired traits. (For fun, you could ask other teachers to watch their proposal videos and give feedback.) Tell students that a good research proposal is like a good sales pitch! It should be clear, concise, and compelling. In other words, their research proposal should be easy to understand, informative, and contain many supporting details. It should not be too long, and it should be original and interesting.

RECORDING A RESEARCH PROPOSAL

Pass out the Animals in Space Research Proposal and the Animals in Space Research Rubric. Review the requirements for each component. You may want to re-read the SpaceLab experiment description on jumping spider predation as an example of a strong start to a proposal. Have students brainstorm different animals, choose appropriate subjects for experiments in space, research their subject animals and then complete their proposals using the rubric as a guideline.

Animals in Space Research Proposal Presentations

> **Connecting to the Common Core**
> **Speaking and Listening**
> Presentation of Knowledge and Ideas: 3.4, 4.4, 5.4

Have students share their completed videos with the class, or invite other classes to attend. Use the rubric to evaluate their video presentations.

STEM at Home

Have students complete the "I learned that …" and "My favorite part of the lesson was …" portions of the STEM at Home student page as a reflection on their learning. They may choose to do the following at-home activity with an adult helper and share their results with the class. If students do not have access to the internet or these materials at home, you may choose to have them complete this activity at school.

"At home, we can watch a video from Wild Kratts called 'Make a Web,' which shows how some spiders make orb webs and how spiders use different kinds of silk for different jobs."

Search for "Wild Kratts Make a Web" on www.pbslearningmedia.org *to find the video at* www.pbslearningmedia.org/resource/37 1aa09e-46d0-4665-8a80-5ca94f6cf423/make-a-web-wild-kratts.

"After we watch the video, we can write the steps a spider takes to build an orb web [answers shown in parentheses]:

1. First, (the spider makes a frame).
2. Then, (the spider drops a silk line down the center).
3. Next, (the spider puts in the spokes).
4. Finally, (the spider adds the spiral)."

For Further Exploration

This section is provided to help you encourage your students to use the science and engineering practices in a more student-directed format. This box lists questions and challenges related to the lesson that students may select to research, investigate, or innovate. Students may also use the questions as examples to help them generate their own questions. After selecting one of the questions in the box or formulating their own questions, students can individually or collaboratively make predictions, design investigations or surveys to test their predictions, collect evidence, devise explanations, design solutions, or examine related resources. They can communicate their findings through a science notebook, at a poster session or gallery walk, or by producing a media project.

Research

Have students brainstorm researchable questions:

? What is spider silk made of, and how strong is it? Which spiders make the strongest silk?

? What kinds of animal research projects have been done in space?

? How do chimpanzees learn sign language?

Investigate

Have students brainstorm testable questions to be solved through science or math:

? Survey your friends and family: Do you think spiders are "Ooh!" or "Ick!"? Graph the results, then analyze your graph. What can you conclude?

? Survey your family and friends: What inherited traits do they have? (widow's peak or straight hairline, attached earlobes or detached earlobes, can roll tongue or can't roll tongue, left handed or right handed) Graph the results, then analyze your graph. What can you conclude?

? Can you find the ISS in the night sky? (go to *spotthestation.nasa.gov* and enter your location)

Innovate

Have students brainstorm problems to be solved through engineering:

? If scientists could make artificial spider silk, what would be some potential uses for it?

? Can you design a habitat for a tarantula in space?

? Can you design a habitat for a pet to keep astronauts on the ISS company?

Reference

Ezkurdia, I., D. Juan, J. M. Rodriguez, A. Frankish, M. Diekhans, J. Harrow, J. Vazquez, A. Valencia, and M. L. Tress. 2014. Multiple evidence strands suggest that there may be as few as 19,000 human protein-coding genes. *Human Molecular Genetics* 23 (22): 5866–5878.

YouTube Space Lab. 2016. ISS science for everyone. NASA. www.nasa.gov/mission_pages/station/research/experiments/208.html.

Websites

"Jumping Spider, Nefertiti, Onboard the International Space Station" (video)
www.youtube.com/watch?v=EPPGQeZ4aw4

"Meet Amr From Egypt" (video)
www.youtube.com/watch?v=2YV1WHjNs4E

Orb Web Time Lapse: "Why Do Spiders Spin Webs?" (video)
http://wonderopolis.org/wonder/why-do-spiders-spin-webs

Spiderlings: "Young Garden Spiders Emerging and Spinning Webs" (video)
www.arkive.org/garden-spider/araneus-diadematus/video-09c.html

More Books to Read

Bardoe, C. 2015. *Gregor Mendel: The friar who grew peas.* New York: Abrams.
Summary: Watercolor illustrations and clear text explain the theory of heredity in simple-to-understand language and examples. Regarded as the world's first geneticist, Gregor Mendel discovered one of the fundamental aspects of genetic science: Animals, plants, and people all inherit and pass down traits through the same process.

Berger, M. 2003. *Spinning spiders*. New York: Harper-Collins.
Summary: From the *Let's-Read-and-Find-Out Science* series, this book featuring remarkably realistic artwork by S. D. Schindler, teaches about the silk spiders produce, the webs they spin, and the prey they capture.

Bishop, N. 2007. *Spiders*. New York: Scholastic.
Summary: Nic Bishop's signature up-close, stop-action photographs show spiders larger than life. Amazing images show the beauty and otherworldliness of spiders. Simple, engaging text conveys basic information about spiders as well as cool and quirky facts.

Boothroyd, J. and B. Silverman. 2011. *What traits are in your genes? Lightning Bolt Books* series featuring volumes on body parts, eye color, facial features, hair traits, unusual traits, and vison. Minneapolis, MN: Lerner.
Summary: Simple text, vivid photographs, and a colorful design illustrate how we inherit genetic traits, while familiar examples make abstract concepts easy to grasp. Back-matter special features, including glossaries and further reading pages, bolster these informative texts.

Green, J. 2014. *Inheritance of traits: Show me science: Why is my dog bigger than your dog?* North Mankato, MN: Raintree.
Summary: This book teaches children about genes and the inheritance of traits through the engaging topic of dogs.

Heos, B. 2013. *Stronger than steel: Spider silk DNA and the quest for better bulletproof vests, sutures, and parachute rope*. Boston: HMH Books for Young Readers.
Summary: From the *Scientists in the Field* series, this book takes older readers into the lab of Dr. Randy Lewis, where goat embryos are injected with genes from golden orb weaver spiders. When the goats mature, some of the females will produce spider silk proteins in their milk. This project aims to produce threads of varying degrees of strength and flexibility typical of spider silk.

Simon, S. 2007. *Spiders: All about their web-building skills, bodies, diets, and more!* New York: Harper Collins.
Summary: Stunning full-color, up-close photographs and interesting text introduce the physical characteristics, behaviors, and life cycles of different kinds of spiders.

Spiderweb Cards

Cob Web

Sheet Web

Funnel Web

Orb Web

Name : _____

Spiderweb Design Challenge

1. Brainstorm possible solutions. Choose a solution and sketch it below.

> Type of web: _____

2. Describe how you will build your spiderweb model.

_____

_____

_____

_____

3. Teacher checkpoint ☐

Now build your model!

Name : _____

Inherited Versus Acquired Traits

Sorting Cards

A wolf cub learning to hunt in a pack with other wolves	A trout swimming in a stream
A parrot saying "Hello!"	A girl speaking English and Spanish
A dalmatian puppy being born with spots	A beagle eating so many dog treats that it gets obese
A sunflower plant in a window growing toward the sun	A garden spider hatching from an egg and spinning a web
A bean plant in a dark closet turning yellow	A robin laying an egg in a nest
A jumping spider losing a leg to a centipede	A goldfish swimming to the top of a bowl when it sees your hand holding food
A tiger cub being born with stripes	A boy having brown eyes

Name: _____

Animals in Space Research Proposal Rubric

Component	Description
Title and Introduction *6 Points*	• Choose an animal that you think could survive for a long period of time in orbit around Earth on the International Space Station (ISS). • Research the traits of the animal. • Describe a normal behavior of this animal on Earth. • Identify whether this behavior is **inherited** (instinctive) or **acquired** (learned) and how it is inherited or acquired. • Use supporting details when describing the animal's normal Earth behavior. • Give your proposal a catchy and descriptive title.
Problem Statement *2 Points*	Describe what you want to find out about the animal through a microgravity experiment in orbit ("This experiment seeks to determine if … ").
Prediction *2 Points*	Write a prediction about what you think will happen. The prediction should be based on your observations and research of your animal subject ("If a _____ is sent into orbit, it will/will not be able to _____.").
Methods *6 Points*	• Describe the animal's temporary habitat on the ISS. • Draw a labeled diagram of the animal's habitat. • Describe how your prediction will be tested on the ISS. Explain how scientists will perform the experiment in microgravity. • Describe how you will collect the data (video, photos, tallying movements or behaviors, etc.). • Include a drawing of the animal or experiment in action. • Describe what you will test or observe once the animal returns to Earth's gravity.
Video *4 Points*	Record your presentation. Use a clear speaking voice and look directly at the camera. Props, pictures, and diagrams are helpful! Keep your presentation to 2 min. or less.
Total Score _____/20	Reviewer comments:

Picture-Perfect STEM Lessons, 3–5

Name: _____

Animals in Space Research Proposal

Title: _____

Title:

Introduction:

Problem Statement:

Prediction:

Methods:

Name: _____

STEM at Home

Dear _____,

At school, we have been learning about **spiders, spiderwebs,** and **inherited traits versus acquired traits**.

I learned that: _____

_____

_____

My favorite part of the lesson was:

_____

_____

_____

At home, we can watch a video from Wild Kratts called "Make a Web," which shows how some spiders make orb webs and how spiders use different kinds of silk for different jobs.

Search for "Wild Kratts Make a Web" on ***www.pbslearningmedia.org*** to find the video at ***www.pbslearningmedia.org/resource/371aa09e-46d0-4665-8a80-5ca94f6cf423/make-a-web-wild-kratts.***

After we watch the video, we can write the steps a spider takes to build an orb web.

1. First, _____.

2. Then, _____.

3. Next, _____.

4. Finally, _____.

Picture-Perfect STEM Lessons, 3–5

Bionic Animals

Description

After hearing the true story of a young bottlenose dolphin who injures her tail and is helped by a team of dedicated prosthetists, students are challenged to design, build, and test a prosthetic part for a wind-up toy animal. In the process, they learn how biomedical engineers design everything from life-saving medical devices to life-changing prosthetic limbs. They also learn how the fast-growing field of three-dimensional (3-D) printing technology is helping both people and animals with limb differences.

Suggested Grade Levels: 3–5

LESSON OBJECTIVES Connecting to the *Framework*		
Science and Engineering Practices	**Disciplinary Core Ideas**	**Crosscutting Concept**
Engaging in Argument From Evidence Obtaining, Evaluating, and Communicating Information	**LS1.A:** Structure and Function **ETS1.A:** Defining and Delimiting Engineering Problems **ETS2.B:** Influence of Engineering, Technology, and Science on Society and the Natural World	Structure and Function

Featured Picture Books

- TITLE: **Winter's Tail: How One Little Dolphin Learned to Swim Again**
- AUTHORS: **Juliana Hatkoff, Isabella Hatkoff, Craig Hatkoff, and David Yates**
- PUBLISHER: **Scholastic**
- YEAR: **2011**
- GENRE: **Narrative Information**
- SUMMARY: *This book recounts the remarkable true story of Winter, a young bottlenose dolphin who injures and then loses her tail after being mangled in a crab trap off the coast of Florida. After Winter is rescued and taken to the Clearwater Marine Aquarium, a group of prosthetists design and build a tail that enables her to once again swim like a dolphin.*

- TITLE: **Biomedical Engineering and Human Body Systems**
- AUTHOR: **Rebecca Sjonger**
- PUBLISHER: **Crabtree**
- YEAR: **2016**
- GENRE: **Non-Narrative Information**
- SUMMARY: *This fascinating book emphasizes the design process that biomedical engineers use to define problems, brainstorm solutions, and build and test models as they develop technologies to diagnose, treat, and prevent medical problems in human body systems.*

Time Needed

This lesson will take several class periods. Suggested scheduling is as follows:

Day 1: **Engage** with *Winter's Tail* Read-Aloud, Part 1, and Structures and Functions T-Chart

Day 2: **Explore** with Animal Prosthesis Design Challenge, "Wild About Animals" Video Segment 1, and Animal Locomotion Videos

Day 3: **Explain** with *Winter's Tail* Read-Aloud, Part 2; "Wild About Animals" Video Segment 2; and Comparing the Design Process

Day 3: **Elaborate** with *Biomedical Engineering and Human Body Systems* Read-Aloud and "Giving the World a Helping Hand" Video

Day 4: **Evaluate** with Derby the Bionic Dog

Materials

For Animal Prosthesis Design Challenge (teacher use only)

- Multipurpose snips
- Pliers
- Hot-glue gun
- Hot-glue sticks

For Animal Prosthesis Design Challenge (per team of 4–5 students)

Battat Wind Up Tubbies (dolphin, turtle, and fish) can be purchased at Amazon.com. You will need 4 sets (12 animals) for a class of 24–30 students. The toys are reusable if handled with care.

In advance, prepare two sets (6 animals total) of the wind-up bath toys by removing most of the tail (dolphin and fish) or most of a flipper (turtle). To do this, grip the tail or flipper with the pliers and gently snap or cut it off with the snips. Use caution to avoid damaging the base of the moving part. Leave a small stump so that you have a base on which you can hot glue students' prosthetic prototypes.

> **SAFETY**
> - Have students wear safety glasses or goggles during this activity.
> - Handle pliers, paint scrapers, scissors and other tools with care. They can pinch and injure skin.
> - Handle hot-glue guns and melted glue with care. They can burn skin.
> - Be sure that the classroom has sufficient ventilation. Hot-glue sticks can release irritating vapors.
> - Immediately wipe up any spilled water to avoid a slip-and-fall hazard.
> - To avoid being shocked, never work with water near a wall with an electrical outlet.

- Two of the same wind-up animal bath toy per team (one with a fin or tail pre-cut by the teacher). A sample distribution is as follows:
  - Team 1: 2 dolphins
  - Team 2: 2 turtles
  - Team 3: 2 fish
  - Team 4: 2 dolphins
  - Team 5: 2 turtles
  - Team 6: 2 fish

WIND-UP TUB TOYS

- Assortment of flexible plastic lids of varying thickness (e.g., plastic coffee can lids, food-storage container lids, chip can lids)
- Scissors
- Plastic paint scraper
- Paper towels
- Large, shallow plastic storage container or wallpaper trough for testing toys, filled with approximately 3 in. of water (or use a small baby pool or an extra-large, shallow container for testing all of the toys instead of giving each team a separate container)

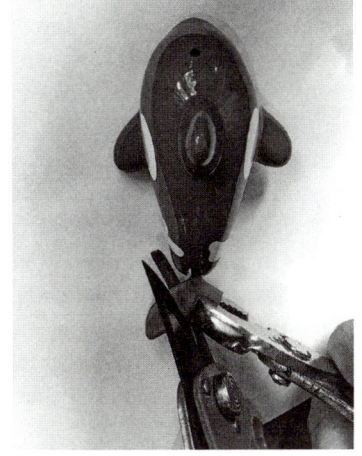

PREPARING THE MODELS

Student Pages

- Animal Prosthesis Design Challenge
- Comparing the Design Process
- Derby the Bionic Dog
- STEM at Home

DAMAGED MODELS

Background for Teachers

Biomedical engineering is the application of engineering principles to the fields of biology and health care. As do other kinds of engineers, biomedical engineers use math, science, technology, and creativity to design solutions to problems, but they also need to understand medicine and body systems. Their work involves designing solutions—from life-saving medical devices to life-changing prosthetic limbs—for detecting, treating, and preventing all sorts of medical problems. Biomedical engineering has led to the invention or improvement of devices such as heart-lung machines, incubators for newborns, dialysis machines, cochlear implants, pacemakers, MRI (magnetic resonance imaging) machines, dental implants, insulin pumps, gear for high-performance athletes, and prosthetic legs, feet, arms, and hands.

In this lesson, students are engaged in disciplinary core idea ETS2.B: Influence of Engineering, Technology, and Science on Society and the Natural World as they learn how prosthetic limbs, or *prostheses*, can improve lives by replacing human and animal limbs lost through accident, illness, or congenital conditions. Biomedical engineers (along with healthcare professionals called *prosthetists*, such as Kevin Carroll who famously designed a new tail for a dolphin named Winter) design prostheses for people and animals with amputated limbs or congenital limb differences. Prosthetic limbs must be comfortable to wear and aesthetically pleasing and must function efficiently and accurately. To meet these criteria, biomedical engineers carefully study the structure and function of a healthy hand, for example, to design a prosthetic hand that works in the same way. Traditionally, prosthetic limbs have been very expensive, but 3-D technology is making them more affordable. In addition, charitable organizations such as Enabling the Future are connecting children with limb differences to a global community of passionate volunteers who are building upper-limb assistive devices using 3-D printers. *3-D printing* is a process for making a physical object from a 3-D digital model, typically by laying down many successive thin layers of a material. If you have access to a 3-D printer in your school, this lesson is a great vehicle for increasing your students' interest in this printing technology!

Chapter 14

Students are first introduced to biomedical engineering and the sensitive subject of prosthetic limb replacement through the remarkable true story of Winter, a young bottlenose dolphin that injures and then loses its tail after being mangled in a crab trap off the coast of Florida. After Winter is rescued and taken to the Clearwater Marine Aquarium, a group of dedicated prosthetists led by Kevin Carroll spend months devising a way to enable Winter to once again swim like a dolphin. Students are engaged in the crosscutting concepts of structure and function as well as systems and system models as they explore how prosthetists and biomedical engineers study structures, functions, and body systems to design comfortable and effective prostheses. They learn that studying *locomotion*, movement that results in progression from one place to another, is important when designing prostheses for swimming or running. After watching videos of marine animal locomotion and making observations about structures used for swimming (fins, flippers, and flukes), students are challenged to design, build, and test a prosthetic tail or flipper for a wind-up bath toy. They learn that it is helpful to build a *prototype*, a working model made to be tested in real-life situations (in this case, a tub of water!)

Note: If you have access to 3-D printing technology, you may want to have students use a simple online 3-D design and modeling tool, such as Tinkercad, to design and 3-D print their prototypes. Go to *www.tinkercad.com* for more information.

After this design challenge, they compare the steps they went through with the design process used by Kevin Carroll and his team as they invented the first prosthetic to ever be used by a marine animal. Through this design challenge, students explore the disciplinary core idea ETS1.A: Defining and Delimiting Engineering Problems by identifying the criteria and constraints for designing animal prostheses and identifying the steps of the design process. They are also introduced to the science and engineering practice of engaging in argument from evidence by using a model (an animal prosthesis) to test the locomotion of an animal with and without a limb injury. They use the evidence gained from the test to make a claim about the merit of their solution by citing relevant evidence about how the solution meets the criteria and constraints of the problem.

Then, students apply what they have learned about animal prosthetic design to human prosthetic design by reading about the development of a "mind-controlled" prosthetic arm and by learning how 3-D printing technology and volunteerism is making a difference for children with limb differences. Finally, their understandings are evaluated as they respond to an article and video about a special dog named Derby, whose life is changed forever through the power of 3-D printing.

engage

Winter's Tail Read-Aloud, Part 1

 **Making Connections: Text to World**

Show students the cover of the book *Winter's Tail: How One Little Dolphin Learned to Swim Again*. Ask

? How many of you have heard about Winter from TV shows, news reports, or perhaps from even visiting the Clearwater Marine Aquarium in Florida? (Answers will vary.)

? How many of you have seen the movie *Dolphin Tale*? (Answers will vary.)

? What do you know about Winter? (Answers will vary, but some students will know that Winter is a dolphin who injured her tail and was rescued and fitted with an artificial tail, or prosthesis.)

Tell students that if they have already seen the movie or read the book, they are going to use the information they obtained in a different way. Instead of merely being entertained by this remarkable true story, they are going to learn how science, medicine, math, technology, and creativity come together to help animals and people in the field of biomedical engineering. Biomedical engineers use their knowledge of body structures (parts) and body systems to help doctors treat and prevent medical problems. More and more, biomedical engineers are figuring out ways to use their knowledge and skills to help pets and wild animals such as dolphins.

Structures and Functions T-Chart

 Determining Importance

Tell students that as you read the story aloud, you would like them to listen for any structures and functions a biomedical engineer might need to study to help an injured dolphin. Make a T-chart on the board titled "Bottlenose Dolphin Structures and Functions," which you will fill in later. Tell students to enjoy the story, and explain that you will revisit the chart after you read. Then, read pages 5–9 of the book aloud.

Making Connections: Text to Self and Turn and Talk

After reading page 9, which describes how Jim Savage freed the baby dolphin from the crab trap, have students share with a partner any experiences they have had rescuing a wild animal or pet in a difficult situation. Be sure to add that they should never touch a wild animal and must get an adult's help before trying to rescue an animal. Some animals can be dangerous or carry diseases that can be transmitted to humans. Also explain that animals' actions can be unpredictable when they are in a life-threatening situation. In many cases, attempting to care for a sick, injured, or orphaned animal is illegal unless you are a licensed wildlife rehabilitator. That is why the fisherman did the right thing by calling the Florida Fish and Wildlife Conservation Commission for help.

Then, finish reading the first half of the book aloud (pp. 10–19), stopping after you read the paragraph about Kevin Carroll at the bottom of page 18: "Kevin was not only a dolphin lover, he was also a premier creator of *prostheses*—special devices that can help replace a body part such as an arm or a leg. Kevin believed he could help." (You will read the rest of the book aloud during the Explain phase of the lesson.)

As a class, generate data for the T-chart and briefly discuss each dolphin structure or function (mentioned in the book) that a biomedical engineer might need to study to help an injured dolphin. Then, discuss how each of these structures serves various functions in a dolphin's growth, survival, behavior, and reproduction. For example, the tail helps the dolphin swim to catch fish so it can grow, swim to find a mate so it can reproduce, and swim to the surface to breathe air so it can survive. Table 14.1 (p. 228) shows a sample completed T-chart, with page numbers for the first mention of the structure or description noted for your reference.

Animal Prosthesis Design Challenge

> Connecting to the Common Core
> **Reading: Informational Text**
> KEY IDEAS AND DETAILS: 3.1, 4.1, 5.1

Questioning

After completing the chart together, *ask*

? Do you know what structures dolphins use to make their clicking and whistling sounds? (Students will most likely not know because the information is not provided in the book, but you can explain that these sounds are made in the larynx as well as in air sacs in their heads. Dolphins don't have vocal cords like humans do.)

Picture-Perfect STEM Lessons, 3–5

Chapter 14

Table 14.1. Sample Structures and Functions T-Chart

Bottlenose Dolphin Structures and Functions		
Structure	**Function**	**Description**
Tail (p. 5)	Swimming	Up-and-down motion facilitates movement.
Mouth (p. 6)	Eating, drinking	Adults eat fish and babies drink milk until age 2.
Blowhole (p. 6)	Breathing	Dolphin must raise this hole out of water to breathe.
Throat (p. 10)	Swallowing	Throat helps food go from the mouth to the stomach.
Head (not in book)	Communicating	Clicks and whistles form the dolphin's vocabulary. (p. 14)
Backbone (p. 14)	Moving	Backbone gives structure and support and aids in moving.

? What is a dolphin's form of *locomotion*, or way of moving from place to place? (swimming)

? What structure on the chart do dolphins use for swimming? (tail—you may want to mention that dolphins also produce sounds by slapping their tails against the surface of the water, so a tail can also be used for communication)

? What is the normal swimming motion of a dolphin's tail? (up and down—have students make an up-and-down motion with one hand)

? How did Winter learn to swim without her tail? (She swished her tail stump from side to side, like a fish or shark—have students make a side-to-side motion with one hand.)

? Why did this new motion concern her trainers? (They worried that she might damage her backbone by swimming the wrong way.)

? What was Kevin Carroll's job? (creating prostheses)

? What are prostheses? (special devices that can help replace a body part such as an arm or leg)

? What structures and functions would he need to study to design a prosthesis for Winter? (her tail, her backbone, the way a dolphin typically swims)

"Wild About Animals" Video Segment 1

 Making Connections: Text to Text

Next, tell students that they are going to watch part of a TV show about Winter. As they watch, ask them to listen for any new information or perspectives that were not in the book. Then, show the first part of the video "Wild About Animals—Winter, the Dolphin With a Prosthetic Tail," which shows actual footage of Winter's rescue and her unusual side-to-side swimming motion never before seen in dolphins. Stop at minute mark 4:06, after the narrator says, "This is where Kevin Carroll entered the picture. As vice-president of Hanger Prosthetics, he had been helping people with a range of debilitating injuries for years. But late in 2006, Kevin Carroll and his team had been formulating a ground-breaking idea—to design a prosthetic dolphin tail for Winter."

> Connecting to the Common Core
> **Reading: Informational Text**
> INTEGRATION OF KNOWLEDGE AND IDEAS: 3.9, 4.9, 5.9

 Questioning

After watching the video from the beginning to minute mark 4:06, *ask*

? What new information did you learn, or what different perspectives did you hear that were not in the book? (Answers will vary, but students may mention the following: The lagoon where Winter was found was famous for having different types of fish. Rescuers also came from the Harbor Branch Oceanographic Institute and were perhaps contacted by the Florida Fish and Wildlife Conservation Commission. Head dolphin trainer Abby Stone provided her first-person perspective. They actually had to cut the flukes off of Winter's tail.)

Then, announce the design challenge for the next class period, "You are going to learn how biomedical engineers and prosthetists such as Kevin Carroll help animals … by designing a prosthesis for an animal yourselves!" (*Note:* In advance, prepare two sets (6 animals total) of the wind-up toys by removing most of the tail (dolphin and fish) or most of a flipper (turtle). To do this, grip the tail or flipper with pliers and gently snap or cut it off with snips. Use caution to avoid damaging the base of the moving part. Leave a small stump so that you have a base on which you can hot glue students' prosthetic prototypes.)

Distribute the Animal Prosthesis Design Challenge to each student, and read the design challenge together:

> **Challenge:** *Using teamwork, your knowledge of the animal's structures and functions, your imagination, and your artistic skills, design a* **prosthesis** *for a wind-up bath toy animal that is missing its tail or a flipper. Before you begin, be sure to research the problem.*

Animal Locomotion Videos

Next, show students the wind-up bath toys (but do not pass them out yet), and tell them which real-life marine animal each toy is going to represent in the design challenge. Explain that, like real biomedical engineers, they will need to be very familiar with the body structures, functions, and locomotion of the animal they are working with to design a suitable prosthesis for it. Read the research instructions together:

> **Research:** *Biomedical engineers study body structures and functions to design solutions. As you watch the videos of marine animals, study their* **locomotion** *(or how they move from place to place). Sketch the shape of their flukes, flippers, or fins, and observe whether they move their tails or flippers up and down or side to side.*

As students watch the Arkive videos (see "Websites" section), have them sketch each animal's swimming structures (dolphin tail flukes, green sea turtle flippers, or clownfish fins) and make observations of its locomotion in the water. They can watch the videos at computer stations, or you can project the videos for the whole class to see. The videos are very short, so students may want to watch them more than once to have sufficient time to make notes. Table 14.2 (p. 230) will help you match each wind-up bath toy to its corresponding, real-life marine animal and to the Arkive.org video that shows it swimming in its natural habitat.

After students have watched the videos and made their sketches and observations on the Animal Prosthesis Design Challenge student page, form teams of four to five students at tables with the following supplies:

- Assortment of flexible plastic lids of varying thickness (e.g., plastic coffee can lids, food-storage container lids, chip can lids)
- Scissors
- Plastic paint scraper
- Paper towels

Table 14.2. Animal and Video Chart for Design Challenge Research

Wind-Up Bath Toy	Corresponding Marine Animal	Search Terms for Arkive	Length of Video (seconds)
Dolphin	Bottlenose dolphin	"Bottlenose dolphin feeding"	40
Turtle	Green sea turtle	"Green turtle swims over reef towards surface"	37
Fish	Clownfish	"Common clownfish in anemone habitat"	28

- Large, shallow plastic storage container or wallpaper trough for testing toys, filled with approximately 3 in. of water (or use a small baby pool or an extra-large, shallow container for testing all of the toys instead of giving each team a separate container)

OBSERVING LOCOMOTION

Before passing out the wind-up bath toys, explain that when designing a solution building a prototype is helpful. A *prototype* is a working model made to be tested in real-life situations. Their prototypes will be made out of plastic and tested in a tub of water (which is the natural habitat of a bath toy!) Then, review the procedure together.

Next, demonstrate, using one of the toys, how to gently wind the toy up until it will not wind any further. (*Helpful hint:* Hold the moving part—the tail or the flipper—still with one hand while you wind with the other. Be careful not to overwind!)

By this point, students will be very excited to receive their wind-up bath toys and begin the design challenge! You can either pass out the toys randomly, or devise a system that allows each team to choose the type of toy it wants. Every team will need one undamaged toy and one toy of the same type with a missing tail fluke, fin, or flipper. Set up a hot-glue station (for teacher use only) and give students a reasonable amount of time to test at least two prototypes and fill out the student page.

explain

Winter's Tail Read-Aloud, Part 2

 Questioning

After all teams are satisfied with their model prostheses and have completed the Animal Prosthesis Design Challenge student page, have them explain what they learned. *Ask*

? What were the *criteria* for your design? In other words, what were the desired features for the prosthetic part? (It had to work in the water, it had to mimic the natural movement of the swimming toy, and it had to handle the force of each thrust without coming unglued.)

? What were the *constraints* for your design? In other words, what limits did your team have on time or materials? (We had a set amount of time, and we could use only plastic and hot glue.)

? What was your final solution? (Answers will vary.)

Next, ask students to listen for how Kevin Carroll and his team designed a prosthetic tail for Winter. Read the rest of *Winter's Tail* (pp. 20–29) aloud.

> Connecting to the Common Core
> **Reading: Informational Text**
> Key Ideas and Details: 3.1, 4.1, 5.1

 Questioning

After reading, *ask*

? What were the *criteria* for the design? In other words, what were the desired features for Winter's prosthetic tail? (It had to work in the water, it had to mimic the up-and-down movement of a swimming dolphin, and it had to handle the force of each thrust of her tail without coming off.)

? What were the *constraints* for the design? In other words, what limits do you think they had on time, money, or materials? (They probably needed to work quickly before Winter permanently damaged her backbone. It is unclear how much money they had to spend, but perhaps the team donated time or people donated money. They were also constrained by the lack of materials designed specifically for making a dolphin prosthesis.)

? What obstacles did they have to overcome? (Winter did not have a tail joint or any other place for a prosthesis to attach to her body, and dolphins have especially sensitive skin.)

? What new technology did they design to overcome the problem of sensitive skin? (a special silicone gel that would be smooth against her skin and would add a cushion to make the prosthesis more comfortable)

? How long did it take to develop a prosthesis that worked? (several months)

? How many designs did they try? (several)

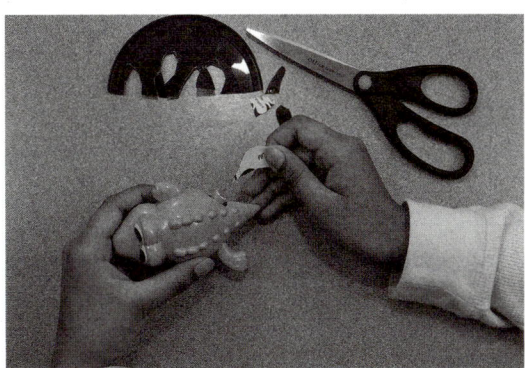

DESIGNING A PROSTHETIC LIMB

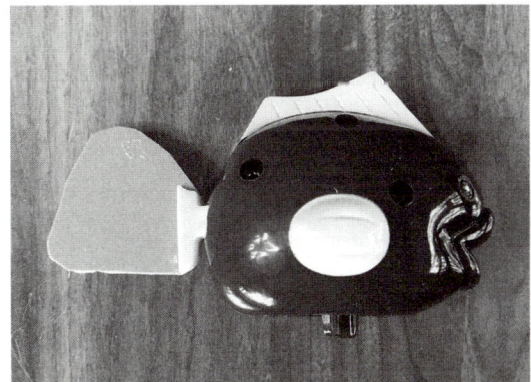

TAIL PROTOTYPE

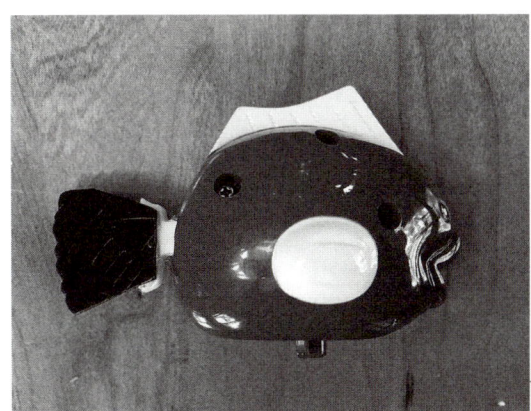

REDESIGNED TAIL PROTOTYPE

? What did the final solution consist of? (a sleeve made of silicone that fits right on Winter's tail stump and a second sleeve that fits on top of the first sleeve to hold the tail and its brace in place)

"Wild About Animals" Video Segment 2

Making Connections: Text to Text

After reading, show the second part of the "Wild About Animals" video (minute mark 4:07–6:03). Tell students that as they watch, you would like them to listen for any new information or perspectives that were not in the book.

> **Connecting to the Common Core**
> **Reading: Informational Text**
> INTEGRATION OF KNOWLEDGE AND IDEAS: 3.9, 4.9, 5.9

Questioning

After watching the second segment of the video, *ask*

? What new information did you learn, or what different perspectives did you hear that were not in the book? (Answers will vary, but students may mention the following: The team thought the task would be difficult. We heard the first-person perspective of Kevin Carroll. We were able to see more about how they built the tail. Winter became the first sea creature in history to use a prosthetic tail. New prostheses have to be made constantly to fit her growing body. The prosthetic tail serves as therapy by strengthening Winter's muscles.)

Comparing the Design Process

Pass out the Comparing the Design Process student page, and have teams work together to describe the steps that Kevin Carroll and his team went through to design a prosthetic tail for Winter and then compare those steps with what their team went through to design a prosthetic part for a wind-up bath toy.

Synthesizing

Ask

? How did your team's design process compare with that of Kevin Carroll's team?

? What was similar?
? What was different?
? What did you learn about designing a prosthesis for an animal?
? How do you think this process compares with designing a prosthesis for a human?

elaborate

Biomedical Engineering and Human Body Systems Read-Aloud

Tell students that the book *Biomedical Engineering and Human Body Systems* by Rebecca Sjonger will give them more insight into the real-world design process that biomedical engineers use to define problems, brainstorm solutions, and build and test prostheses. Because the book is nonfiction and has a lot of information that doesn't pertain to prosthetics design, you will be skipping some pages and "chunking" the information into manageable bits as you read. You may want to use sticky notes in advance to mark the pages you will be reading.

What Is Biomedical Engineering?

 Chunking and Questioning

Read page 4, including captions and insets; the "Scientists and Engineers" inset on page 5; and the top of page 10.

 Turn and Talk

Then, ask students to turn to a partner and discuss the following question:

? From what we've learned from *Winter's Tail*, what we explored when we designed animal prostheses, and what we just read, how would you describe *biomedical engineering*? (Answers will vary, but they should include some understanding of the idea that biomedical engineering is a field of engineering that involves designing solutions to medical problems, in both humans and animals. Biomedical engineering is the application of engineering principles and design concepts to medicine and biology for healthcare purposes.)

Next, have pairs discuss the following question:

? What subjects and skills do biomedical engineers need to know or use to solve challenges? (They need to know the subjects of math, science, and health; they need to use computer, chemical, electrical, and mechanical engineering skills; and they need to be creative people with good communication skills, who can work well in teams.)

What Do Biomedical Engineers Design?

Picture Walk

Next, show the picture of the artificial toe on page 9, and read the caption. Explain that people have been trying to figure out ways to design prosthetic limbs for thousands of years. Tell students that you are going to do a "picture walk" through the book to show them the wide variety of innovations that biomedical engineers have designed or improved on—from life-saving medical devices to life-changing prosthetic limbs. As you flip through the book, share some of the following pictures and their captions with students:

- Page 8—Ophthalmoscope
- Page 9—Incubator
- Page 10—Artificial kidney
- Page 11—Cochlear implant
- Page 12—Prosthetic leg for cycling
- Page 13—Prosthetic arm
- Page 15—MRI machine
- Page 19—Dental implant
- Page 21—Mouth guard
- Page 22—Prosthetic hand
- Page 23—Defibrillator
- Page 25—Camera pill, insulin pump

Then, show the picture on page 24, and explain that innovations such as these are shared at conferences for biomedical engineers around the world.

How Are Prostheses Designed?

Tell students that biomedical engineers are always looking for new ways to make life better for people. For example, after Kevin Carroll designed the new silicone gel for Winter's tail, he realized that it could also help people who wear prostheses. He put the gel to the test on a war veteran whose artificial legs were uncomfortable. The gel created an extra cushion that helped reduce his discomfort. This discovery was a big breakthrough, making life a little easier for people needing prostheses.

Show the picture of Max Ortiz Catalan on page 16. Tell students that in Sweden, a team led by this biomedical engineer set out to improve the kind of prosthetic arm that is controlled by electrodes attached to the surface of the skin. Sometimes those electrodes don't work well, and the prosthetic arm doesn't move the way the wearer wants it to. Have students listen for how Max Ortiz Catalan's team used a variation of the design process to solve this problem as you read aloud the light yellow insets. There is quite a bit of text, so you may want to paraphrase or summarize. (Note that the design process terms used in the Comparing the Design Process student page are in parentheses.) Read the following insets aloud:

- Page 13—Investigating the problem (IDENTIFY PROBLEM)
- Page 14—Exploring ideas (BRAINSTORM)
- Page 16—Choosing a design (DESIGN)
- Page 18—Constructing a model (BUILD)
- Page 20—Test and assess (TEST AND EVALUATE)
- Page 22—Refine the design (REDESIGN)
- Page 24—Telling the story (SHARE SOLUTION)

After reading, you may want to show all or part of the 4:36 min. "Mind-Controlled Neuroprosthetics" video (see "Websites" section), which shows the actual test subject from the book wearing his amazing "mind-controlled" prosthetic arm! The video isn't narrated, so when it shows the man being unable to complete a task explain that he can't

Picture-Perfect STEM Lessons, 3–5

complete the task because he is wearing the original prosthesis with the electrodes attached to the surface of the skin. When the video shows him completing a task, explain that he can complete the task because he is wearing the improved prosthesis with the electrodes surgically implanted.

"Giving the World a Helping Hand" Video

Tell students another exciting new technology has become a real game-changer in the world of biomedical engineering. This technology is helping people with limb differences all over the world. *Ask*

? Have you ever heard of a 3-D printer?

Then, read the inset on the top of page 29 titled "Powerful Printers." Next, have students listen for how 3-D printing technology, free software, and passionate volunteers are changing the lives of children around the world as you show the 7:18 min. video "Giving the World a Helping Hand."

 **Synthesizing**

> Connecting to the Common Core
> **Reading: Informational Text**
> INTEGRATION OF KNOWLEDGE AND IDEAS: 3.9, 4.9, 5.9

After watching, *ask*

? Why do you think the video is titled "Giving the World a Helping Hand"? (Max Ortiz Catalán's idea is helping people all over the world who need prosthetic hands.)

? Why do you think the video ends with this quote: "Never underestimate the power of your ideas. They just might change the world"? (One idea led to a whole movement to help people all over the world.)

? How do the title and the quote relate to the work of Max Ortiz Catalán, the biomedical engineer who designed the "mind-controlled" arm, or Kevin Carroll, the prosthetist who designed Winter's tail? (Max Ortiz Catalán and Kevin Carroll are both helping others and changing the world with their ideas.)

? Would you like to be a biomedical engineer? (Answers will vary.)

evaluate

Derby the Bionic Dog

Tell students that they are going to apply what they have learned about structures, functions, prosthetics, and the design process by watching a video, reading an article, and responding to questions about the text. The article is about a special dog named Derby. Before passing out the Derby the Bionic Dog student page, have students watch the first 36 sec. of a 3:03 min. video titled "Derby the Dog: Running on 3D Printed Prosthetics" (see "Websites" section).

Sketch to Stretch

At the 36 sec. mark of the video, where Tara Anderson says, "I had to try and help this dog," stop the video and *ask*

? What could you design to help Derby get around better?

Have students sketch their ideas and then share and discuss with others. Next, show the rest of the video, and have students compare their solution with the solution in the video. Then, pass out the article found on the Derby the Bionic Dog student page. Have students read it and respond to the questions. Answers are as follows:

? What structures on Derby's body are deformed or missing? (His front legs are deformed, and he has no front paws.)

? What functions do these structures serve in a dog? (sitting, standing, walking, running, digging, playing, etc.)

? What problems did this cause for Derby? (He couldn't do any of those functions normally.)

? What was the solution to the problems? (prosthetic front legs)

? Think about the steps that Tara Anderson and her team went through to design and build prosthetic legs for Derby the dog. Then, fill in the chart on the student page. (A sample completed chart is shown in Table 14.3.)

Table 14.3. Answers to the Design Process for Derby's New Legs

Derby's New Legs
What problem was Tara's team trying to solve? **They wanted to design prosthetic legs so Derby could move like other dogs.**
What two things did they study about Derby before they started designing? **They studied how he moved and the shape of his deformed legs.**
What were some of the ideas the team brainstormed? **They talked about designing the legs in the shape of a loop so they wouldn't dig into the dirt. They thought that using treads would possibly give Derby better traction. They thought that if the legs were low to the ground, maybe Derby wouldn't hurt himself as he got used to them.**
What shape were the first prostheses they designed, and what technology helped them design the legs? **They used digital 3-D scanning on a computer to design oval-shaped legs.**
What technology did they use to build the legs? **They used a 3-D printer.**
When they tested Derby's first set of new, taller legs, what did they discover? How did they improve their design? **His elbows flopped out of the sides, he looked clumsy, and he seemed uncomfortable. To improve the design, they changed the shape and made the legs out of nylon for more bounce.**
How did the final solution improve Derby's health and quality of life? **He can sit like other dogs, he can stand tall and straight without hurting his back, and he can run and play like other dogs.**

STEM at Home

Have students complete the "I learned that …" and "My favorite part of the lesson was …" portions of the STEM at Home student page as a reflection on their learning. They may choose to do the following at-home activity with an adult helper and share their results with the class. If students do not have access to the internet or these materials at home, you may choose to have them complete this activity at school.

"At home, we can watch the trailer for a documentary called *My Bionic Pet** to explore the technologies that biomedical engineers, prosthetists, and veterinarians are using to help pets."

🔍 *Search "My Bionic Pet PBS Trailer" on YouTube to find the video at* www.youtube.com/watch?v=VkJFLHEBFaI.

"After we watch the video together, we can fill in the chart below [see chart and answers in Table 14.4, p. 236]."

**The full documentary is currently free with an Amazon Prime membership. You can also check your local library for a copy.*

Table 14.4. Completed STEM at Home Chart

Animal	Missing Structure(s)	Function(s) of Structure(s)	Solution
Golden retriever	Front legs	Walking, running	Prosthetic legs
Border collie	Back legs	Walking, running, jumping	Prosthetic legs
Pig	Back legs	Walking, running	Wheelchair
Swan	Beak	Eating, preening	Prosthetic beak
Alligator	Tail	Swimming	Prosthetic tail
Horse	Front leg	Walking, running	Prosthetic leg

For Further Exploration

This section is provided to help you encourage your students to use the science and engineering practices in a more student-directed format. This box lists questions and challenges related to the lesson that students may select to research, investigate, or innovate. Students may also use the questions as examples to help them generate their own questions. After selecting one of the questions in the box or formulating their own questions, students can individually or collaboratively make predictions, design investigations or surveys to test their predictions, collect evidence, devise explanations, design solutions, or examine related resources. They can communicate their findings through a science notebook, at a poster session or gallery walk, or by producing a media project.

Research

Have students brainstorm researchable questions:

? What were the first prosthetic legs for humans made of?

? What kinds of animals have benefited from prosthetics?

? How is Winter doing with her prosthetic tail? Go to *www.seewinter.com* to view a live webcam!

Investigate

Have students brainstorm testable questions to be solved through science or math:

? What is the relationship between the number of turns and the length of time a wind-up bath toy swims in a tub of water? Graph the results, then analyze your graph. What can you conclude?

? What happens to swimming motion as you increase the length of a prosthetic tail on a wind-up dolphin? Is there a limit to how long the tail can be?

? What happens to swimming motion when you increase the length of a prosthetic flipper on a wind-up turtle? Is there a limit to how long the flipper can be?

Innovate

Have students brainstorm problems to be solved through engineering:

? Can you design a device to help someone with a prosthetic leg swim?

? Can you invent a "bionic toy" by replacing one part of it with something else to make it move differently?

? Can you build a model of a prosthetic hand? Go to www.sciencebuddies.org and search "build a helping hand" to try it!

Websites

Arkive: "Bottlenose Dolphin Feeding" (video)
www.arkive.org/bottlenose-dolphin/tursiops-truncatus/video-08a.html

Arkive: "Common Clownfish in Anemone Habitat" (video)
www.arkive.org/common-clownfish/amphiprion-ocellaris/video-03a.html

Arkive: "Green Turtle Swims Over Reef Towards Surface" (video)
www.arkive.org/green-turtle/chelonia-mydas/video-01.html

"Derby the Dog: Running on 3D Printed Prosthetics" (video)
www.3dsystems.com/media/derby-dog-running-3d-printed-prosthetics

"Giving the World a 'Helping Hand'" (video)
https://vimeo.com/152492035

"Mind-Controlled Neuroprosthetics" (video)
www.youtube.com/watch?v=ZuJu_ulpvq4

Wild About Animals: Winter, the Dolphin With a Prosthetic Tail" (video)
www.youtube.com/watch?v=jGC_s3OjNZI

More Books to Read

Hatkoff, C., and D. Yates. 2014. *Hope for Winter: The true story of a remarkable dolphin friendship.* New York: Scholastic.

Summary: Exactly five years and one day after Winter was rescued, something amazing happened. Just feet from where Winter was found, another orphaned and injured dolphin appeared. The Clearwater Marine Aquarium team quickly went to work to nurse this new dolphin back to health. The little dolphin named Hope now lives with Winter, and they are best friends.

Marquardt, M. 2016. *Bioengineering in the real world.* North Mankato, MN: Abdo.

Summary: Part of the *Stem in the Real World* series, this book for older readers describes historical innovations in bioengineering, the evolution of prosthetic devices, the future of bioengineering, and ethical issues.

Name: _____

Animal Prosthesis Design Challenge

Challenge: Using teamwork, your knowledge of the animal's structures and functions, your imagination, and your artistic skills, design a **prosthesis** for a wind-up bath toy animal that is missing its tail or a flipper. Before you begin, be sure to research the problem.

Research: Biomechanical engineers study body structures and functions to design solutions. As you watch the videos of marine animals, study their **locomotion** (or how they move from place to place). Sketch the shape of their flukes, flippers, or fins, and observe whether they move their tails up and down or side to side.

Dolphin Tail Flukes (Sketch)	Observations
Green Sea Turtle Flippers (Sketch)	Observations
Clownfish Tail Fins (Sketch)	Observations

National Science Teachers Association

Name: _____

Procedure

1. Study the structures and locomotion of the undamaged wind-up bath toy in water.
   Observations: _____

2. Study the structures and locomotion of the damaged wind-up bath toy in water.
   Observations: _____

3. Brainstorm possible solutions for a prosthesis with your team.

4. Possible solutions: _____

5. Design and sketch a solution. Type of plastic: _____

   Sketch

6. Build a prototype by cutting it out of plastic.

7. Take your prototype to the teacher, who will attach it to the toy with hot glue.

8. Let the hot glue set on the toy for at least 2 minutes before touching it.

9. Test the prototype by winding the toy up and observing its locomotion in water.
   Observations: _____

10. Evaluate the solution by comparing its locomotion in water with the locomotion of the undamaged toy in water.
    Observations: _____

11. If you are not satisfied with your solution, carefully peel off the prototype and use a paint scraper to remove any remaining glue. Then go back to step 4 and redesign a solution. If you are satisfied, sketch your solution below and be ready to share your team's solution with others. Type of plastic: _____

Our Solution

Sketch

Comparing the Design Process

Think about the steps that prosthetics designer Kevin Carroll and his team followed to design and build a prosthetic tail for Winter the dolphin. Compare this process to the one you used to design and build a prosthetic tail or flipper for a wind-up bath toy animal.

Kevin Carroll's Process for Winter's Tail	THE DESIGN PROCESS	Your Process for the Toy Animal's Limb
1. What was the problem he was trying to solve? What were the criteria?	**Identify Problem**	1. What was the problem you were trying to solve? What were the criteria?
2. What kinds of experts came together to share ideas?	**Brainstorm** 	2. What were some of the ideas your team brainstormed?
3. What was the unique design his team came up with?	**Design** 	3. What was the shape of your first design? Sketch it here.
4. How did he test the designs he built? How did he evaluate whether they worked?	**Build**  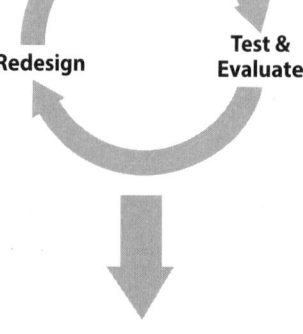 Redesign — Test & Evaluate	4. How did you test the designs you built? How did you evaluate whether they worked?
5. What was the final solution? Is the work finished?	**Share Solution**	5. What was the shape of your final design? Sketch it here.

Derby the Bionic Dog

Derby is a friendly husky mix who was born with no front paws and small, deformed front legs. He moved using awkward little hops, and he couldn't sit up straight. His owners realized that they couldn't take care of Derby's special needs, so they took him to an animal shelter. There, he was taken in by a rescue group called Peace and Paws. Luckily, that's how he met Tara Anderson, who happens to work for a 3-D printing company. Tara felt sorry for Derby, so she agreed to foster him.

Tara wanted to help Derby move the same way other dogs move. At first, she tried a small, two-wheeled cart for Derby. The cart supported Derby's body, but he wasn't able to run and play with other dogs when he was using the cart. It also got stuck easily. Tara knew there had to be a better way to help Derby get around.

That's when Tara decided to try 3-D printing technology to help Derby. With the help of veterinary prosthetics specialist Derrick Campana at a company called Animal Orthocare, Tara and her team went to work designing a pair of prosthetic legs for Derby. First, they studied how Derby moved. They studied the shape of his deformed legs. Then, they brainstormed ideas. If the legs were designed in the shape of a loop, maybe they wouldn't dig into the dirt. Maybe treads on the bottoms would give Derby better traction. If the new legs were low to the ground, maybe Derby wouldn't hurt himself as he got used to them.

Next, Derrick made molds of Derby's legs using plaster of Paris. He then hand-sculpted the plaster molds to look just like the dog's uniquely shaped front legs. From there, he measured and took pictures of the molds using digital 3-D scanning and used this data to create digital models of the legs on a computer. Digital 3-D scanning made it much easier to get the shape and the dimensions of the prosthetic legs right, without building a lot of prototypes first.

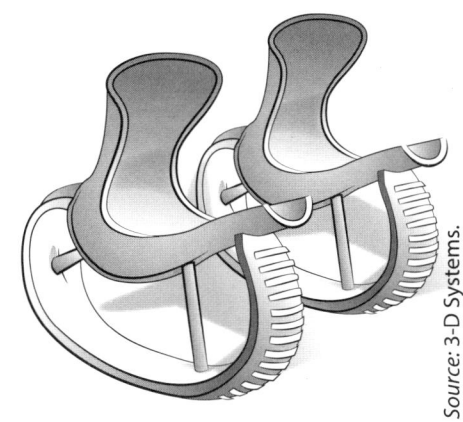

Source: 3-D Systems.

DERBY'S FIRST PROSTHETIC LEGS

After several design attempts, the team printed Derby's first set of custom prosthetic legs using a 3-D printer. The legs were oval-shaped, made of hard plastic, and were short enough that Derby could get used to them gradually without hurting himself. The first time Derby tried them on, he took off running! His new owners, Sherry and Dom, were amazed. They loved Derby despite his differences, but they were thrilled that he could now move more like other dogs.

Tara and her team planned to make taller prostheses later, gradually increasing their height until Derby's back was straight when he walked. But when it came time to try a taller version of the original legs on Derby, Tara found out that they didn't work very well. Derby walked with his elbows flopped out to the sides. He looked clumsy and seemed uncomfortable. So the team decided to try a different design. They wanted to create prosthetic legs that were taller than the original, but had a little bit more bounce to them. They would have to use a different material. Designing the new legs took many attempts. Each time they would test the legs on Derby, and each time they would have to redesign them.

Finally, they built a new pair of prosthetic legs using a different type of 3-D technology. The new legs had a completely different shape than the original prostheses. They were made of nylon instead of plastic; which gave them just the right amount of bounce. They were straighter and tall enough that Derby could stand on them without leaning forward. Now, Derby can sit just like other dogs, he can stand tall and straight without hurting his back, and he can even run and play wearing his new prosthetic legs. The best part of all is that Derby's story continues to inspire people everywhere. Thanks to science, technology, engineering, math, teamwork, and a lot of love, Derby is now a happy and healthy pup.

Questions After Reading

1. What structures on Derby's body are deformed or missing?

_____

2. What functions do these structures serve in a dog?

_____

3. What problems did this cause for Derby?

_____

4. What was the solution to the problems?

_____

Name : _____

5. Think about the steps that Tara Anderson and her team went through to design and build prosthetic legs for Derby the dog. Then, fill in the chart below.

THE DESIGN PROCESS	Derby's New Legs
Identify Problem	What problem was Tara's team trying to solve?
Brainstorm	What two things did they study about Derby before they started designing?
	What were some of the ideas the team brainstormed?
Design	What shape were the first prostheses they designed, and what technology helped them design the legs?
Build / Redesign / Test & Evaluate	What technology did they use to build the legs?
	When they tested Derby's first set of new, taller legs, what did they discover? How did they improve their design?
Share Solution	How did the final solution improve Derby's health and quality of life?

Picture-Perfect STEM Lessons, 3–5

Name : _____

STEM at Home

Dear _____,

At school, we have been learning about how **biomedical engineers** study structures and functions to design creative solutions to medical problems.

I learned that: _____
_____

My favorite part of the lesson was:
_____
_____

At home, we can watch the trailer for a documentary called *My Bionic Pet** to explore the technologies that biomedical engineers, prosthetists, and veterinarians are using to help pets.

 Search "My Bionic Pet PBS Trailer" on YouTube to find the video at ***www.youtube.com/watch?v=VkJFLHEBFaI.***

After we watch the video together, we can fill in the chart below.

Animal	Missing Structure(s)	Function(s)	Solution
Golden retriever	Front legs	Walking, running	Prosthetic legs
Border collie			
Pig			
Swan			
Alligator			
Horse			

*The full documentary is currently free with an Amazon Prime membership. You can also check your local library for a copy.

Chapter 15

From Seed to Tree

Description

Students explore the "magic" of seeds by investigating maple samaras and reading both fiction and non-fiction picture books about tree seeds. They discover that plants get their material for growth chiefly from air and water through the remarkable process of photosynthesis. Then, they learn how plants are grown on the International Space Station (ISS).

Suggested Grade Levels: 3–5

LESSON OBJECTIVES Connecting to the *Framework*		
Science and Engineering Practices	**Disciplinary Core Ideas**	**Crosscutting Concept**
Engaging in Argument From Evidence	**LS1.C:** Organization of Matter and Energy Flow in Organisms	Energy and Matter
	ETS2.A: Interdependence of Science, Engineering, and Technology	

Featured Picture Books

TITLE: ***If You Hold a Seed***
AUTHOR: **Elly MacKay**
ILLUSTRATOR: **Elly MacKay**
PUBLISHER: **Running Press**
YEAR: **2013**
GENRE: **Story**
SUMMARY: *Illustrated with layered paper collages, this beautiful book depicts a little boy discovering the magic inside a seed.*

TITLE: ***Next Time You See a Maple Seed***
AUTHOR: **Emily Morgan**
PUBLISHER: **NSTA Press**
YEAR: **2014**
GENRE: **Non-Narrative Information**
SUMMARY: *Clear text and stunning, full-color photographs reveal the secrets held inside a maple seed and encourage children and adults to discover the extraordinary in these ordinary springtime flyers.*

Chapter 15

Time Needed

This lesson will take several class periods. Suggested scheduling is as follows:

Day 1: Engage with *If You Hold a Seed* Read-Aloud and From Seed to Tree Probe

Day 2: Explore with Exploring Maple Seeds

Day 3: Explain with *Next Time You See a Maple Seed* Read-Aloud and Photosynthesis Video

Day 4: Elaborate with "Plants in Space" Pairs Read

Day 5: Evaluate with Revisiting the From Seed to Tree Probe

Materials

- Maple seeds (If not in season, maple seeds can be purchased at *www.treeseeds.com* and many other online vendors.)
- Maple or other tree log
- Hand lenses
- Clipboards
- Pencils

Student Pages

- From Seed to Tree Probe
- Exploring Maple Seeds Journal
- Plants in Space
- STEM at Home

Background for Teachers

The transformation from seed to plant is one of the most amazing processes in nature. To sprout, most seeds need only water and the appropriate temperature. As a seed begins to germinate, the tiny embryo uses the material in the rest of the seed for food and grows roots to soak up water. When the leaves appear, they begin to capture light energy from the Sun with *chlorophyll*, a special green pigment. Chlorophyll is contained in a plant's *chloroplasts*, which are like tiny factories in which plants make their food. Using the Sun's energy, carbon dioxide from the air is combined with water and transformed through a chemical reaction inside the chloroplasts into a kind of sugar called *glucose*. This simple sugar provides the energy the plant needs to survive and grow. The chemical reaction also produces oxygen, which is released into the air as a waste product. The entire remarkable process by which plants make food is called *photosynthesis*.

$$\text{water + carbon dioxide} \xrightarrow[\text{(energy from the Sun)}]{\text{Photosynthesis}} \text{sugar + oxygen}$$

A Framework for K–12 Science Education suggests that by the end of grade 5, students are able to support an argument that "plants acquire their material for growth chiefly from air and water" (NRC 2012, p. 148). This lesson begins with a read-aloud to capture students' interest in seeds, followed by a formative assessment probe that addresses this disciplinary core idea. The probe is adapted from the "Giant Sequoia Tree" probe in *Uncovering Student Ideas in Science, Volume 2* (Keeley, Eberle, and Tugel 2007). The authors of the probe suggest that you might show a tree that is more familiar to students than the giant sequoia, for example a maple tree. In the probe adapted for this lesson, students are asked to think about how a small maple seed could grow into a towering maple tree. They select their choices and explain their thinking. You can collect the probes and use them to assess preconceptions. The probes will also help the students focus in on the essential concept in this lesson: The materials plants need for growth come mainly from air and water. At the end of the lesson, students have the opportunity to revise their answer and explain their thinking.

The probe used in this lesson is similar to the "seed-and-log" question in the *Private Universe* film series (Harvard-Smithsonian Center for Astrophysics 1995), where Harvard and MIT graduates' misconceptions about photosynthesis were revealed. (See "Websites" section for a link to the video.) Most of the Harvard and MIT graduates who appear in the video attributed the mass of the tree to soil and water, but, in fact, most of the matter that makes up the structure of a tree can be traced back to carbon dioxide in the air.

This idea of carbon dioxide being the main building material for plants is abstract and counterintuitive because carbon dioxide is a gas and might seem like "nothing" to students. A smaller amount of the matter in a plant can be attributed to water. The authors of *Uncovering Student Ideas in Science* explain that, "The atomic mass of one molecule of carbon dioxide is approximately 44 atomic mass units; one molecule of water is approximately 18 atomic mass units" (p. 123). So most of the "stuff" the tree is made of comes from carbon dioxide and a smaller portion comes from water. The authors also suggest that "the unscientific word *stuff* can be used intentionally in this probe to explore students' ideas without being hindered by their understanding of the concept of matter or mass" (p. 124).

We have taken that recommendation and have used the word *stuff* in some of the discussion questions in the explain phase of the lesson. It is important to note that soil is not part of the process of photosynthesis. The soil does contain nutrients that help the plant differentiate tissues, but the nutrients from the soil do not provide food that the plant needs to live and grow.

For the technology component of this lesson, students explore how plants are grown on the International Space Station. They learn how science and technology are interdependent. For example, to grow plants in space, scientific knowledge about how plants grow must be applied and new technologies must be invented. In this way, science and technology work together to solve a problem. At the end of the lesson, students revisit the From Seed to Tree Probe, revise their choices, and provide evidence for their answers. This gives students the opportunity to employ the science and engineering practice of engaging in argument from evidence.

Chapter 15

engage

If You Hold a Seed Read-Aloud

Show students the cover of *If You Hold a Seed,* and introduce the author and illustrator, Elly MacKay (pronounced Mack-EYE). Give each student a maple seed (which they will later find out is technically called a *samara*). As you read the book aloud, have students hold the seeds in their hands.

MAPLE TREE SAMARA

 Determining Importance

Connecting to the Common Core
Reading: Literature
KEY IDEAS AND DETAILS: 3.2, 4.2, 5.2

Tell students that as you read, you would like them to think about the author's purpose. What message is she trying to get across to the reader? After reading, *ask*

? What do you think the author's purpose was in writing this book? (Answers will vary.)

? What about the book makes you think that? (Answers will vary.)

? What does the author mean by the *magic* in a seed? (That a seed can grow into something.)

Have students look closely at a few of the illustrations. *Ask*

? How do you think MacKay made the art for this book? (Answers will vary.)

Tell students that they can get some insight into the author's purpose, as well as the fascinating way she creates her art, by watching a video of her introducing the book. Show the 2:13 min. video of MacKay explaining the message she wanted to convey with the book and demonstrating how she created the art (see "Websites" section). Afterward, *ask*

? Did watching this video change what you thought about the author's purpose? (Answers will vary.)

Next, ask students to think about the seed they are holding in their hands. *Ask*

? What will the seed grow into? (Students will likely recognize it as a maple seed that will grow into a maple tree.)

? What would need to happen for it to grow into a maple tree? (You have to plant it, water it, and make sure it gets sunlight.)

Show students a log cut from a tree. Hold a maple seed in one hand and hold the log in the other. *Ask*

? How does this seed turn into a log like this one? Where does the "stuff" come from that makes up a tree? (Answers will vary.)

From Seed to Tree Probe

Give each student a copy of the From Seed to Tree Probe student page (adapted from Keeley, Eberle, and Tugel 2007). Have them circle the response they agree with and explain their thinking. Collect their completed probes, and use them to assess their preconceptions. Write this question on the board: *Where does the "stuff" come from that makes up a maple tree?*

Tell students that you would like them to think about this question as they explore and read more in the coming days. Tell students that they will also have a chance to respond to the probe again later.

explore

Exploring Maple Seeds

Give each student a copy of the Exploring Maple Seeds Journal. (Be sure to copy page 7 and the cover back-to-back with pages 1 and 6; copy pages 3 and 4 back-to-back with pages 5 and 2.) Also give students a clipboard, a pencil, a hand lens, and a handful of maple seeds. (If possible, allow students to collect their own maple seeds.) Have students write their names on the cover of the journal. Take students to an outdoor area where they can do the activities described in the journal. Allow them to complete the activities in pairs or small groups, with each student recording observations and findings in his or her own journal. This exploration journal is intended to be self-guided, but here are a few things to check as you are visiting each group:

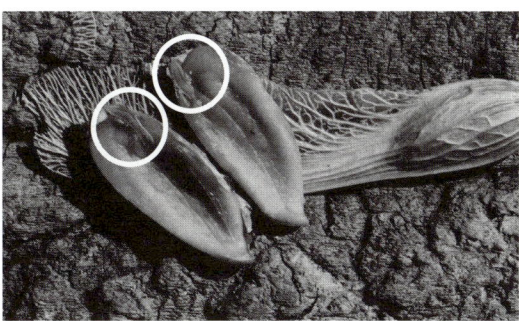

MAPLE TREE EMBRYOS

> **SAFETY**
> Have students wash their hands with soap and water after completing this activity.

- **Page 1—Maple Seed Observations.** Encourage students to label any parts they know, even if they are not sure of the names at this point (e.g., wing and seed).
- **Page 2—Maple Seed Names.** Students will learn later from the book that what we commonly call *maple seeds* are actually the fruits of the maple tree. These fruits are called *samaras*. At this point, encourage students to use their imaginations to come up with their own name for these whirling wonders.
- **Pages 3 and 4—How Do They Fly?** Encourage students to try these activities and record their findings in their journals. Ask them what the motion of the falling maple seed reminds them of.
- **Page 5—What's Inside?** Visit groups and demonstrate how to gently pull off the wing and remove the seed. Then, show students how the seed splits into two sections. If you look *very* closely, you can see the part of the seed that will grow into a maple tree! The bigger the seed, the easier it is to see.
- **Page 6—What Do You Wonder About Maple Seeds?** If students are having trouble coming up with questions, share some of your own wonderings about maple seeds (e.g., Why do they have a wing? Why do they spin? How does a tiny seed grow into a towering tree?).

explain

Next Time You See a Maple Seed Read-Aloud

 Determining Importance

At this point, gather the students together (either indoors or outdoors) and have them share some of the wonderings they wrote on page 6 of the journal. Tell them that you have a book to share with them that might answer some of their questions. Show the cover of *Next Time You See a Maple Seed*. Tell them that the author, Emily Morgan, created the *Next Time You See* series to inspire kids and adults to experience the wonders of nature. Have them signal when they hear something that relates to one of their wonderings, then pause the reading and discuss.

> **Connecting to the Common Core**
> **Reading: Informational Text**
> KEY IDEAS AND DETAILS: 3.1, 4.1, 5.1

After reading the book aloud, have students share some things that they learned from the book. Share some of your new learnings, too. Then, ask students to record some of their learnings on page 7 of the journal ("What Did You Learn About Maple Seeds?"). Be sure that students also fill in the blank at the bottom of page 2.

Next, ask students if they have any other questions about maple seeds (or *samaras*, as they know to call them now). Remark that new learning often leads to new questions, and then invite them to add any new questions they have about maple seeds to their list on page 6. Encourage students to share their new wonderings and refer to the text for the information that inspired those new questions.

After reading the book, students may want another opportunity to open up a maple seed to see the embryo, or baby tree, inside. Give them the chance to do this, and invite them to add the word *embryo* to their drawing on page 5 of the journal ("What's Inside?").

Encourage students to take their completed journal home, along with a few maple seeds, so they can share what they learned with someone at home.

 Rereading

Refer to the question you wrote on the board earlier: *Where does the "stuff" come from that makes up a maple tree?* Reread page 21 of *Next Time You See a Maple Seed*. Then, write out the process of photosynthesis on the board:

Photosynthesis
water + carbon dioxide ⟶ sugar + oxygen
(energy from the Sun)

Students might be wondering how water and carbon dioxide can turn into sugar. Hold up a balloon with CO_2 written on it in marker, and tell students that the balloon represents invisible carbon dioxide gas. Then, hold up a cup of water with H_2O written on it to represent water. (Although chemical formulas are not part of the standards for elementary students, they will likely already know that CO_2 is the formula for a carbon dioxide molecule and H_2O is the formula for a water molecule.) *Ask*

? How can this "stuff" (carbon dioxide gas) and this "stuff" (water) make up a tree?

Tell students that it might be helpful for them to see the process written using chemical symbols:

Photosynthesis
$$CO_2 + H_2O \xrightarrow[\text{(energy from the Sun)}]{} C_6H_{12}O_6 + O_2$$

Seeing the process in this way might help students understand that sugar is made up of the same "ingredients" (elements) as water and carbon dioxide—carbon, hydrogen, and oxygen—and that those ingredients are rearranged to make sugars. The Sun provides the energy for this change to take place. Oxygen is released as a waste product. So most of the "stuff" that makes up a plant (in this case, a maple tree) comes from a gas in the air (carbon dioxide) and water.

Students might also be wondering what the nutrients from the soil have to do with plant growth. Explain that they can think of nutrients from the soil as similar to vitamins that people take. Vitamins do not provide our bodies with energy; we need food for that. Nutrients from the soil do not provide plants with the energy they need to grow; plants need food for that. A plant's food is the sugar that it makes out of water and carbon dioxide through the process of photosynthesis.

Photosynthesis Video

 Turn and Talk

Tell students that people used to think plants ate soil because they grew up from the ground. Ask students to discuss the following question:

? How could you prove that plants don't get food from the soil?

Connecting to the Common Core
Reading: Informational Text
KEY IDEAS AND DETAILS: 3.1, 4.1, 5.1

Next, show students the 2.25 min. photosynthesis video from NOVA (see "Websites" section), which explains how plants get their food chiefly from carbon dioxide and water. Before watching, ask the students to listen for how a Belgian scientist in the 1600s proved that plants don't get food from the soil. After watching the video, *ask*

? How did the Belgian scientist's experiment prove that plants don't get food from soil? (He planted a tree in a pot of soil and then weighed it. For two years, he gave the tree only water. Then, he weighed the tree and the soil separately. The tree gained a lot of weight—164 lb.—but the soil lost less than 1 lb.)

? The scientist then concluded that the tree got its food from the water he had given it. Why was his conclusion that plants get food from water only partly correct? (He couldn't see carbon dioxide, which is the other raw material needed for plants to make food. Carbon dioxide is an invisible gas.)

? The video talks about things that are like tiny factories in which photosynthesis takes place inside a plant cell. What are these "factories" called? (chloroplasts)

Explain that *chloroplasts* are the food producers of the plant cell. They contain tiny green molecules called *chlorophyll*. Animal cells do not have chloroplasts. Chloroplasts work to convert light energy from the Sun into sugar that can be used by cells. So all of the "stuff" that makes up a maple tree, or any plant for that matter, comes from the remarkable process of photosynthesis taking place in those tiny chloroplasts!

elaborate

"Plants in Space" Pairs Read

Connecting to the Common Core
Reading: Informational Text
Range of Reading and Level of Text Complexity 3.10, 4.10, 5.10

Source: NASA/Frank Ochoa-Gonzales.

PLANT BIOLOGIST MEASURING RADISHES GROWN UNDER LED LIGHTS

Tell students that scientists and engineers at NASA are taking what they have learned about photosynthesis and using it in an innovative way. Give each student a copy of the article on the Plants in Space student page, and have students do a pairs read. In a pairs read, one student reads a paragraph while the other listens and then makes comments (I think …"), asks questions ("I wonder …"), or shares new learning ("I didn't know …). After the reading, *ask*

? What problems would astronauts have if they tried to grow plants in space? (A weightless environment has no up or down, so roots would grow in all directions. Water and any materials used to anchor the plants and allow for root growth would tend to float away.)

? How did astronauts try to solve these problems? (They invented *plant pillows* [bags of substrate containing controlled-release fertilizer] as part of a vegetable production system called *Veggie*. Wicks are implanted in the pillows to draw water into the substrate and provide a place to glue the seeds. The seeds are oriented so roots will grow "down" into the substrate and shoots that emerge will push out of the pillow. LEDs [light emitting diodes] furnish light energy for photosynthesis and give the shoots a sense of direction so they keep growing upward.)

? If you were going to design a greenhouse to be built on Mars, what materials would you need to include? (Answers will vary, but students should realize that to grow plants, they will need to supply light, water, and carbon dioxide. They should understand that soil is not necessary.)

After the students read the article, show the NASA video "Historic Vegetable Moment on the Space Station" (see "Websites" section) so that students can see the astronauts trying the lettuce on the ISS. The video gives them a better idea of how the Veggie system works.

Astronaut harvesting lettuce grown on the ISS

evaluate

Revisiting the From Seed to Tree Probe

Connecting to the Common Core
Writing
Text Types and Purposes: 3.2, 4.2, 5.2

Give students their first probe from the beginning of the lesson. Have them read what they wrote and think about how they might change their choice or explain their thinking differently. Then, give students a new copy of the probe. Have them circle the correct response and, again, explain their thinking. The correct response is Jack's: "I think plants get the materials they need to grow mainly from water and air."

Students should be able to explain that a plant combines air (or carbon dioxide from the air) and water to make food (sugars) for the plant to grow. Plants get nutrients from soil, but not food. Sunlight provides the energy for the process of photosynthesis, but does not constitute a "material."

STEM at Home

Have students complete the "I learned that …" and "My favorite part of the lesson was …" portions of the STEM at Home student page as a reflection on their learning. They may choose to do the following at-home activity with an adult helper and share their results with the class. If students do not have access to the internet or these materials at home, you may choose to have them complete this activity at school.

"At home, we can watch a video about the Robotic Samara, a one-winged helicopter that mimics the movements of a maple seed."

Search "World's first controllable MAV monocopter" on YouTube to find the video at www.youtube.com/watch?v=u23Hqq8QbeE.

"After we watch the video, we can discuss these questions:

1. How is the Robotic Samara like a maple seed?
2. How is it different?
3. NASA is considering something similar technologies for exploring the surface of Mars. What other uses for the Robotic Samara can you think of?"

For Further Exploration

This section is provided to help you encourage your students to use the science and engineering practices in a more student-directed format. This box lists questions and challenges related to the lesson that students may select to research, investigate, or innovate. Students may also use the questions as examples to help them generate their own questions. After selecting one of the questions in the box or formulating their own questions, students can individually or collaboratively make predictions, design investigations or surveys to test their predictions, collect evidence, devise explanations, design solutions, or examine related resources. They can communicate their findings through a science notebook, at a poster session or gallery walk, or by producing a media project.

Research

Have students brainstorm researchable questions:

? What is hydroponic technology?

? What challenges are there to growing plants on Mars?

? What other plants make seeds that have unique ways to disperse?

Investigate

Have students brainstorm testable questions to be solved through science or math:

? How long does it take for a maple seed to sprout? Try planting one!

? What happens when sunlight or water is taken away from a plant for an extended period of time?

? Can you estimate how many maple seeds a particular maple tree produces? (*Hint*: Count the seeds on one branch, estimate the number of branches, and then multiply.)

Innovate

Have students brainstorm problems to be solved through engineering:

? Can you design a system for growing plants without soil?

? Can you design a schoolyard garden?

? Can you design a plant experiment to be performed on the ISS?

References

Harvard-Smithsonian Center for Astrophysics. 1995. *Private Universe Project: Workshop 2. Biology: Why are some ideas so difficult?* Videotape. Burlington, VT: Annenberg/CPB Math and Science Collection.

Keeley, P., F. Eberle, and J. Tugel. 2007. Giant sequoia tree. In *Uncovering student ideas in science, volume 2: 25 more formative assessment probes,* P. Keeley, F. Eberle, and J. Tugel, 121–128. Arlington, VA: NSTA Press.

National Research Council (NRC). 2012. *A framework for K–12 science education: Practices, crosscutting concepts, and core ideas.* Washington, DC: National Academies Press.

Websites

"A Private Universe: Minds of Our Own—2: Lessons From Thin Air" (video)
www.learner.org/resources/series26.html

"Historic Vegetable Moment on the Space Station" (video)
http://science.nasa.gov/science-news/science-at-nasa/2016/historic-vegetable-moment-space-station

"NOVA: Photosynthesis" (video)
www.pbslearningmedia.org/resource/tdc02.sci.life.stru.photosynth/photosynthesis

Author and Illustrator Elly MacKay Discussing *If You Hold a Seed* (video)
www.youtube.com/watch?v=XvPqxFhKyE0

More Books to Read

Gibbons, G. 2002. *Tell me, tree: All about trees for kids*. New York: Little Brown.

Summary: This book is a kid-friendly guide to the wide variety of trees around us. It includes an explanation of photosynthesis, diagrams of tree parts, and information on how to identify trees.

Robbins, K. 2005. *Seeds*. New York: Atheneum Books for Young Readers.

Summary: Readers will learn how seeds grow and how they vary in shape, size, and dispersal patterns. Ken Robbins's signature photographs and simple explanations illuminate how wondrous nature springs up throughout the year before our very eyes.

Salas, L. P. 2008. *From seed to maple tree: Following the life cycle*. Mannkato, MN: Picture Window Books.

Summary: Provides a simple, straightforward description of the life cycle of a maple tree.

Name: _____

From Seed to Tree Probe

Ashley was hiking in the woods with three of her friends when she stopped to pick up a maple seed on the ground. Looking up at a towering maple tree overhead, she asked, "How can a seed this small grow into a tree that big?"

The friends all had different ideas about where plants get the materials they need for growth. This is what they said:

Maria: I think plants get the materials they need to grow mainly from soil.

Flavio: I think plants get the materials they need to grow mainly from soil and water.

Jack: I think plants get the materials they need to grow mainly from water and air.

Ashley: I think plants get the materials they need to grow mainly from sunlight.

Which friend do you agree with and why? Explain your thinking.

_____

_____

_____

_____

_____

_____

_____

Exploring Maple Seeds Journal

Name: _____

What Did You Learn About Maple Seeds?

After reading, *Next Time You See a Maple Seed*, record some of your new learnings below.

_____
_____
_____
_____
_____
_____

National Science Teachers Association

How Do They Fly?

1. Throw a handful of maple seeds into the air. Watch them fall to the ground. How would you describe their motion?

   _____

2. Hold one seed high and drop it. Watch it fall to the ground. Repeat several times. Draw a picture to represent its path as it falls.

---

3. Find two seeds. Remove the wing from one and keep the wing on the other. Drop the seeds from the same height at the same time and compare the way they fall. Draw or write your observations.

4. Which seed hit the ground first, the one with the wing or the one without?

   _____

5. Which seed landed the farthest from you, the one with the wing or the one without?

   _____

Picture-Perfect STEM Lessons, 3–5

Maple Seed Names

Maple seeds are known by many different names such as *helicopters* and *whirligigs*. In fact, *maple seed* is not even the real name! You'll find out later what the real name is.

Imagine you were the first person to see a maple seed and you had to give it a name. What name would you give it? Think about the way the seed looks and moves. Then, list your ideas.

_____
_____
_____
_____
_____

I learned that the real name is _____.

What's Inside?

Gently open up the hard, round end of the maple seed. Use a hand lens to get a closer look. Draw and label what you find inside.

What Do You Wonder About Maple Seeds?

List some questions you have about maple seeds.

_____
_____
_____
_____
_____
_____
_____
_____
_____
_____
_____

6

Maple Seed Observations

Use a hand lens to closely observe a maple seed. Draw and label any parts you know.

[ ]

What words would you use to describe the maple seed?

_____
_____
_____

1

Picture-Perfect STEM Lessons, 3–5

Plants in Space

Why Grow Plants in Space?

As we set our sights on extended space travel, including a mission to Mars, we need to figure out how to provide enough food for long space journeys. One way to do so is by growing plants. Imagine the challenges of growing plants in a spacecraft. How would you keep the soil from floating around? How would you water the plants without making a mess? How would the plant shoots know to grow "up" and the roots "down" without gravity?

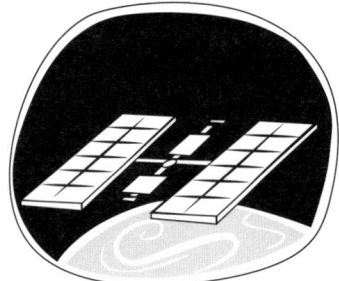

Plant Pillows

Astronauts on the International Space Station (ISS) are experimenting with a "veggie plant growth chamber" as they orbit 200 miles above Earth's surface. Instead of growing the plants in pots, scientists and engineers have developed a new technology called a *plant pillow*. The pillows are much lighter and less messy than soil. Because plants get their material for growth chiefly from carbon dioxide and water, soil is not necessary. The pillow contains a substrate in which the roots of the plant can grow, along with some fertilizer.

NASA's Veggie

The plant pillows are part of a vegetable production system from NASA that has been nicknamed *Veggie*. It uses the cabin environment for the source of carbon dioxide, and the astronauts periodically place water on wicks that draw water into the pillows.

Which Way Is Up?

The system includes LEDs (light emitting diodes) to provide the energy needed for photosynthesis and give the plant shoots a sense of direction so they keep growing upward. One question scientists had when first growing plants on the ISS is whether plants would know which way to grow without gravity. Early experiments show that in space, the plants grow toward the lights that are placed above them, and the roots grow in the opposite direction.

One Small Bite for Man

In 2015, astronauts on the ISS ate the first lettuce grown using the plant pillow technology. It took 33 days for the lettuce to grow from a seed to an edible plant. On the day they ate the lettuce for the first time, 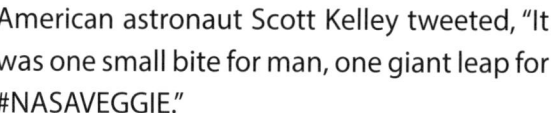 American astronaut Scott Kelley tweeted, "It was one small bite for man, one giant leap for #NASAVEGGIE."

More Than Just Food

Scientists believe that growing plants in space will not only provide fresh vegetables for the astronauts to eat but also improve astronauts' morale. Growing a garden has been proven to reduce stress and will give the astronauts some of the sights, smells, and tastes of Earth. Plans are being made to experiment with growing more kinds of vegetables on the ISS in the future.

Name : _____

STEM at Home

Dear _____,

At school, we have been learning about **maple seeds** and how plants make food through **photosynthesis**.

I learned that: _____

_____

_____

My favorite part of the lesson was: _____

_____

_____

_____

At home, we can watch a video about the Robotic Samara, a one-winged helicopter that mimics the movements of a maple seed.

🔍 Search "World's first controllable MAV monocopter" on YouTube to find the video at *www.youtube.com/watch?v=u23Hqq8QbeE*.

After we watch the video, we can discuss these questions:

How is the Robotic Samara like a maple seed _____

_____

How is it different?

_____

NASA is considering similar technologies for exploring the surface of Mars. What other uses for the Robotic Samara can you think of?

_____

_____

_____

Picture-Perfect STEM Lessons, 3–5

Chapter 16

Hurricane!

Description

After reading the heartwarming true story of a cat and a dog that survived the devastation following Hurricane Katrina, students learn why this storm was one of the worst natural disasters in U.S. history—many of the levees built to protect New Orleans failed. Students then learn more about levees and are challenged to build and test a model of one.

Suggested Grade Levels: 3–5

LESSON OBJECTIVES Connecting to the *Framework*		
Science and Engineering Practices	**Disciplinary Core Ideas**	**Crosscutting Concept**
Constructing Explanations and Designing Solutions Engaging in Argument From Evidence	**ESS3.B:** Natural Hazards **ETS.1.B:** Developing Possible Solutions **ETS2.B:** Influence of Engineering, Technology, and Science on Society and the Natural World	Cause and Effect

Featured Picture Books

TITLE: **Two Bobbies: A True Story of Hurricane Katrina, Friendship, and Survival**
AUTHORS: **Kirby Larson and Mary Nethery**
ILLUSTRATOR: **Jean Cassels**
PUBLISHER: **Walker**
YEAR: **2008**
GENRE: **Story**
SUMMARY: *This book tells the remarkable true story of a dog and cat that survived the aftermath of Hurricane Katrina by sticking together.*

TITLE: **Building Dikes and Levees**
AUTHOR: **Rebecca Stefoff**
PUBLISHER: **Cavendish Square**
YEAR: **2015**
GENRE: **Non-Narrative Information**
SUMMARY: *This book from the Great Engineering series explains how dikes and levees protect areas from flooding and describes what happens when they break. Photographs and diagrams accompany the text.*

Picture-Perfect STEM Lessons, 3–5

Time Needed

This lesson will take several class periods. Suggested scheduling is as follows:

Day 1: Engage with *Two Bobbies* Read-Aloud

Day 2: Explore with Hurricane Katrina Investigation Stations and **Explain** with Putting It All Together

Day 3: Explain with *Building Dikes and Levees* Read-Aloud and New Vocabulary List

Day 4: Elaborate with Model Levee Design Challenge

Day 5: Evaluate with Evaluate Your Design

Materials

For Hurricane Katrina Investigation Stations

- Station 1: Where Is New Orleans?
  - Station 1 info card
  - Globe or U.S. map
  - Computer or tablet with Google Maps (optional)
- Station 2: What Is a Hurricane?
  - Station 2 info card
  - Computer or tablet with NOAA video
- Station 3: How Are Hurricanes Measured?
  - Station 3 info card
- Station 4: How Do We Monitor Hurricanes?
  - Station 4 info card
  - Computer or tablet with NOAA website and hurricane researcher video from PBS Learning Media
- Station 5: What Happened to the Levees?
  - Station 5 info card
  - Photos of the levees in New Orleans after Hurricane Katrina
- Station 6: What Were Some of the Rescue Efforts During Hurricane Katrina?
  - Station 6 info card
- Station 7: How Do Scientists Predict Hurricanes?
  - Station 7 info card
  - Computer or tablet with weather scientist video from PBS Learning Media

For Model Levee Design Challenge (per group of 3 students)

- Model Levee Design Challenge Team Cards
- Clear, plastic shoebox
- Measuring cup
- LEGOs or small building blocks for making models of buildings (not to be used for building the levee)

- Play sand (about 10 lb.)
- Pea gravel (about 10 lb.)
- Straws (box of 100)
- Craft sticks (bag of 100)
- Roll of masking tape
- Roll of duct tape
- Box of 50 sandwich-size plastic zipper bags
- Bag of cotton balls
- $10 in play money (precut)

Student Pages

- Hurricane Katrina Investigation Journal
- New Vocabulary List
- Model Levee Design Challenge
- Evaluate Your Design
- STEM at Home

Background for Teachers

Hurricane Katrina was one of the worst natural disasters in U.S. history. On August 29, 2005, this 350 mile-wide hurricane spun through the Gulf Coast of the United States with winds of more than 150 miles per hour. Katrina was a category 5 hurricane, the highest level on the Saffir-Simpson Hurricane Wind Scale (see Table 16.1). This scale rates hurricanes based on wind speed and storm surge (the height the water rises above normal in a storm). Although Hurricane Katrina affected much of the southeastern United States, the area most devastated was New Orleans, Louisiana. New Orleans is surrounded by water—lakes, rivers, canals, wetlands, and the ocean. Levees were built throughout the city to keep the water out. However, the levees were not high enough or strong enough to handle the massive storm surges produced by Hurricane Katrina. Levees and floodwalls failed in more than 50 places that day. Some levees were *overtopped*, meaning waves poured over the tops, and others were *breached*, which means cracked or broken, under the enormous weight of the water pressing on them (Figure 16.1, p. 266). Hurricane Katrina took an estimated 1,836 lives and cost billions of dollars. Tens of thousands of people from all over the United States volunteered to help. People donated food and water, clothes, first

Table 16.1. The Saffir-Simpson Hurricane Wind Scale

Category	Wind speed (mph)	Storm surge (feet)
5	157 or higher	More than 18
4	130–156	13–18
3	111–129	9–12
2	96–110	5–8
1	74–95	4–5
Additional Classifications		
Tropical Storm	39–73	0–3
Tropical Depression	0–38	0

Chapter 16

Figure 16.1. Levee Overtopping Versus Levee Breaching

HURRICANE KATRINA

aid, clean-up assistance, building help, and even "music aid" to replace the musical instruments lost by the many musicians who lived in New Orleans.

The morning of August 29, the mayor of New Orleans announced a mandatory evacuation. People piled into cars and took as much of their belongings with them as they could. But others were not able to leave and became stranded in their homes. According to a Harris Interactive survey of people affected by Hurricane Katrina, nearly half of the people who stayed behind did so because they did not want to leave their pets (Fritz Institute 2006). Some people left their pets with food and water, thinking they would be able to come back to get them in a day or two. Rescue workers were not instructed on whether they could rescue pets. After the disaster, the public flooded elected officials with letters about the pets who were left behind in Katrina. In response, Congress passed the Pets Evacuation and Transportation Standards (PETS) Act in 2006 with near unanimous support. The law requires rescue agencies to save pets as well as people during natural disasters.

In the engage phase of this lesson, students listen to the heartwarming true story of a cat and dog that survived Hurricane Katrina together. The story is used to engage their emotions about the disaster in order to arouse their curiosity, leading them into an exploration of the levee failures that devastated

New Orleans. This lesson addresses disciplinary core idea ESS3.B: Natural Hazards, which suggests that by the end of grade 5, students understand that "a variety of hazards result from natural processes (e.g., earthquakes, tsunamis, volcanic eruptions, severe weather, floods, coastal erosion) [and that] [h]umans cannot eliminate natural hazards but can take steps to reduce their impacts" (NRC 2012, p. 193). The lesson concludes with an activity during which students build a model levee, test it, and then make a claim about the merit of the levee. This design challenge ties into the engineering design standards for grades 3–5 as well as the science and engineering practices of constructing explanations and designing solutions and engaging in argument from evidence. In these activities, students also explore the crosscutting concept of cause and effect in the context of a natural disaster.

engage

Connecting to the Common Core
Reading: Literature
KEY IDEAS AND DETAILS: 3.1, 4.1, 5.1

Two Bobbies Read-Aloud

 Questioning

Show students the cover of *Two Bobbies* and introduce the authors, Kirby Larson and Mary Nethery, and the illustrator, Jean Cassels. Read the subtitle of the book: *A True Story of Hurricane Katrina, Friendship, and Survival*, and *ask*

? Have you ever heard of Hurricane Katrina?
? What do you think this book might be about?

Read the brief introduction on the page opposite the copyright page. Then, read the book aloud. After reading, *ask*

? How did the two Bobbies help each other? (Bobbi the dog was a seeing-eye dog for Bob Cat. Bobbi protected Bob Cat. They both gave each other companionship.)
? How did people help them? (Rich, the man at the construction site, fed them and took them to an animal shelter; the people at the animal shelter took care of them; Anderson Cooper had them on his news show; and Melinda, who saw the show, adopted them.)
? Where did most of this story take place? (New Orleans)
? What happened in New Orleans that put the two Bobbies in danger? (Hurricane Katrina)
? What is a hurricane? (Answers will vary.)

Next, read the back flap of the book about the authors' friendship and how they are donating a portion of the proceeds from this book to the Best Friends Animal Society. Also, read the paragraph about the illustrator, Jean Cassels, who lives in New Orleans and had to evacuate during Hurricane Katrina. Tell students that Hurricane Katrina happened in August 2005. *Ask*

? Were you born before or after 2005?
? Have you ever heard people talk about Hurricane Katrina?

Tell students that they are going to learn much more about Hurricane Katrina, which was one of the worst natural disasters in U.S. history.

explore

Hurricane Katrina Investigation Stations

Connecting to the Common Core
Reading: Informational Text
INTEGRATION OF KNOWLEDGE AND IDEAS: 3.7, 4.7, 5.7

Tell students that Hurricane Katrina affected a large portion of the southeastern United States, but New Orleans, Louisiana (where the two Bobbies were found), was the most devastated place of all. Tell students that they are going to become "hur-

Chapter 16

HURRICANE KATRINA INVESTIGATION STATIONS

ricane investigators" and put the clues together to find out why New Orleans was destroyed during the storm. Turn your classroom into a Hurricane Investigation Center by setting up the seven Hurricane Investigation Stations as outlined in the "Materials" section (p. 264).

Give each student a copy of the Hurricane Katrina Investigation Journal and have groups of three to four students rotate through each station (for about 5 min. per station). The stations do not need to be visited in order, so each group can begin at a station and then rotate. Visit groups as they are investigating the stations.

explain

Putting It All Together

 Synthesizing

After students have visited all the stations, discuss the answers to the questions in their Hurricane Katrina Investigation Journals. The questions and answers are as follows:

Station 1: Where Is New Orleans?

- Find New Orleans on a globe or map. What bodies of water surround the city? (Gulf of Mexico, Mississippi River, Lake Pontchartrain, Lake Borgne)

Station 2: What Is a Hurricane?

- How fast does the wind need to blow to be considered a hurricane? (at least 74 mph)

Station 3: How Are Hurricanes Measured?

- Hurricane Katrina was a category 5 hurricane. What is the wind speed of a category 5 hurricane? (157 mph or higher)
- What is the storm surge of a category 5 hurricane? (greater than 18 ft.)

Station 4: How Do We Monitor Hurricanes?

- What does NOAA stand for? (National Oceanic and Atmospheric Administration)
- Are there any tropical storms or hurricanes being tracked right now at *www.nhc.noaa.gov*? If so, describe them. (If this lesson is done in summer or fall, there may be some tropical storms or hurricanes on the map.)
- Watch the video of the hurricane researcher. Where is the National Hurricane Center? (Miami, Florida)
- How are hurricanes different from tornadoes? (Hurricanes are much larger than tornadoes, and hurricanes spin only counterclockwise. Tornadoes can spin either clockwise or counterclockwise.)

Station 5: What Happened to the Levees?

- What is a levee? (a wall or barrier that protects against high water)
- What happened to the levees around New Orleans during Hurricane Katrina? (Water went over many of the levees, and many of the levees broke.)

Station 6: What Were Some of the Rescue Efforts During Hurricane Katrina?

- Who helped during the rescue efforts? (volunteers, FEMA [Federal Emergency Management Agency], U.S. Navy, firefighters, Coast Guard, and animal shelters)
- How did the relationship between people and their pets affect the rescue efforts? (Some

people chose not to evacuate because they did not want to leave their pets.)

Station 7: How Do Scientists Predict Hurricanes?

- Watch the video of the weather scientist. How does she get a firsthand look at how hurricanes start? (She flies into them on an airplane.)
- How are satellites useful for studying weather? (With satellites, scientists can see the whole Earth.)

Next, help students summarize the events that made Hurricane Katrina one of the worst natural disasters in American history. *Ask*

? How did the location of New Orleans affect the storm's impact? (New Orleans is on the Gulf of Mexico. Hurricanes form over the ocean, so being on the Gulf put New Orleans in the hurricane's path. New Orleans is also surrounded by water—lakes, rivers, and the ocean. The hurricane caused those bodies of water to rise.)

? How powerful was Hurricane Katrina? (a category 5 on the Saffir-Simpson Hurricane Wind Scale, which is the most powerful hurricane on the scale)

? How were the people of New Orleans affected during Katrina? (Many evacuated, some stayed, and lives were lost.)

? What other places were affected? (the entire Gulf Coast of the United States)

? What did Hurricane Katrina teach rescue agencies about people and their pets? (The bond between people and their pets is very strong. Some people chose not to evacuate because they did not want to leave their pets behind.)

? What was the primary cause of the flooding in New Orleans during Katrina? (The levees were not able to hold the water back from the Gulf, lakes, and rivers surrounding New Orleans.)

Tell students that they will be learning more about the levee failures that caused so much destruction in New Orleans during Hurricane Katrina.

explain

Building Dikes and Levees Read-Aloud

> Connecting to the Common Core
> **Reading: Informational Text**
> CRAFT AND STRUCTURE: 3.4, 4.4, 5.4

Features of Nonfiction

Show students the cover of *Building Dikes and Levees*. Tell them that this book can help them understand more about what happened to the levees in New Orleans during Hurricane Katrina. Explain that *dike* and *levee* are closely related terms and often used interchangeably, both meaning a high wall or barrier against high water or floods. For this lesson, we will refer to theses structures as *levees*. Tell student that you would like to know if this book has any information on Hurricane Katrina. Model how to use the index in the back of the book to look for specific information. Find "Hurricane Katrina" in the index, and turn to the pages listed (pp. 24–25). Read the section titled "Historic Floods," which describes how the levees that were built to protect New Orleans were overtopped and breached during Hurricane Katrina. *Ask*

? Why was the flooding in Hurricane Katrina so bad? (The levees did not hold up.)

New Vocabulary List

> Connecting to the Common Core
> **Reading: Informational Text**
> CRAFT AND STRUCTURE: 3.4, 4.4, 5.4

Tell students that *overtopping* and *breach* are important words to know when learning about levees. Give students a copy of the New Vocabulary List student page. A new vocabulary list is a "guess-and-check" type of visual representation. Students develop vocabulary as they draw and write predictions about a new word's meaning, read the word

Picture-Perfect STEM Lessons, 3–5

269

in context, and draw and write their new definitions of the word. Have them fill in the "What I Think It Means" column for each word. Students will need to use these words in the elaborate portion of the lesson.

Tell students that they are going to be challenged to design a model of a levee to protect a city from flooding. Show them the clear, plastic shoebox. Tell them that the city will be on one side (they will build a city out of LEGOs or small plastic blocks), and the body of water will be on the other side. Explain that they will need to construct something in between to hold the water back. Tell students that you are going to read the other sections of the book *Building Dikes and Levees* so that they can learn more about how levees are made. They will need to listen for the new vocabulary words and for information about how levees are built. Read the book aloud, stopping when you reach each word on the new vocabulary list.

- Page 2—Overtopping
- Page 24—Breach
- Page 13—Civil engineer, hydrologist
- Page 14—Geologist

After reading each word in context, discuss the word's meaning and model how to use the glossary in the back to of the book. Allow students to compare the glossary definitions with the ones they developed. The glossary definitions are as follows:

- *Overtopping*—When water rises high enough to spill over the top of a levee
- *Breach*—A hole, crack, or other break
- *Civil engineer*—An engineer who designs bridges, dams, roads, and other structures the public uses
- *Geologist*—A scientist who studies geology, the subject of the earth and what it is made of
- *Hydrologist*—A scientist who studies water and the movement of water

elaborate

Model Levee Design Challenge

Note: The following activity was adapted from "Protecting Our City With Levees," an activity from *www.teachengineering.org*.

Give each student the Model Levee Design Challenge sheet, and give each group a set of Model Levee Design Challenge Team Cards; a clear, plastic shoebox; and some LEGOs or small plastic blocks. Have team members divide up the building blocks

> **SAFETY**
> - Have students wear goggles for this activity.
> - Be sure to immediately clean up spilled sand, gravel, or water to avoid a slip-and-fall hazard.
> - Have students wash their hands with soap and water after completing this activity.

evenly, and then have each student construct a building to put on one side of the plastic shoebox. Explain that these buildings will represent the city. Tell students that the opposite end of the box will represent a body of water next to the city. Their challenge is to build a levee that will keep the water from flooding the city. The Model Levee Design Challenge student page explains the criteria and constraints of the challenge:

Criteria (desired features or outcomes):

1. Must form a barrier between the body of water and the city.
2. Must hold two cups of water without breaching or overtopping.

Constraints (materials, time, and cost):

1. Materials: You may use only materials that your teacher provides or approves.
2. Time: You must build your model within the time limit your teacher sets.

3. Cost: You can spend only $10 on materials.

Connecting to the Common Core
Mathematics
NUMBER AND OPERATIONS IN BASE TEN: 3.NBT.1

Describe the different supplies listed on the Model Levee Design Challenge student page, and go over how much each one costs. Then, give each team $10 in play money, and explain how to use the cost chart to record the team's spending on the project (Table 16.2). That is, Cost per Unit × Number of Units = Total Cost for Each Material. Sum the Total Cost column to get the Grand Total for the Levee Project (not to exceed $10).

Then, give each team member a team card. One student will share the information on his or her card from the viewpoint of a *geologist*, one an *engineer*, and the third a *hydrologist*. Explain that in a real levee project, those are three of the key people who might work together to design the levee, but they would likely be part of a much larger team. Other people (construction workers and other tradespeople) do the actual building of a levee. But for this challenge, students will be both the designers and the builders.

Next, teams can design a plan for the project and record the plan on the student page. After getting your sign off at the teacher checkpoint, students can purchase their materials from you using their play money. Then, they can build and test their models.

After students have made their first levee-building attempt, have groups share their designs and their results. Explain that communicating with peers about proposed solutions in an important part of the design process. Sharing ideas can lead to improved designs. *Ask*

? Did your levee solve the problem?
? Did it meet the criteria within the given constraints?
? If your levee did not solve the problem, what did you think was the cause?

MODEL LEVEE DESIGN CHALLENGE

Table 16.2. Cost Chart

Material	Cost per Unit	Number of Units	Total Cost
1 cup of sand	$1		
1 cup of gravel	$1		
2 straws	$1		
2 craft sticks	$1		
1 ft. of masking tape	$1		
1 ft. of duct tape	$2		
1 plastic bag	$2		
10 cotton balls	$2		
Grand Total for the Levee Project			

? What materials seem to work best?
? How would you improve your design?

Be sure to tell students that they should not get discouraged if their levee did not work. The teams of engineers that design real levees build and test many different models. Instead of being discouraged by what didn't work, they learn from it to make a better model, and eventually a better levee. Next, show students the video "Rebuilding the Levees" from the Science Channel series *Build It Bigger* (see "Websites" section). After the video, *ask*

? How has the city of New Orleans prepared for future hurricanes? (They have built a surge barrier 8 miles away from the city that is bigger and stronger than any of the barriers they had before.)

? Are there any ideas from the video that you can incorporate into your design? (Answers will vary.)

Give students time to improve their levee designs. You may consider increasing the budget by giving each team $5 in play money and letting them add or switch out materials.

Evaluate Your Design

Connecting to the Common Core
Writing
Production and Distribution of Writing: 3.4, 4.4, 5.4

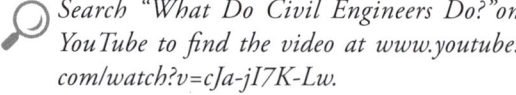

Give students the Evaluate Your Design student page. Using their best levee design, have them review the problem, criteria, and constraints and make a claim about whether their design solution worked. They should use evidence from their test to support the claim. For example, students might say, "Our levee solved the problem, meeting all of the criteria and constraints. The levee was not overtopped or breached. The city was protected from the water." Or students might say, "Our levee did not solve the problem because it did not meet the constraint of not overtopping. The levee did not breach, but it was not tall enough to keep the water out." Be sure to explain to students that they are not being evaluated on whether their levee worked;

rather, you are evaluating how well they can use evidence from their tests to support a claim about whether their design solution worked.

STEM at Home

Have students complete the "I learned that …" and "My favorite part of the lesson was …" portions of the STEM at Home student page as a reflection on their learning. They may choose to do the following at-home activity with an adult helper and share their results with the class. If students do not have access to the internet or these materials at home, you may choose to have them complete this activity at school.

"We learned how civil engineers help design and build levees to protect people from flooding, but civil engineers also design and build many other kinds of structures. Together, we can watch a video called 'What Do Civil Engineers Do?' and then answer the questions below. In this video, we will meet three civil engineers whose work is having a big influence on people's lives."

Search "What Do Civil Engineers Do?" on YouTube to find the video at www.youtube.com/watch?v=cJa-jI7K-Lw.

"What are some of the projects these civil engineers were involved in designing and building? [*Answer:* a huge underground tunnel to capture storm water, a system for bringing clean water to people in Haiti, and a hurricane-proof retractable roof for a stadium]

If you were a civil engineer, what would YOU like to design? Sketch it below!"

For Further Exploration

This section is provided to help you encourage your students to use the science and engineering practices in a more student-directed format. This box lists questions and challenges related to the lesson that students may select to research, investigate, or innovate. Students may also use the questions as examples to help them generate their own questions. After selecting one of the questions in the box or formulating their own questions, students can individually or collaboratively make predictions, design investigations or surveys to test their predictions, collect evidence, devise explanations, design solutions, or examine related resources. They can communicate their findings through a science notebook, at a poster session or gallery walk, or by producing a media project.

Research
Have students brainstorm researchable questions:

- How do hurricanes get their names?
- How much does it cost to build a real levee?
- How far from you is the nearest large body of water? Are there levees near it? Are the levees manmade or natural?

Investigate
Have students brainstorm testable questions to be solved through science or math:

- Which materials from our levee design challenge are water absorbent?
- Which materials from our levee design challenge are water resistant?
- How much water can a small bag of sand hold before leaking?

Innovate
Have students brainstorm problems to be solved through engineering:

- Can you build a model to test how sandbags can be used to create a water barrier?
- Can you build a model to test how different types of roofs resist wind?
- Can you build a model to test how buildings can be made earthquake resistant?

References

Fritz Institute. 2006. Harris Interactive Katrina survey reveals inadequate immediate relief provided to those most vulnerable. www.fritzinstitute.org/prsrmPR-FI-HIKatrinaSurvey.htm.

National Research Council (NRC). 2012. *A framework for K–12 science education: Practices, crosscutting concepts, and core ideas.* Washington, DC: National Academies Press.

Websites

National Hurricane Center
www.nhc.noaa.gov

PBS Learning Media
www.pbslearningmedia.org

"Protecting Our City With Levees" (activity)
www.teachengineering.org/activities/view/cub_weather_lesson05_activity1

"Rebuilding the Levees Video"
www.sciencechannel.com/tv-shows/build-it-bigger/videos/build-it-bigger-rebuilding-the-levees

"What Is a Hurricane?" (video)
http://oceanservice.noaa.gov/facts/hurricane.html

More Books to Read

Stewart, M. 2015. *Hurricane watch*. New York: HarperCollins.
Summary: From the *Let's-Read-and-Find-Out Science* series, this book explains how hurricanes form, how scientists predict and study hurricanes, and how to prepare for a hurricane.

Callery, S. 2015. *Hurricane Katrina: Day by day in one of the world's deadliest storms*. New York: Scholastic.
Summary: This book from Scholastic's *Discover More* series takes the reader day by day through Hurricane Katrina in words and pictures.

London, J. 1998. *Hurricane!* New York: Scholastic.
Summary: This beautifully illustrated picture book tells the story of two young brothers surviving a hurricane in Puerto Rico.

Hurricane Katrina Info Cards

Station 1: Where Is New Orleans?

Locate New Orleans, Louisiana, on a map of the United States or on a globe. Notice the bodies of water that surround New Orleans.

Station 2: What Is a Hurricane?

Watch the NOAA video "What Is a Hurricane?" at **http://oceanservice.noaa.gov/facts/hurricane.html**.

Station 3: How Are Hurricanes Measured?

The Saffir-Simpson Hurricane Wind Scale

Category	Wind speed (mph)	Storm surge (feet)
5	157 or higher	More than 18
4	130–156	13–18
3	111–129	9–12
2	96–110	5–8
1	74–95	4–5
Additional Classifications		
Tropical Storm	39–73	0–3
Tropical Depression	0–38	0

Wind speed is how fast the wind is blowing, measured in miles per hour.

Storm surge is the height of the water above normal levels during a storm, measured in feet or meters.

Station 4: How Do We Monitor Hurricanes?

NOAA, the National Oceanic and Atmospheric Administration, is a government agency that focuses on the oceans and the atmosphere.

NOAA's National Hurricane Center has the most current information on tropical storms, depressions, and hurricanes.

Visit the National Hurricane Center's website at ***www.nhc.noaa.gov.***

Next, meet hurricane researcher Jason Dunion at the National Hurricane Center. Go to ***www.pbslearningmedia.org,*** and search "hurricane researcher Jason Dunion."

Station 5: What Happened to the Levees?

New Orleans is surrounded by water. Two large bodies of water that surround the city are the Mississippi River and Lake Pontchartrain. Structures called **levees** were built to keep those bodies of water from flooding New Orleans. Levees are walls or barriers that protect against high 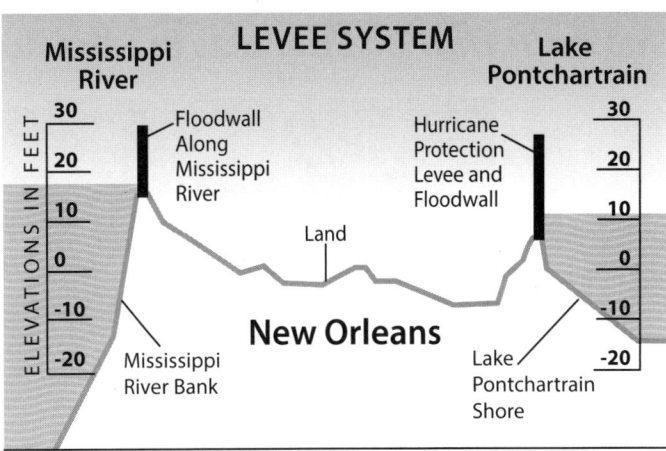 water. When Hurricane Katrina happened, the water went over many of the levees, and in many places the levees broke! The city was completely flooded. In some places, the floodwater was more than 10 feet high.

Station 6: What Were Some of the Rescue Efforts During Hurricane Katrina?

- Tens of thousands of people from all over the United States volunteered to help.
- The Coast Guard rescued approximately 33,000 people.
- FEMA (Federal Emergency Management Agency), the U.S. Navy, firefighters, and many others groups organized rescue efforts.
- Many people did not evacuate because they had nowhere else to go, did not have a car, or could not leave their homes without help.
- Some people chose to stay behind because they did not want to leave their homes or loved ones, including pets.
- Many people left their pets with food and water, thinking that they would be able to come back in a day or two to get them.
- Animal shelter volunteers from across the country worked hard to reunite people with their pets.

Station 7: How Do Scientists Predict Hurricanes?

Meet NASA atmospheric scientist Robbie Hood, who studies hurricanes to improve the ability to predict when one will occur. She has even flown over and through hurricanes to collect data!

Go to *www.pbslearningmedia.org,* and search "weather scientist Robbie Hood."

Hurricane Katrina Investigation Journal

Investigator: _____

Station 6: What Were Some of the Rescue Efforts During Hurricane Katrina?

Who helped during the rescue efforts?
_____

How did the relationship between people and their pets affect the rescue efforts?
_____

Station 7: How Do Scientists Predict Hurricanes?

Watch the video of the weather scientist. How does she get a first-hand look at how hurricanes start?
_____
_____

How are satellites useful for studying weather?
_____
_____

Station 1: Where Is New Orleans?
Find New Orleans on a globe or map. What bodies of water surround the city?

Station 2: What Is a Hurricane?
How fast does the wind need to blow to be considered a hurricane?

Station 3: How Are Hurricanes Measured?
Hurricane Katrina was a category 5 hurricane. What is the wind speed of a category 5 hurricane?

What is the storm surge in a category 5 hurricane?

Station 4: How Do We Monitor Hurricanes?
What does NOAA stand for?

Are there any tropical storms or hurricanes being tracked right now at *www.nhc.noaa.gov*? If so, where?

Watch the video of the hurricane researcher. Where is the National Hurricane Center?

How are hurricanes different from tornadoes?

Station 5: What Happened to the Levees?
What is a levee?

What happened to the levees throughout New Orleans during Hurricane Katrina?

National Science Teachers Association

Name : _____

New Vocabulary List

Word	What I Think It Means (Draw and Write)	What It Means (Draw and Write)
Breach		
Overtopping		
Civil engineer		
Geologist		
Hydrologist		

Picture-Perfect STEM Lessons, 3–5

Name : _____

Model Levee Design Challenge

Problem: Design a model levee to protect a city from flooding.

Criteria (desired features or outcomes):
1. Must form a barrier between the body of water and the city
2. Must hold two cups of water without breaching or overtopping

Constraints (limits on available resources and time):
1. Materials: You may use only materials that your teacher provides or approves.
2. Time: You must build your model within the time limit your teacher sets.
3. Cost: You can spend only $10 on materials.

Brainstorm ideas and sketch them below.

Name : _____

Chapter 16

Choose a Solution and Make a Plan

Circle the supplies you plan on using, determine how many units of each you will need, and calculate the grand total for your levee project.

Material	Cost per Unit	Number of Units	Total Cost
1 cup of sand	$1		
1 cup of gravel	$1		
2 straws	$1		
2 craft sticks	$1		
1 foot of masking tape	$1		
1 foot of duct tape	$2		
1 plastic bag	$2		
10 cotton balls	$2		
Grand Total for Levee Project			

How will you know if your design was successful?

_____

_____

_____

_____

Teacher Checkpoint ☐

Model Levee Design Challenge
Team Cards

Geologist

Share the following information with your team before planning the levee:

- Heavy materials work best for building a levee.
- Light materials can easily wash away.
- Sand and gravel can be moved by water if they are not protected by a waterproof barrier.

Civil Engineer

Share the following information with your team before planning the levee:

- Creating a barrier that water cannot seep through is important when building a levee.
- The levee must be tall enough to prevent overtopping (water flowing over the top of the levee).
- The levee must be strong enough to prevent breaching (cracking or breaking).

Hydrologist

Share the following information with your team before planning the levee:

- The higher the water level, the more force there is pushing on the levee.
- Flowing water can easily carry away small or light objects.
- Water can seep underneath a levee if the levee has any holes or openings.

Name : _____

Evaluate Your Design

Problem: Design a model levee to protect a city from flooding.
Did your design meet the criteria and stay within the constraints?
Criteria (✓):

- ☐ Must form a barrier between the body of water and the city
- ☐ Must hold two cups of water without breaching or overtopping

Constraints (✓):

- ☐ Materials: You may use only materials that your teacher provides or approves.
- ☐ Time: You must build your model within the time limit your teacher sets.
- ☐ Cost: You can spend only $10 on materials.

1. Did your model levee solve the problem and meet the criteria, within the given constraints? Explain, using evidence from your test.

_____

_____

_____

2. If you were given a chance to build another levee, what changes would you make?

_____

_____

3. If you had a bigger budget ($15), do you think you could have built a better levee? What would you have purchased?

_____

_____

National Science Teachers Association

Name: _____

STEM at Home

Dear _____,

At school, we have been learning about **hurricanes and levees**. I learned that:

_____

_____

My favorite part of the lesson was:

_____

_____

We learned how civil engineers help design and build levees to protect people from flooding, but civil engineers also design and build many other kinds of structures. Together, we can watch a video called "What Do Civil Engineers Do?" and then answer the questions below. In this video, we will meet three civil engineers whose work is having a big influence on people's lives.

🔍 Search "What Do Civil Engineers Do?" on YouTube to find the video at *www.youtube.com/watch?v=cJaRjI7K-Lw.*

What are some of the structures these civil engineers were involved in designing and building?

_____

_____

_____

If you were a civil engineer, what would *you* like to design? Sketch it below!

Picture-Perfect STEM Lessons, 3–5

Chapter 17

Solving the Puzzle Under the Sea

Description

After hearing the inspiring story of Marie Tharp, who helped create the first scientific map of the ocean floor, students learn that the ocean floor has many interesting features, including large mountain ranges. Students use a model to explore how sonar measurements help us map undersea mountains. Then, they learn how mountains form and observe patterns in mountain locations. Finally, students summarize how Tharp's maps changed the way we look at Earth.

Suggested Grade Levels: 3–5

LESSON OBJECTIVES Connecting to the *Framework*		
Science and Engineering Practices	**Disciplinary Core Ideas**	**Crosscutting Concept**
Analyzing and Interpreting Data Obtaining, Evaluating, and Communicating Information	**ESS2.B:** Plate Tectonics and Large-Scale System Interactions **ETS2.A:** Interdependence of Science, Engineering, and Technology	Patterns Stability and Change

Featured Picture Books

- **TITLE:** ***Solving the Puzzle Under the Sea: Marie Tharp Maps the Ocean Floor***
- **AUTHOR:** **Robert Burleigh**
- **ILLUSTRATOR:** **Raúl Colón**
- **PUBLISHER:** **Simon & Schuster**
- **YEAR:** **2016**
- **GENRE:** **Narrative Information**
- **SUMMARY:** *This book tells the inspiring story of how geologist Marie Tharp, in partnership with Bruce Heezen, created the first scientific map of the entire ocean floor.*

- **TITLE:** ***How Mountains Are Made***
- **AUTHOR:** **Kathleen Weidner Zoehfeld**
- **ILLUSTRATOR:** **James Graham Hale**
- **PUBLISHER:** **HarperCollins**
- **YEAR:** **1995**
- **GENRE:** **Non-Narrative Information**
- **SUMMARY:** *From the Let's-Read-and-Find-Out Science series, this book explains various types of mountains (folded, fault block, dome, and volcanic) and how movements in the lithosphere cause mountains to form.*

Picture-Perfect STEM Lessons, 3–5

Chapter 17

Time Needed

This lesson will take several class periods. Suggested scheduling is as follows:

Day 1: **Engage** with Mystery Map and *Solving the Puzzle Under the Sea* Read-Aloud

Day 2: **Explore** with "How to Map the Ocean Floor" Video and Sounding Boxes

Day 3: **Explain** with *How Mountains Are Made* Anticipation Guide and Read-Aloud and Looking for Patterns

Day 4: **Elaborate** with Google Earth Field Trip: Mountain Ranges of the World

Day 5: **Evaluate** with Comparing Maps

Materials

For Mystery Map (per pair or projected on a screen)

- Color copy of the "Heezen–Tharp World Ocean Map" (downloadable from the Library of Congress; see "Websites" section)

For Sounding Boxes

- Straw (1 per box)
- Permanent marker
- Ruler
- Sounding box (prepared in advance; 1 per group of 3–4 students)
- Plaster of Paris
- Shoebox
- Screwdriver
- Permanent marker

Instructions for Making a Sounding Box

1. Find an empty shoebox with a lid.
2. Mix a tub of plaster of Paris, following the instructions on the package.
3. Create a model of an ocean mountain range by sculpting the plaster of Paris high in the center of the box, perpendicular to the long edge of the box. (You can use cardboard to make forms to reduce the amount of plaster used.)
4. Make a ridge down the center of the mountain range by scooping out some of the plaster with one finger.
5. The lowest point of the box (ridge) should be no more than 15 cm from the top of the box.
6. Using a screwdriver, poke a hole in the center of the lid. Equally space a total of 10 holes down the

SAFETY

- Have students wear safety goggles, gloves, and aprons during this activity.
- Review the Safety Data Sheet information for plaster of Paris.
- Tell students to use caution when working with skewers, scissors, screw drivers, and other sharps to avoid puncturing skin and eyes.
- Have students wash their hands with soap and water after completing this activity.

center of the lid, parallel to the long edge of the box. Label the holes A–J.

7. Seal the box lids on with tape.

For Google Earth Field Trip: Mountain Ranges of the World

- Mountain Ranges Cards
- Tablets, laptops, computers, or whiteboard with Google Earth or Google Maps

INSIDE OF A SOUNDING BOX

OUTSIDE OF A SOUNDING BOX

For Looking for Patterns (per pair or projected on a screen)

- Color copy of the Seismicity of the Earth, 1960–1980, map (downloadable from the Library of Congress, see "Websites" section)
- Color copy of the Heezen–Tharp map (from the engage phase)

For Comparing Maps (per pair or projected on screen)

- Color copy of the Seismicity of the Earth map (from the explain phase)
- Color copy of the Heezen–Tharp map (from the engage and explain phases)

Student Pages

- Sounding Box Graph
- *How Mountains Are Made* Extended Anticipation Guide
- Mountain Ranges on Land and Under the Ocean
- Comparing Maps
- STEM at Home

Background for Teachers

Although five oceans are labeled on our world maps—the Atlantic Ocean, Pacific Ocean, Indian Ocean, Arctic Ocean, and Southern Ocean—all of these ocean regions actually flow into each another, forming one "world ocean." Until relatively recently, much of what lies beneath the surface of the ocean has been a mystery. Long ago, people attempted to measure the depth of the ocean by dropping ropes with weights attached. During World War I, sonar was developed to detect submarines. Since then, ships have been using sonar to measure the depth of the ocean floor. A *sonar device* sends a sound wave, or "ping," down through the water and records the time it takes for the sound wave to reach the ocean floor and echo back. Because scientists know how fast sound travels through water, they can calculate the exact depth of water in that spot. These measurements are called *soundings*.

In 1957, scientists Marie Tharp and Bruce Heezen (HAYZ-in) began using soundings to create a comprehensive map of the ocean floor. The soundings revealed great mountain ranges, deep trenches, and vast plains. Heezen and others went on many ocean trips to collect the sonar data. Tharp, who was not allowed on the research ships at the time (because she was a woman), used the information

Chapter 17

to create a series of maps that revealed the features of the seafloor. But these were not ordinary maps; these maps would be merged to create something quite extraordinary. The *New York Times Magazine* describes her discovery in this way:

> As details of the ocean floor emerged, Tharp noticed a fascinating feature. A well-known mountain range running down the Atlantic, known as the Mid-Atlantic Ridge, appeared as expected. But as Tharp's careful drafting made clear, there was also a valley that ran down through the middle of the mountain range. It was a hugely important geophysical feature; this "rift valley" marked a dynamic seam in the crust of the planet, the boundary of huge continent-size plates where new portions of crust rose from the interior of the earth to the surface like a conveyor belt and then, in a geological creep known as "drift," moved outward in both directions from the midocean ridge." (Hall 2006)

Tharp's discovery led to the theory of seafloor spreading. It also supported the theories of plate tectonics and continental drift, which scientists were debating at the time. Although more advanced sonar and satellite technologies have been developed to map the ocean floor, the Heezen–Tharp map, completed in 1977, was groundbreaking and remains remarkably accurate. The map is also known for its aesthetic beauty. It was painted by a famous Austrian landscape artist named Heinrich Berann, who captured the dramatic mountain ranges and other ocean floor features in exquisite detail. It hung in museums, libraries, schools, and even homes. Marie Tharp's work helped change the way people saw the world, revealing some of the mysteries of the sea and supporting the idea that our planet is composed of moving plates. Tharp, who died in 2006 at the age of 86, is considered one of the 20th century's most important scientists.

Solving the Puzzle Under the Sea provides a context in which students can explore a core idea about plate tectonics and large-scale system interactions. Students look closely at maps made by Tharp and learn how Earth's features, specifically mountain ranges, form in patterns because of Earth's moving plates. Students read about how different kinds of mountains are made, but this lesson does not evaluate the details and specific vocabulary pertaining to mountain formation. Instead, this lesson is about conveying the general idea that the patterns of Earth's features can be mapped and that those maps provide evidence that Earth is composed of moving plates. Students will later build on this idea in middle school as they learn more specifically about tectonic processes. This lesson also gives students the opportunity to explore disciplinary core idea ETS2.A: Interdependence of Science, Engineering, and Technology, as students learn how advancements in technology have provided more detailed views of the ocean floor. Students are involved in the science and engineering practice of analyzing and interpreting data as they use a model to get "soundings" of the ocean floor and graph the soundings to create a profile of the terrain. They are also engaged in the practice of obtaining, evaluating, and communicating information as they use a variety of resources—books, maps, and an app—to gain more information about mountains under the sea and on land.

engage

Mystery Map

Tell students that you have a mysterious-looking document to share! Give pairs of students a color copy of the Heezen–Tharp map titled, "World Ocean Floor," or project the map on a large screen. (See "Websites" section for a downloadable map from the Library of Congress.) Have them study the map carefully and make observations. *Ask*

? Does this map look like any map you have seen before?

- ? What do you notice about the map?
- ? What are you wondering about the map?
- ? How do you think the map was made?
- ? How old do you think the map might be?
- ? Who might have made the map?

Solving the Puzzle Under the Sea Read-Aloud

> **Connecting to the Common Core**
> **Reading: Informational Text**
> Key Ideas and Details: 3.1, 4.1, 5.1

Next, tell students that you have a book that will answer these questions and many more. Show students the cover of *Solving the Puzzle Under the Sea*, and introduce the author, Robert Burleigh, and illustrator, Raúl Colón. Read the book aloud, stopping to ask the questions provided in the next sections. (You may want to write the questions on sticky notes and place them in the book ahead of time as reminders.)

 Making Connections: Text to Self

After reading page 3, *ask*

- ? Do you like to look at maps?
- ? Do you ever use maps—printed or digital—to help you find your way?
- ? Have you ever made a map?

After reading page 6, *ask*

- ? Have you ever seen the ocean with your own eyes?
- ? What do you think the bottom of the ocean might look like?

 Questioning

After reading page 9, *ask*

- ? How do you think Marie felt when her first boss told her that "Having a woman on a ship is bad luck?" (confused, angry, upset)

After reading page 11, *ask*

- ? How do you think a map could be made of the ocean floor? (Answers will vary.)

After reading page 19, *ask*

- ? How was Marie's work like "piecing together an immense jigsaw puzzle"? (The soundings being measured were each like puzzle pieces. She was putting them together to get a full picture of the ocean floor.)
- ? How did she feel about this work? (She found it thrilling.)

After reading page 21, *ask*

- ? Why did she add colors to the map? (to show depths)

After reading page 25, *ask*

- ? How did Marie and Bruce's map support the new scientific theory of plate tectonics or continental drift? (They found a deep trench at the center of the Atlantic Ocean's mountain chain, which they thought might be a boundary between two plates.)

> **Connecting to the Common Core**
> **Reading: Informational Text**
> Integration of Knowledge and Ideas: 3.7, 4.7, 5.7

Next, have students refer to the Heezen–Tharp map, and *ask*

- ? The book says that the map appeared in museums, school, and even on the walls of many homes. Why do you think people hung this map in their homes? (It is beautiful and represented exciting, new findings about Earth.)
- ? What features do you see on the map? (land, sea, mountain ridges clearly defined in the ocean and on land, and so on)
- ? Can you find the north–south rift in the center of the Atlantic Ocean mentioned in the book? Why was this rift an important discovery? (It supported the new theory of plate tectonics.)

Share the following quote from Marie Tharp: "The whole world was spread out before me. I had a blank canvas to fill with extraordinary possibilities, a fascinating jigsaw puzzle to piece together: mapping the world's vast hidden seafloor. It was a once-in-a-lifetime—a once-in-the-history-of-the-world—opportunity for anyone, but especially for a woman [at that time]." *Ask*

? Why did Marie view mapping the ocean floor as such an amazing opportunity for a woman? (At the time, very few women were scientists. She was confronted with opposition because she was a woman.)

? How long do you think it took to create the map? (Answers will vary.)

Explain that Marie and her research partner, Bruce Heezen, started on this project in 1957 and finished in 1977. It was a 20-year long undertaking! Heezen and others collected the sonar data on many ocean trips, and Marie used the information to create a series of maps that revealed the features of the seafloor. *Ask*

? Marie had college degrees in geology, mathematics, music, and English. How did these subjects come together in her map-making? (Geology is the study of Earth's structure and processes. Her knowledge of geology helped her understand the features of the ocean floor. Mathematics helped her understand and represent the sonar measurements. Music may have helped her understand sound. English made her a good communicator as she worked in a team and shared her discoveries.)

? What were the sonar measurements called? (soundings)

? How does sonar work? (Answers will vary.)

explore

"How to Map the Ocean Floor" Video

> **Connecting to the Common Core**
> **Reading: Informational Text**
> INTEGRATION OF KNOWLEDGE AND IDEAS: 3.7, 4.7, 5.7

Tell students that you are going to explore how sonar works. Have students make observations as they watch a short video on sonar. Show the 43 sec. animated video "How to Map the Ocean Floor." (You may want to show it more than once).

 Turn and Talk

After watching the video, have students share their observations with a partner and describe how sonar works in their own words. Then, *ask*

? How does sonar work? (An instrument on a ship emits a sound wave and "listens" for how long the sound wave takes to return. Multibeam sonar systems pick up these echoes from various locations.)

Explain that because scientists know how fast sound travels through water, they can calculate the exact depth of water in that place. Today, computers can stitch all this information together to give us detailed maps of the ocean floor.

Sounding Boxes

Ahead of time, prepare one sounding box for each team of three to four students. (See instructions in the "Materials" section.)

> **Connecting to the Common Core**
> **Mathematics**
> MEASUREMENT AND DATA: 4.MD.4

Tell students that they are going to use a model to explore how sonar helped Marie get a picture of

MAKING A SONAR STICK

MAKING A SOUNDING BOX GRAPH

what the seafloor looked like. Show them one of the sounding boxes you have prepared in advance. Tell them that inside of the sealed box is a model of a feature of the ocean floor. They are not going to be able to look at it directly; instead, they will need to take measurements, graph the data, and then use the graph to infer what the inside of the box looks like. They will work with a team of three to four students.

First, students should create a measuring device. Have each team create a "sonar stick" out of a straw by placing a ruler next to the straw and marking a line for each centimeter with a permanent marker. They do not need to write the numbers on it. If using bendable straws, be sure they start making their marks on the end of the straw that doesn't bend.

Remind students that the measurements taken by sonar are called *soundings*, so the boxes will be referred to as *sounding boxes*. Explain to students that they will use the sonar stick to determine the shape of the terrain by placing the sonar stick in each hole as far as it will go and measuring the depth at each location. Model how to place the sonar stick in one of the holes on a sounding box and hold your fingers where the stick meets the top of the box. Pull the stick out and count the number of centimeters deep that hole was. Tell students that they will be recording that number on a graph.

Next, distribute a copy of the Sounding Box Graph student page to each student, and give each team a sounding box. Explain that the *y*-axis of this graph is a little unusual—the zero point is at the top of the graph instead of the bottom. The graph looks this way because their model will show how sonar soundings are measured in terms of distance from the surface of the water to the bottom of the ocean floor. Have students begin with the first hole and measure its depth using the sounding stick. Then, have them plot the measurement on their graph. Students should repeat the procedure until they are finished with the graph.

After all groups have completed the activity, have students take a look at their graphs. *Ask*

What do you think the bottom of your sounding box looks like? (a mountain with a crack down the middle or two mountains) Explain that the graph they made is a profile of the terrain. A *profile* means a side-view. Revisit page 16 of *Solving*

Chapter 17

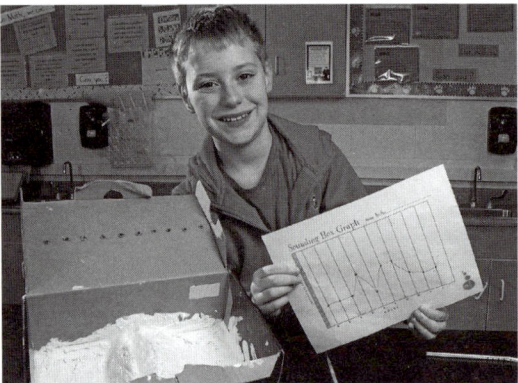

COMPARING THE GRAPH TO THE MODEL

the Puzzle Under the Sea, which shows a profile of the seafloor like the ones Marie Tharp created. *Ask*

? How does this profile compare with the one you made? (It has a peak rising upward and some drop-offs.)

? What can profiles tell us? (the depths and shapes of the ocean floor)

? What *can't* profiles tell us? (what lives there, what things are made of, the color of things, temperature, etc.)

? How is the activity we did like the "sounding" that Marie's team did? (We took measurements without ever seeing what we were measuring.)

? How is it different? (Sonar devices use sound waves, but we used a straw marked in centimeters.)

Next, allow students to open up the boxes to reveal what is inside. *Ask*

? How does your Sounding Box Graph compare with the terrain inside the box? (They are similar.)

? What ocean feature from the book do you think this model represents? (Mid-Atlantic Ridge)

? How do mountains form in the ocean? (Answers will vary.)

? How do mountains form on land? (Answers will vary.)

? Do mountain locations on Earth have a pattern? (Answers will vary.)

explain

How Mountains Are Made Anticipation Guide and Read-Aloud

Connecting to the Common Core
Reading: Informational Text
KEY IDEAS AND DETAILS: 3.1, 4.1, 5.1

 ## Anticipation Guide

Show students the cover of *How Mountains Are Made* and introduce the author, Kathleen Weidner Zoehfeld and illustrator, James Graham Hale. Have students complete the "Before Reading: Agree or Disagree" side of the *How Mountains Are Made* Extended Anticipation Guide. Tell them that you will be reading the book to find the answers to the questions on the anticipation guide.

Read the book aloud. As you read, have students signal (raise their hands) when they hear evidence for or against each statement on the anticipation guide. After you read, they can fill in the "After Reading" column and for each statement fill in the "Evidence from the Text" section of the anticipation guide with evidence that supports each answer. When students finish, go over each question and ask students to share their answers. Answers are as follows:

1. Agree
2. Agree
3. Disagree
4. Disagree
5. Agree

Explanations from the reading are as follows:

1. Fossil seashells have been found on Mt. Everest (p. 10).
2. Mt. Everest was once a flat plain under an ocean (p. 11).

3. Earth's outer shell (lithosphere) is broken into plates that float on top of partially melted rock (pp. 14–15).
4. Some mountains form by volcanic eruptions, but some form in other ways (pp. 16–23)
5. Mt. Everest is growing as much as 2 in. per year. (p. 29)

Looking for Patterns

> Connecting to the Common Core
> **Reading: Informational Text**
> INTEGRATION OF KNOWLEDGE AND IDEAS: 3.7, 4.7, 5.7

Next, turn back to the map on page 14, which shows the eight major tectonic plates. Hold that page side by side with the Heezen–Tharp map so students can compare the two maps. *Ask*

? What similar patterns do you notice between these two maps? (Most of the ocean mountain ranges on the Heezen–Tharp map are along the plate boundaries on the map in the book.)

? How is this similarity described in *How Mountains Are Made*? ("In a few places on earth, plates are moving away from one another. This usually happens under the oceans. Magma pushes up through the space between the plates. When it hits the cool ocean water, the magma becomes solid. Great underwater mountain ranges are built up," p. 21; refer to the diagrams as you read.)

? How did the Heezen–Tharp map help support the new theory of plate tectonics? (It showed a deep, narrow valley that divided the Mid-Atlantic into two parts, which supported the theory that Earth's surface is divided into moving plates.)

Explain that in 1982, Marie Tharp released another map that showed seismic (earthquake) activity. She used red dots to represent major seismic events (earthquakes) that occurred between 1960 and 1980. Give each pair of students a color copy of the Seismicity of the Earth, 1960–1980, map or project the map (see "Websites" section). *Ask*

COMPARING MAPS

? Compare this map with the Heezen–Tharp map. What pattern do you notice about the seismic, or earthquake, activity on this map? (A lot of seismic activity occurred on or near the ocean mountain ranges.)

? Compare the Seismicity of the Earth map with the map on page 14 of *How Mountains Are Made*. What patterns do you notice? (Most of the seismic [earthquake] activity takes place on the edges of the plates.)

? Why do you think Marie Tharp made this second map? (She wanted to see if there was a relationship between seismic [earthquake] activity and the ocean features she mapped years earlier.)

? How does this map support the idea of plate tectonics? (It supports the idea that Earth is made of moving plates and that these moving plates cause changes in Earth's surface.)

elaborate

Google Earth Field Trip: Mountain Ranges of the World

Note: Depending on the availability of technology, the following activity could be done as a whole group, with the application projected on a screen, or students can work in pairs using computers, laptops, or tablets.

Tell students that they are going on a field trip! This is not a real field trip; it is a *virtual* field trip to explore some of Earth's mountain ranges using the computer application, or app, Google Earth. If students have not had previous experience with this app, you can show them the Google Earth tutorial on searching for places (see "Websites" section). The search feature of Google Earth allows students to enter the name of a place and "fly" there. Point out that sometimes the red place marker that shows up is not located exactly on the mountain range they have entered in the search box, so they will need to zoom to get a better view. Be sure students know how to use a "two-finger pinch" to zoom in and zoom out once they arrive at a place on the globe. Tell students that the images on the Google Earth app are made with satellite cameras that orbit Earth, airplanes that fly above Earth, and even cars that take photographs when being driven. Google takes all of these photographs and puts them together so that we can fly around our planet and see what different places actually look like.

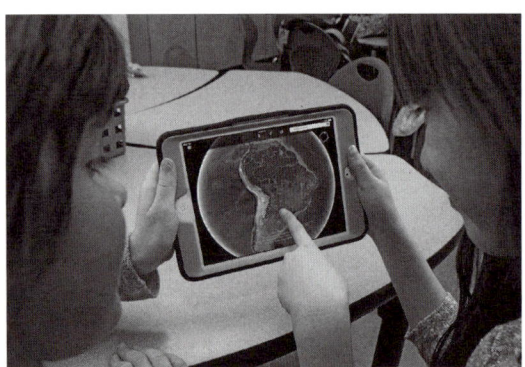

Using Google Earth

Give each student a copy of the Mountain Ranges Cards student page and the Mountain Ranges on Land and Under the Ocean student page. In pairs or as a whole class, students should use Google Earth (or Google Maps) to fly to all of the different mountain ranges listed on the student page, labeling the locations by cutting out the cards and gluing or taping them on the map. Explain that the cards list only some of the world's mountain ranges. Answers are as follows:

1. Appalachian Mountains
2. Alps
3. Ural Mountains
4. Rocky Mountains
5. Mid-Atlantic Ridge
6. Himalayas
7. East Pacific Rise
8. Andes
9. Pacific-Antarctic Ridge
10. Great Dividing Range

evaluate

Comparing Maps

 **Synthesizing**

Connecting to the Common Core
Writing: Text Types and Purposes
INTEGRATION OF KNOWLEDGE AND IDEAS: 3.2, 4.2, 5.2

Give students a copy of the Comparing Maps student page, and provide them with color copies of (or project) both the Heezen–Tharp map and the Seismicity of the Earth map. Remind students that *seismicity* means earthquake activity.

Based on the two maps, have students answer the questions on the student page. Discuss the answers together.

1. Study the Heezen–Tharp World Ocean Floor map. What have scientists learned from this map? (Scientists have learned that mountains and rifts are on the ocean floor.)
2. Study Marie Tharp's Seismicity of the Earth, 1960–1980, map. What have scientists learned from this map? (Scientists learned that the locations where most earthquakes occur on Earth form a pattern.).
3. Compare the two maps (Heezen–Tharp World Ocean Floor and Seismicity of the Earth, 1960–1980). What have scientists learned from combining the information from these two maps? (A lot of seismic activity [earthquakes] occurs along the mountain ranges on the ocean floor.)

4. How do these two maps support the idea that Earth is made of moving plates? (The mountain ranges and earthquakes occur in a pattern that shows the edges of Earth's plates.)

Read the following quote from Mike Purdy, the director of the Lamont-Doherty Earth Observatory, where Marie Tharp made her maps.

The significance of Tharp's achievement and the importance of the maps cannot be overstated. She was a pioneer in her science, playing a crucial role in the early days of sea-floor spreading research, and she was a pioneer in her profession, succeeding as a woman in a field dominated for decades by men. But most of all she was a wonderful person, a great colleague and a happy friend to so many of us.

5. A *pioneer* is a person who is the first to discover or settle a new country or area. Why was Marie Tharp referred to as a *pioneer*? (She was a woman leading in a field that had been led by men. She was the first person to map the ocean floor.)

STEM at Home

Have students complete the "I learned that …" and "My favorite part of the lesson was …" portions of the STEM at Home student page as a reflection on their learning. They may choose to do the following at-home activity with an adult helper and share their results with the class. If students do not have access to the internet or these materials at home, you may choose to have them complete this activity at school.

"At home, we can watch a video about the Okeanos Explorer, a NOAA ship dedicated to exploring the ocean floor."

Search "Okeanos Explorer Exploration Video" on your web browser to find the video at http://oceanexplorer.noaa.gov/okeanos/explorations/ex1605/dailyupdates/media/video/0510-exploration/0510-exploration.html.

"What is the difference between *research* and *exploration*? [*Answer:* The purpose of research is to answer a question. The purpose of exploration is to generate questions.]

Okeanos explores the ocean floor with ROVs (remotely operated vehicles). We can watch videos taken by the ROVs that reveal parts of the ocean floor that had never been seen before!"

Search "Okeanos Photo and Video Log" on your web browser to find the videos at http://oceanexplorer.noaa.gov/okeanos/explorations/ex1402/logs/photolog/welcome.html#clips.

"Choose a few dive highlight videos to watch. What interesting things did you see in the videos?"

For Further Exploration

This section is provided to help you encourage your students to use the science and engineering practices in a more student-directed format. This box lists questions and challenges related to the lesson that students may select to research, investigate, or innovate. Students may also use the questions as examples to help them generate their own questions. After selecting one of the questions in the box or formulating their own questions, students can individually or collaboratively make predictions, design investigations or surveys to test their predictions, collect evidence, devise explanations, design solutions, or examine related resources. They can communicate their findings through a science notebook, at a poster session or gallery walk, or by producing a media project.

Chapter 17

Research
Have students brainstorm researchable questions:

- ? What is the longest submarine (undersea) mountain range? What is the longest above-ground mountain range?
- ? How is sonar similar to echolocation? What kinds of animals use echolocation?
- ? How is the height of a mountain on land measured today? How was it measured in the past?

Investigate
Have students brainstorm testable questions to be solved through science or math:

- ? Can you use a bar graph to compare the heights of the world's tallest mountains (both on land and on the ocean floor)?
- ? Can you calculate the difference between the speed of sound through air and through water?
- ? How does the height of Mt. Everest compare with the tallest building in the world?

Innovate
Have students brainstorm problems to be solved through engineering:

- ? Can you design a map showing your route to school?
- ? Can you design your own sounding box with multiple ocean floor features and have a friend see if she or he can map the terrain without looking?
- ? Can you design models of folded, fault-block, dome, and volcanic mountains to teach others how different types of mountains are formed?

Reference
Hall, S. *New York Times Magazine*. 2006. The Contrary Map Maker. December 31.

Websites
Google Earth
www.google.com/earth

Google Earth Tutorial: Searching for Places
www.google.com/earth/learn/beginner.html#tab=-searching-for-places

Heezen–Tharp World Ocean Floor (map)
www.loc.gov/resource/g9096c.ct003148

"How to Map the Ocean Floor" (video)
www.youtube.com/watch?v=bADFB199Klc

Seismicity of the Earth, 1960–1980 (map)
www.loc.gov/item/85694236/?loclr=blogtea

More Books to Read
Collard, S. B. 2003. *The deep-sea floor*. Watertown, MA: Charlesbridge.
Summary: This book reveals the animal life that lives in the deepest parts of the ocean and examines the technology that allows scientists to conduct research in areas characterized by trenches, vents, and seeps.

Jenkins, S. 2009. *Down, down, down: Journey to the bottom of the sea*. Boston: Houghton Mifflin Books for Children.
Summary: Lively illustrations take the reader down, down, down from the surface to the bottom of the sea.

Locker, T. 2001. *Mountain dance*. New York: Harcourt.
Summary: This book poetically describes various kinds of mountains and how they are formed.

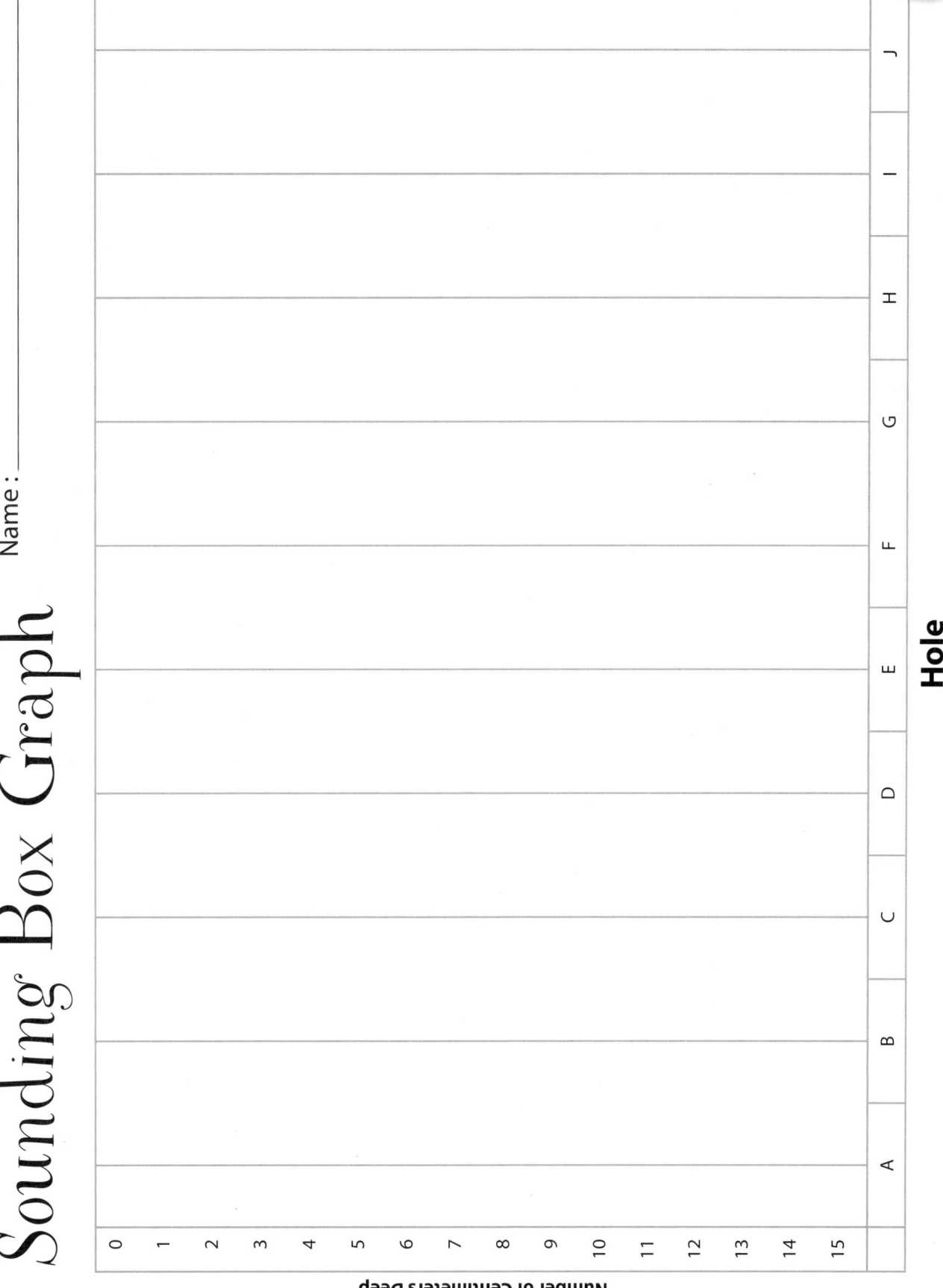

Name: _____

How Mountains Are Made

Extended Anticipation Guide

Before Reading　　　　　　　　　　　　　　　　　　　　**After Reading**
Agree or Disagree　　　　　　　　　　　　　　　　　　　　Agree or Disagree

_____　　1. Fossil seashells have been found on Mt. Everest.　　_____

_____　　2. Over time, flat plains can become mountains.　　_____

_____　　3. The Earth is solid rock all the way to the center.　　_____

_____　　4. All mountains are formed by volcanic eruptions.　　_____

_____　　5. Mt. Everest is getting taller.　　_____

Evidence from the text:

1. _____
   _____

2. _____
   _____

3. _____
   _____

4. _____
   _____

5. _____
   _____

Mountain Ranges Cards

Cut out the names of the mountain ranges below. Some are located on land, and some are under the ocean. Use Google Earth or Google Maps to locate each mountain range. Then, tape or glue each name in the correct location on the map. Note: These are just some of Earth's mountain ranges; there are many more.

Mountain Ranges Cards

Appalachian Mountains	Alps
Andes	Himalayas
Rocky Mountains	Ural Mountains
Great Dividing Range	East Pacific Rise
Mid-Atlantic Ridge	Pacific-Antarctic Ridge

Mountain Ranges Cards

Appalachian Mountains	Alps
Andes	Himalayas
Rocky Mountains	Ural Mountains
Great Dividing Range	East Pacific Rise
Mid-Atlantic Ridge	Pacific-Antarctic Ridge

Picture-Perfect STEM Lessons, 3–5

Chapter 17

Name: _____

Mountain Ranges on Land and Under the Ocean

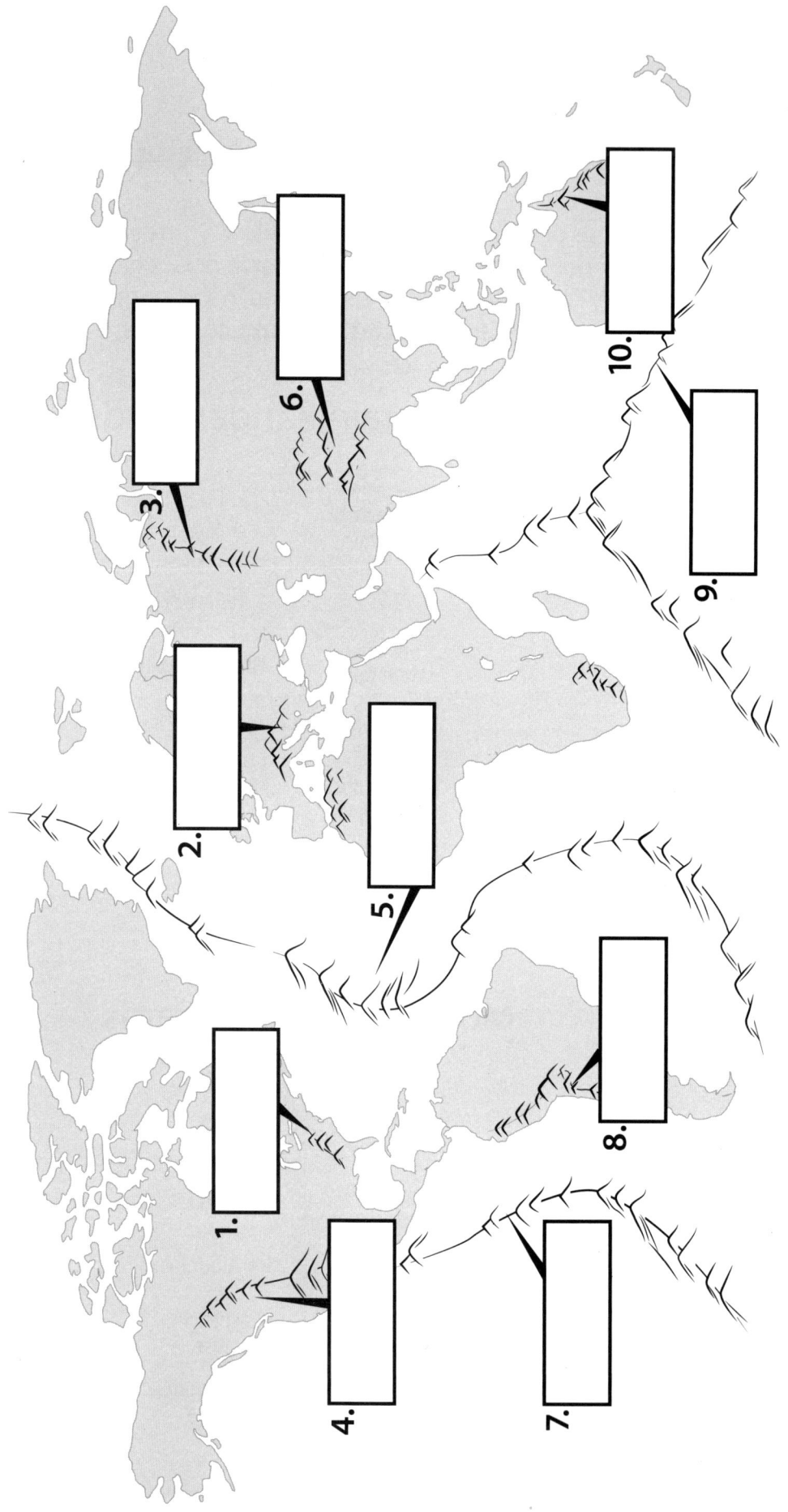

304 National Science Teachers Association

Name : _____

Comparing Maps

1. Study the Heezen–Tharp World Ocean Floor map. What have scientists learned from this map?

   _____
   _____

2. Study Marie Tharp's Seismicity of the Earth, 1960–1980, map. What have scientists learned from this map?

   _____
   _____

3. Compare the two maps (Heezen–Tharp World Ocean Floor and Seismicity of the Earth, 1960–1980, map). What have scientists learned from combining the information from these two maps?

   _____
   _____

4. How do these two maps support the idea that Earth is made of moving plates?

   _____
   _____

Read the following quote from Mike Purdy, the director of the Lamont-Doherty Earth Observatory, where Marie Tharp made her maps.

> *The significance of Tharp's achievement and the importance of the maps cannot be overstated. She was a pioneer in her science, playing a crucial role in the early days of seafloor spreading research, and she was a pioneer in her profession, succeeding as a woman in a field dominated for decades by men. But most of all she was a wonderful person, a great colleague and a happy friend to so many of us.*

5. A *pioneer* is a person who is the first to discover or settle a new country or area. Why was Marie Tharp referred to as a *pioneer*?

   _____
   _____

Picture-Perfect STEM Lessons, 3–5

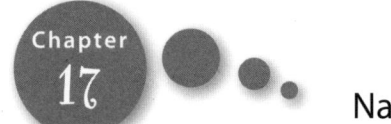

Name : _____

STEM at Home

Dear _____,

At school, we have been learning about scientist **Marie Tharp,** who worked on the first map of the ocean floor and discovered great mountain ranges under the sea.

I learned that: _____

_____

My favorite part of the lesson was: _____

_____

_____

At home, we can watch a video about the Okeanos Explorer, a NOAA ship dedicated to exploring the ocean floor.

Search "Okeanos Explorer Exploration Video in your web browser to find the video at *http://oceanexplorer.noaa.gov/okeanos/explorations/ex1605/dailyupdates/media/video/0510-exploration/0510-exploration.html.*

What is the difference between *research* and *exploration*? _____

_____

Okeanos explores the ocean floor with ROVs (remotely operated vehicles). We can watch videos taken by the ROVs that reveal parts of the ocean floor that had never been seen before!

Search "Okeanos Photo and Video Log" to find the video at ***http://oceanexplorer.noaa.gov/okeanos/explorations/ex1402/logs/photolog/welcome.html#clips.***

Choose a few dive highlight videos to watch. What interesting things did you see in the videos? _____

_____

_____

306

National Science Teachers Association

Chapter 18

Space Exploration

Description

By exploring the history of various scientific discoveries about space, students will recognize a fundamental understanding of the nature of science—scientific knowledge is open to revision as new evidence is discovered. They will also learn about the criteria and constraints involved in developing space exploration technologies by playing a board game.

Suggested Grade Levels: 3–5

LESSON OBJECTIVES Connecting to the *Framework*		
Science and Engineering Practices	**Disciplinary Core Ideas**	**Crosscutting Concept**
Engaging in Argument From Evidence Obtaining, Evaluating, and Communicating Information	**ESS1.B:** Earth and the Solar System **ETS1.A:** Defining and Delimiting Engineering Problems **ETS2.A:** Interdependence of Science, Engineering, and Technology **ETS2.B:** Influence of Engineering, Technology, and Science on Society and the Natural World	Systems and System Models

Featured Picture Books

TITLE: **Boy, Were We Wrong About the Solar System!**
AUTHOR: **Kathleen V. Kudlinski**
ILLUSTRATOR: **John Rocco**
PUBLISHER: **Dutton Books for Young Readers**
YEAR: **2008**
GENRE: **Narrative Information**
SUMMARY: *This book is an entertaining and informative look at how scientific theories change as new evidence is discovered—from the first humans wondering about the night sky to the demotion of Pluto to dwarf planet status.*

TITLE: *Space Exploration*
AUTHOR: **Christine Dugan**
PUBLISHER: **Teacher Created Materials**
YEAR: **2012**
GENRE: **Non-Narrative Information**
SUMMARY: *From the TIME for Kids series, this book describes the history of space exploration, the technologies that have made it possible, and what the future might hold.*

Chapter 18

Time Needed

This lesson will take several class periods. Suggested scheduling is as follows:

Day 1: Engage with "The Known Universe" Video, **Explore** with Eyes on the Solar System

Day 2: Explain with *Boy, Were We Wrong About the Solar System!* Read-Aloud

Day 3: Elaborate with *Space Exploration* Read-Aloud and Space Race Game

Day 4 and beyond: Evaluate with Mission to Space and "The Known Universe" Conclusion

Materials

For Eyes on the Solar System

- Computers with NASA's Eyes on the Solar System app downloaded (*Note:* This app is available only on laptop or desktop computers, not handheld devices.)

For Boy, Were We Wrong About the Solar System! *Read-Aloud*

- The Known Universe: Then and Now Cards (1 set, pre-cut, per group of 2 students)

For Space Race Game (per group of 3–4 students)

- Space Race game board (Tape together pages 320 and 323 and laminate them. A larger, full-color version of the board is available on this book's Extras page at *www.nsta.org/PicturePerfectSTEM3-5*.)
- Die
- 4 different colors of counters or centimeter cubes to use as game markers
- Space Race Game Instructions
- Space Race Game Cards (printed on cardstock and pre-cut)

For Mission to Space (per student)

- Poster board, tape, and markers or a computer or tablet with presentation software

Student Pages

- Eyes on the Solar System
- Mission to Space Info Sheet
- Mission to Space Presentation Rubric
- STEM at Home

Background for Teachers

Throughout history, people have made many fantastic discoveries about the universe. Some of these discoveries stand the test of time, whereas others are replaced by new ideas. But there is always so much more to find out! As new technologies are developed, scientists are able to test and challenge previous findings. For students, understanding space science is not just learning *what* we know about the universe; it is also about learning *how* we know what we know. This lesson focuses on the nature of science, particularly, how our understandings about the universe are based on evidence and how those understandings change as we develop technologies that allow us to gather new evidence.

The lesson opens and closes with a 6 min. video created by the American Museum of Natural History called "The Known Universe." The video begins on Earth and zooms out to show our solar system, galaxy, and many other galaxies—all of the parts of our universe that humans have observed so far. Two of the most astonishing ideas that resonate in this video journey are (1) the enormity of what we have been able to observe and learn about the universe with our advances in technology and (2) the magnitude of how much we still do not know and have not yet observed about the universe.

One of the basic understandings that students should have about the nature of science is the notion that scientific knowledge is open to revision in light of new evidence. Students will explore this understanding through a picture book that explains how our ideas about the solar system change as technologies advance and new evidence surfaces. They also encounter the crosscutting concept of systems and system models as they learn how our models of the solar system have changed over time. Students see several examples of how the science and engineering practice of engaging in argument from evidence plays a critical role in developing new understandings about the universe.

For the technology component of this lesson, students learn how science and technology influence and support each other. For example, they learn that advances in space exploration technologies can lead to scientific discoveries, and in turn scientific discoveries can lead to the development of new technologies. Students also explore disciplinary core idea ETS1.A: Defining and Delimiting Engineering Problems as they learn about the criteria and constraints from four famous space exploration projects: Project Apollo, the Space Shuttle Program, the Hubble Space Telescope, and the International Space Station. Finally, students apply their knowledge of criteria and constraints to their research on other current and future space missions, and they share their findings with others in a "Space Expo."

engage

"The Known Universe" Video

Tell students that you have a 6 min. video from the American Museum of Natural History for them to watch called "The Known Universe" (see "Websites" section). Have them sit back, relax, and take an amazing journey through the known universe as they watch the video!

 Questioning

After watching the video, *ask*

? How do we know what we know about the universe? (Answers will vary. Also write this question on the board or on a poster that can be displayed throughout the lesson)

Discuss students' initial ideas, and tell them that you would like them to think about this question as they learn about space exploration over the next several days.

explore

Eyes on the Solar System

In advance, open NASA's Eyes on the Solar System app (see "Websites" section) and download the app onto student computers. This app is available only on laptop or desktop computers, not handheld devices. You may want to view some of the Eyes on the Solar System tutorials before beginning (see "Websites" section) to familiarize yourself with the app.

Tell students that you are going to give them an opportunity to explore what is known about our

Chapter 18

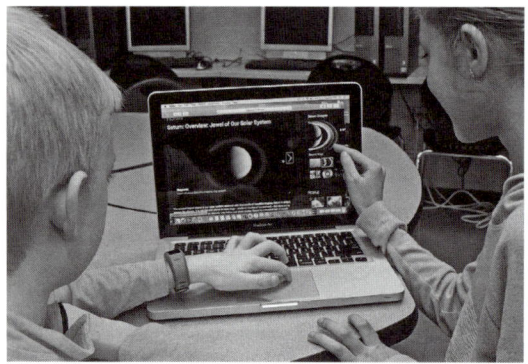

EXPLORING WITH NASA'S EYES ON THE SOLAR SYSTEM APP

solar system with a computer application called Eyes on the Solar System. Give each student a copy of the Eyes on the Solar System student page, and then go through the following steps together:

1. Open the Eyes on the Solar System app by clicking the SIMPLE button on the NASA's Eyes homepage.
2. Listen to the brief introduction, and observe some of the simulations.
3. Click on the DESTINATIONS tab.
4. Click on the picture of Earth. A sidebar will appear.
5. Under the INFO tab, read the first paragraph, which gives a brief description of our planet.
6. Click on the STATS tab to find out who discovered Earth (*Answer:* Known by the ancients).
7. Click on the HOME icon at the top right corner of the screen to get back to the simulation of the solar system.
8. From there, use the DESTINATIONS tab to find pictures of the Sun, the eight planets, Pluto, asteroids, and comets.

Tell students that the celestial bodies they can explore on the app represent a very small portion of the objects in our solar system. Countless objects—asteroids, comets, and so on—orbit our Sun. Any object orbiting the Sun is considered part of our solar system. Students can work in pairs or teams to explore the app. They should choose five more celestial bodies to research, in addition to Earth.

Have them use the Eyes on the Solar System student page to record their findings.

 Questioning

Connecting to the Common Core
Reading: Informational Text
KEY IDEAS AND DETAILS: 3.1, 4.1, 5.1

After students have completed the Eyes on the Solar System student page, *ask*

? Which objects in the solar system were "known by the ancients?" (Students may have noticed that the Sun and the first six planets were "known by the ancients," and the dates of their discovery are unknown.)

? Why do you think those bodies were discovered by the ancients? (They are large, relatively close, and can be seen with the naked eye.)

? Why do you think the other celestial bodies on the app were not known until later? (They are too far away, making them too small to see with the naked eye.)

? What are some of the interesting facts you recorded about the objects in our solar system? (Answers will vary.)

? How do we know all of these things about our solar system? (Answers will vary.)

? Where is the farthest that people have actually traveled in our solar system? (the Moon)

? Do you think scientists know all there is to know about our solar system? (Answers will vary.)

? Do you think scientists have ever made wrong guesses in their quest to understand our solar system? (Answers will vary.)

explain

Boy, Were We Wrong About the Solar System! Read-Aloud

Tell students that author Kathleen V. Kudlinski and illustrator John Rocco have made a fascinat-

THEN AND NOW CARD SORT

ing book that looks at the mistakes, mishaps, and creativity that are part of scientific discovery. Show students the cover of *Boy, Were We Wrong About the Solar System!* Read the first two pages aloud, ending with "it took a long time and a lot of wrong guesses to learn what we now know today."

 Card Sort

Before reading the rest of the book, give each pair of students a pre-cut set of The Known Universe Then and Now Cards. Tell them that some of the statements on the cards represent what people used to think about the universe, and other cards represent what people now understand about the universe. Have pairs read each statement and sort the cards into two columns: what people thought "then" and what people understand "now."

> **Connecting to the Common Core**
> **Reading: Informational Text**
> CRAFT AND STRUCTURE: 3.1, 4.1, 5.1

After students have had a chance to sort the cards, tell them that you are going to read the rest of the book and as you read, you would like them to signal when they hear one of the ideas from the cards. Then, read the book aloud, stopping to discuss each idea from the cards. Make sure students place the corresponding card in the correct column, with the old idea directly across from the new idea. The correct answers are shown in Table 18.1.

Table 18.1. Answers for Then and Now Card Sort

Then	Now
Earth is flat.	Earth is spherical.
Earth is the center of the solar system.	The Sun is the center of the solar system.
The Sun and Moon are perfect, smooth spheres.	The Sun has spots, and the Moon has mountains and craters.
Martians dug canals on Mars.	There are no canals on Mars.
Pluto is considered a planet.	Pluto is not considered a planet.

Students might think some of the statements in the "Then" column were silly for anyone to believe, such as the Earth being the center of the solar system. Have students think about the technologies people had at the time and what daily, monthly, and yearly patterns people must have observed that made them think that way. For example, people would observe the Sun, Moon, and stars "rising" and "setting" each day, so you can see why it made sense to them that the Sun, Moon, and planets were circling Earth. But for one scientist, the idea of an Earth-centered system did not make sense. This astronomer, named Nicolaus Copernicus, had a new idea to explain the apparent motions of the planets, Moon, and Sun. His idea was that the Sun was the center of the system. *Ask*

? Do you remember from the book how people reacted when Copernicus shared his new and controversial idea? (p. 10: They laughed at him.)

? Why did they laugh? (p. 10: Because he had no proof.)

? When did the proof come? (pp. 10–11: When telescopes were invented many years later, people started to see for the first time that Venus appears to change shape just like the Moon, which means that sunlight is shining on Venus from all different angles. That means that Venus has to be circling the Sun. If Venus is circling the Sun, then the other planets must be circling the Sun, too.)

Point out how the scientists in this book had to make scientific *arguments* based on the *evidence* they observed to convince others to accept a new idea or model. Convincing people of a new idea is often very difficult. Much of the evidence came as new technologies were invented. Explain that a *technology* is something made by humans to solve a problem. The telescope is an example of a technology. *Ask*

? What problem did the invention of the telescope solve? (The problem of not being able to get a close look at objects in the sky.)

Explain that an interesting relationship exists between science and technology. New technologies are used to answer scientific questions, and scientific discoveries lead to the development of new technologies. In other words, new technologies are created as science advances, and those new technologies lead to new scientific discoveries. For example, the scientific understanding of how curved glass can magnify light led to the invention of the telescope, which led to more discoveries about the universe. *Ask*

? How has the telescope changed over the years? (Telescopes have gotten bigger and more powerful, and we now have a huge telescope orbiting Earth that can see farther than we've ever seen before—the Hubble Space Telescope.)

Show the illustration of the Hubble Space Telescope on page 28. You may want to visit NASA's Hubblesite to show students some pictures that the Hubble Space Telescope took (see "Websites" section). Then, *ask*

? Do you think there will be more advances in the technology of telescopes? (p. 30: Yes, scientists and engineers keep inventing better instruments.)

Next, look back at the then and now card sort and discuss the technological advances and observations that led people to change each idea about the solar system. For example,

- Early telescopes allowed people to observe that the Sun and Moon are not perfect, smooth spheres and that the Earth is not the center of the solar system.

- Better telescopes and photos taken by Mars rovers revealed that there are no canals on Mars, but rather various surface features that were once mistaken for canals.

- Even better telescopes revealed that many more tiny bodies beyond Neptune orbit the Sun, which led to declassifying Pluto as a planet.

Explain that with astronomy, or any other area of science, as we make more observations, our ideas and models can change. This is the *tentative nature of science*. Conclusions in science are not always final, but that doesn't necessarily mean they are wrong. It means that conclusions can be modified or replaced if new evidence becomes available.

elaborate

Space Exploration Read-Aloud

 Making Connections: Text to World

Reread page 30 of *Boy, Were We Wrong About the Solar System!*, which says, "Our ideas will change as we learn more. Scientists keep inventing better instruments. Every year more advanced probes are sent through the solar system." *Ask*

? What new space technologies have you heard about or seen on the news? (Answers will vary.)

 Questioning

Show students the cover of the book *Space Exploration* and introduce the author, Christine Dugan. Read aloud pages 4–9, which are about early space exploration, NASA, and President John F. Kennedy's challenge to the nation to land a person on the Moon.

Connecting to the Common Core
Reading: Informational Text
KEY IDEAS AND DETAILS: 3.1, 4.1, 5.1

Then, *ask*

- **?** What country launched the first artificial satellite into space? (The Soviet Union)
- **?** What was the name of the satellite? (*Sputnik*)
- **?** In what year was it launched? (1957)

Explain that the "space race" between the United States and the Soviet Union began in 1957 with *Sputnik*. *Sputnik's* launch came as a surprise—and not a pleasant one—to most Americans. People were worried about the Soviets. They saw the Soviet Union as a threat to the United States. In the United States, space was seen as the next frontier, an extension of the great American tradition of exploration. Americans did not want to lose the space race to their rivals. *Ask*

- **?** What agency did the United States form in 1958 in response to Sputnik? (NASA)
- **?** What does NASA stand for? (National Aeronautics and Space Administration)
- **?** What challenge did President Kennedy give Americans in his famous 1961 speech? (He asked the nation to land a person on the Moon by the end of the decade.)
- **?** Did the nation meet the challenge? (Answers will vary.)

Space Race Game

Tell students that you will read the rest of the book later to find out what happened next in space exploration. Then, tell them they are going to take part in a "space race" in the form of a board game.

This game is a fun way to learn about four of the greatest space exploration programs the world has ever known. The object of the game is to be the first player to reach "Mission Accomplished!" at the top of the board. Place students into groups of three or four (four is optimal). To each group of players, distribute a Space Race game board, a die, four different colors of counters or centimeter cubes to use as game markers, a pre-cut set of Space Race Game Cards printed on cardstock, and a copy of the Space Race Game Instructions sheet. Read the following instructions together:

PLAYING THE SPACE RACE GAME

HOW TO SET UP

- Place the game board on a flat surface in the center of the players.
- Place the Space Race Game Cards face down next to the board.
- Each player chooses a different color game piece.
- Players place their pieces on the Launch Pad.
- Each player takes turns rolling the die. The player with the highest roll will go first.
- Player order will continue to the left (clockwise).

HOW TO PLAY

The object of the game is to be the first player to reach "Mission Accomplished!"

- On your turn, roll the die and move forward that number of spaces.
- If you land on a "Draw a Card" space, draw a Space Race Game Card and read it aloud so that all players can hear.
  - If your card says, "CRITERION MET: ROLL AGAIN!," then roll the die and move forward that number of spaces.
  - If your card says, "CRITERION NOT MET: MOVE BACK ONE SPACE!," then move back one space.
- If you land on a black (or purple) space, read aloud what it says and follow the instructions.

Chapter 18

- If you land on a shortcut space, read aloud what it says and follow the arrows to take the shortcut.
- Play continues until all players have reached "Mission Accomplished!" or until time is called.

After all players have reached "Mission Accomplished!" or after time runs out, stop the game. Have students put the game board, die, and game pieces away. Ask them to sort the Space Race Game Cards into four stacks representing each of the four missions: Project Apollo, the Space Shuttle Program, the Hubble Space Telescope, and the International Space Station (ISS). Ask each student to take one stack of his or her group's cards. (If there are only three players, one student can take two stacks.)

Ask students if they had ever heard or used the word *criterion* before they played the game. Explain that the word *criterion* is very important to engineers and other people who develop or invent things. (Be sure to explain that the word *criterion* is singular for the word *criteria*.) Then, have students read through the Space Race Game Cards for their mission, and with their group, develop a definition for the word *criteria*. They should come up with a definition in their own words that captures the idea that criteria are the desired features or outcomes of a project or technology (e.g., safety, accuracy, or other design requirements). Explain that, in the real world, the success of a designed solution is determined by considering the criteria of the solution. *Ask*

? What is an example of a criterion that your mission did *not* meet? The following answers can be found on the Space Race Game Cards (one unmet criterion for each mission).

- **Project Apollo Criterion Not Met:** *To land on the Moon and return safely to Earth* (FACT: Apollo 1 was never launched because of a tragic accident before takeoff.)
- **Space Shuttle Program Criterion Not Met:** *To carry people into orbit and return them safely* (FACT: During the 30 years of the Space Shuttle Program, tragedies happened on the *Challenger* and *Columbia* shuttles.)
- **Hubble Space Telescope Criterion Not Met:** *To take very clear pictures of things in the universe* (FACT: The first pictures Hubble took were blurry. The primary mirror was flawed.)
- **ISS Criterion Not Met:** *To be resupplied by U.S. space shuttles and Russian rockets* (FACT: The Space Shuttle Program ended in 2011, and some Russian rocket failures since then have made reaching the ISS more difficult.)

Explain that despite not meeting every criterion at first, these projects continued. For example, the Apollo 1 mission was not able to meet the criterion of landing a person on the Moon, but NASA learned from its mistakes and persevered. Two years later, the Apollo 11 mission successfully met the criterion of landing a person on the Moon.

Sometimes, however, a mission is a failure when certain criteria are not met. For example, the goal of the Mars Polar Lander Mission was to send back information about the surface of Mars. The spacecraft successfully reached Mars, but NASA has never been able to make contact with it!

Next, explain that possible solutions to a problem are limited by *constraints*. Tell students to look at all the black (or purple if using the color version of the game board) spaces and shortcuts on the game board. These are examples of constraints that people who design technologies must consider. *Ask*

? What are the three categories of constraints from the game? (materials, budget, and time.)

? What happens if you land on a black (or purple) space? (You have to move back.)

? Explain that in the real world, when you can't work within the given constraints, your project is more likely to be delayed or even canceled. Then, *ask*

? What happens if you land on the space that reads, "MATERIALS NOT AVAILABLE"? (You have to return to the Launch Pad.)

Explain that often, the materials (technologies) needed for a space program to get started have not been invented. Sometimes engineers need to develop new technologies for a space program to even begin.

? What happens if you land on a shortcut? (You get to skip ahead.)

Explain that the shortcuts represent examples of a project staying within the constraints. In the real world, your project is more likely to move forward when you can work within the given constraints.

Next, read aloud pages 10–25 of *Space Exploration*, which will give students more information about Project Apollo, the Space Shuttle Program, the Hubble Space Telescope, and the ISS. After you finish reading about each space project, ask the class how successful they think that project was. Students should realize that although none of these projects had a perfect path to success, each contributed greatly to our knowledge of space and space travel.

Students may be upset to learn about the tragic loss of life that occurred during Project Apollo and the Space Shuttle Program. So you may want to take some time to have a discussion about risk. Explain that there is significant risk involved in sending people into space. But as John F. Kennedy said in his famous speech, we do these "things, not because they are easy, but because they are hard … because that challenge is one that we are willing to accept." Astronauts are aware of the dangers, but they believe that their mission is worth the risk. Many people consider astronauts heroes because of the risks they take for the sake of discovery, exploration, and advancement of scientific knowledge. NASA works very hard to limit risk by having astronauts work closely with the engineers who design, build, and test spacecraft. The risk to humans involved in space exploration is one of the reasons we use unmanned spacecraft for many missions today.

evaluate

Mission to Space

Connecting to the Common Core
Writing
RESEARCH TO BUILD AND PRESENT KNOWLEDGE: 3.7, 4.7, 5.7

Ask

? After learning about four important space projects, what do you think the future holds for space exploration? (Answers will vary.)

Read the "Our Future in Space" section of *Space Exploration* (pp. 26–29), ending with the sentence "Only time will tell what the future holds for space exploration." Tell students that they will be researching a current or future mission in space exploration, making a poster or digital presentation explaining the mission, and then sharing their findings with others at a Space Expo.

You can have students choose their own missions or assign each student one. NASA's Jet Propulsion Laboratory has information on current and future NASA missions on their website (see "Websites" section). Students can click on a mission and get basic information, illustrations, and links to NASA sites with more details about the mission. Give each student copies of the Mission to Space Info Sheet and Mission to Space Presentation Rubric student pages. Have students create a poster or digital presentation (e.g., using Educreation or PowerPoint) to share their research about their space exploration technology with the class. Review with students the following information to include:

- Name of mission
- Primary destination
- Launch date
- Criteria (desired features or outcomes of the mission)
- Constraints (limits on available resources for completing the project or building the technology)
- How the mission will help us understand more about the universe
- Other interesting facts
- Photographs, drawings, or both

Students can share their mission posters and presentations at a Space Expo. Use the Mission to Space Presentation Rubric to evaluate student presentations. You may want to invite other classes to visit your Space Expo.

Picture-Perfect STEM Lessons, 3–5

"The Known Universe" Conclusion

To conclude this lesson, revisit the question you posed at the beginning of this lesson. *Ask*

? How do we know what we know about the universe? (Advances in space exploration technologies have allowed us to make more detailed observations of the universe. We know what we know because of evidence collected by these technologies).

Finally, show students "The Known Universe" video again. This time, *ask*

? Why do you think the scientists at the museum decided to call this video "The Known Universe" and not just "The Universe"? (Students should realize that although we have learned a lot about the universe, there is much information we still don't know. Because our technology is limited, our views and explorations of space are limited. Humans still have much to learn.)

Explain to students that scientists at the American Museum of Natural History created this video in December 2009. By the time the students are adults, the "known universe" will likely have expanded greatly. Perhaps future students will also say about some of our current ideas and models "Boy, were we wrong!"

STEM at Home

Have students complete the "I learned that ..." and "My favorite part of the lesson was ..." portions of the STEM at Home student page as a reflection on their learning. They may choose to do the following at-home activity with an adult helper and share their results with the class. If students do not have access to the internet or these materials at home, you may choose to have them complete this activity at school.

"At home, we can use a website called 'Trace Space Back To You' to explore the technologies we use every day that are 'spinoffs' of space exploration."

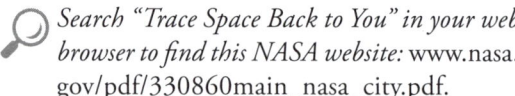 *Search "Trace Space Back to You" in your web browser to find this NASA website:* www.nasa.gov/pdf/330860main_nasa_city.pdf.

"After we explore the website together, we can make a list of things we use that are spinoffs of technologies invented for space exploration."

For Further Exploration

This section is provided to help you encourage your students to use the science and engineering practices in a more student-directed format. This box lists questions and challenges related to the lesson that students may select to research, investigate, or innovate. Students may also use the questions as examples to help them generate their own questions. After selecting one of the questions in the box or formulating their own questions, students can individually or collaboratively make predictions, design investigations or surveys to test their predictions, collect evidence, devise explanations, design solutions, or examine related resources. They can communicate their findings through a science notebook, at a poster session or gallery walk, or by producing a media project.

Research

Have students brainstorm researchable questions:

? How does a telescope work?

? What is space debris or "space junk," and why is it a concern for space exploration?

? What are some things we have learned from New Horizons, NASA's recent mission to Pluto?

Investigate

Have students brainstorm testable questions to be solved through science or math:

? What planets can you see without a telescope?

? Can you see more stars with binoculars than with the naked eye?

? How does what you can observe in the night sky change throughout the year?

Innovate

Have students brainstorm problems to be solved through engineering:

? If you were on the first mission to Mars, what technologies would you need to bring with you?

? If you were going to design a space colony on Mars that could support human life for a year, what would it need to include? Draw your design.

? What questions do you have about the universe? What kind of spacecraft would you design to answer those questions?

Websites

Current NASA Missions
www.jpl.nasa.gov/missions/?type=current

Future NASA Missions
www.jpl.nasa.gov/missions/?search=&type=future&missions_target=&mission_type=&launch_date=#submit

NASA's Eyes on the Solar System Tutorials
http://eyes.nasa.gov/tutorials.html

"The Known Universe" (video)
https://vimeo.com/19568852

More Books to Read

Becker, H. 2015. *National Geographic Kids: Everything space*. Washington, DC: National Geographic Children's Books.
Summary: This fascinating book will pull readers in like gravity. It is full of fun facts, space maps, infographics, reports from explorers, and more than 100 pictures.

Buckley, J. 2015. *Home address: ISS—The International Space Station*. New York: Penguin Young Readers.
Summary: Full-color photographs and clear text describe daily life on the International Space Station.

Caron, M. K. 2016. *Mission to Pluto: The first visit to an ice dwarf and the Kuiper Belt.* Boston: Houghton Mifflin Harcourt.
Summary: A group of planetary scientists, astronomers, engineers, and NASA technicians go where no spacecraft or person has ever been before.

Kudlinski, K. 2006. *Boy, were we wrong about dinosaurs!* New York:
Summary: From the same author as *Boy, Were We Wrong About the Solar System!,* this book explores the tentative nature of science within the context of dinosaurs.

Wooster, P. 2012. *An illustrated timeline of space exploration*. Mankato, MN: Picture Window Books.
Summary: This book highlights the most important events and discoveries in space exploration. Broken up into dates and eras, the reader can follow the events in order or jump back and forth.

Chapter 18

Name: _____

Eyes on the Solar System

Directions: Use NASA's Eyes on the Solar System app to "visit" Earth and five other celestial bodies in our solar system. Be sure to click the SIMPLE button to launch the app. Click the DESTINATIONS button and select a celestial body. Then, use the sidebar tabs to find the information in the chart below.

Object	What Type of Body Is It?	Who Discovered It?	When Was It Discovered?	Interesting Facts
1. Earth				
2.				
3.				
4.				
5.				
6.				

The Known Universe

Then and Now Cards

Earth is the center of the solar system.	The Sun has spots, and the Moon has mountains and craters.
There are no canals on Mars.	Earth is spherical.
Pluto is considered a planet.	Earth is flat.
The Sun and Moon are perfect, smooth spheres.	Martians dug the canals on Mars.
The Sun is the center of the solar system.	Pluto is not considered a planet.

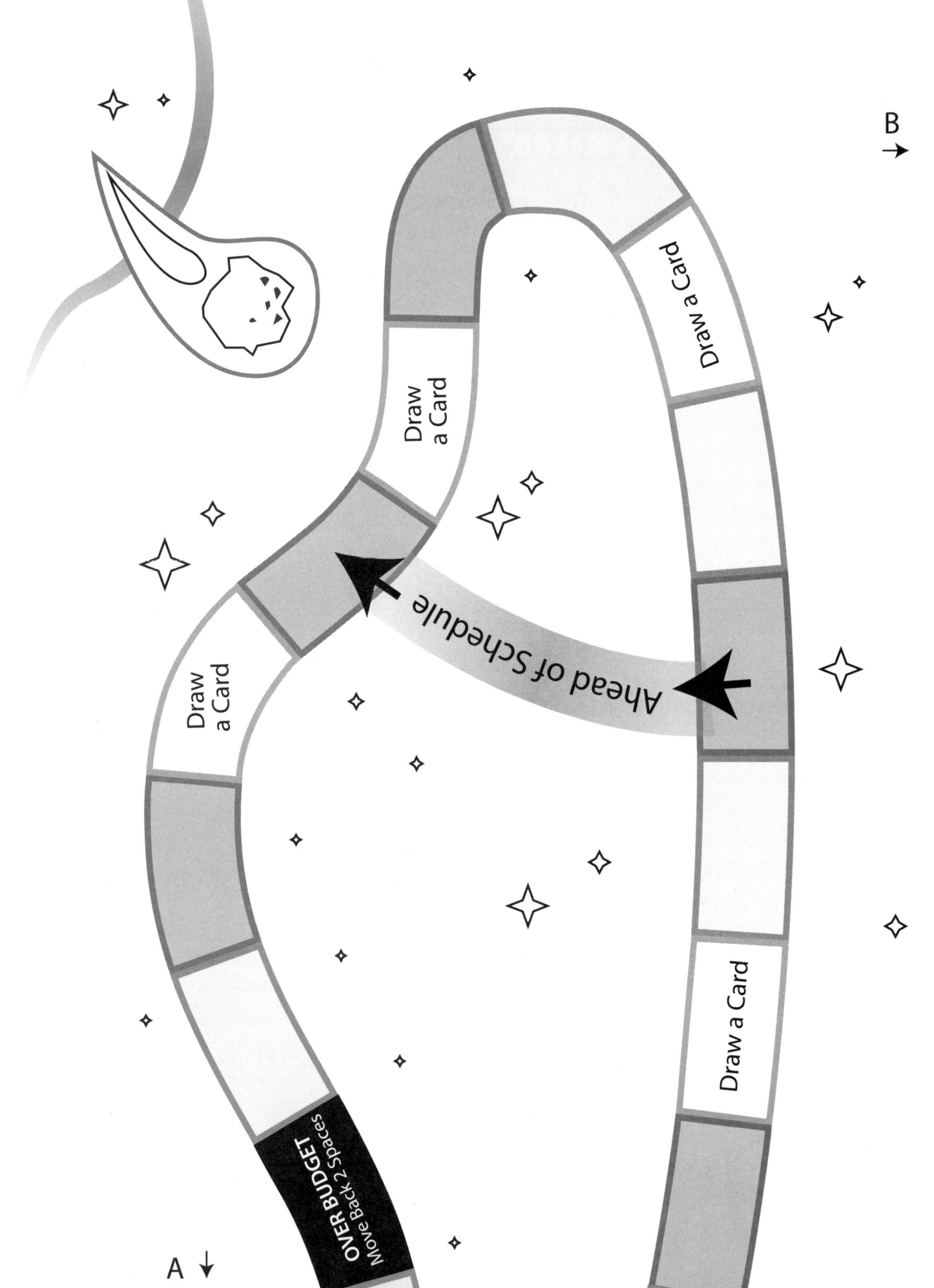

RACE

- Draw a Card
- Draw a Card
- Draw a Card
- Materials Available
- **MATERIALS NOT AVAILABLE** Return to Launch Pad

B ←

C ↓

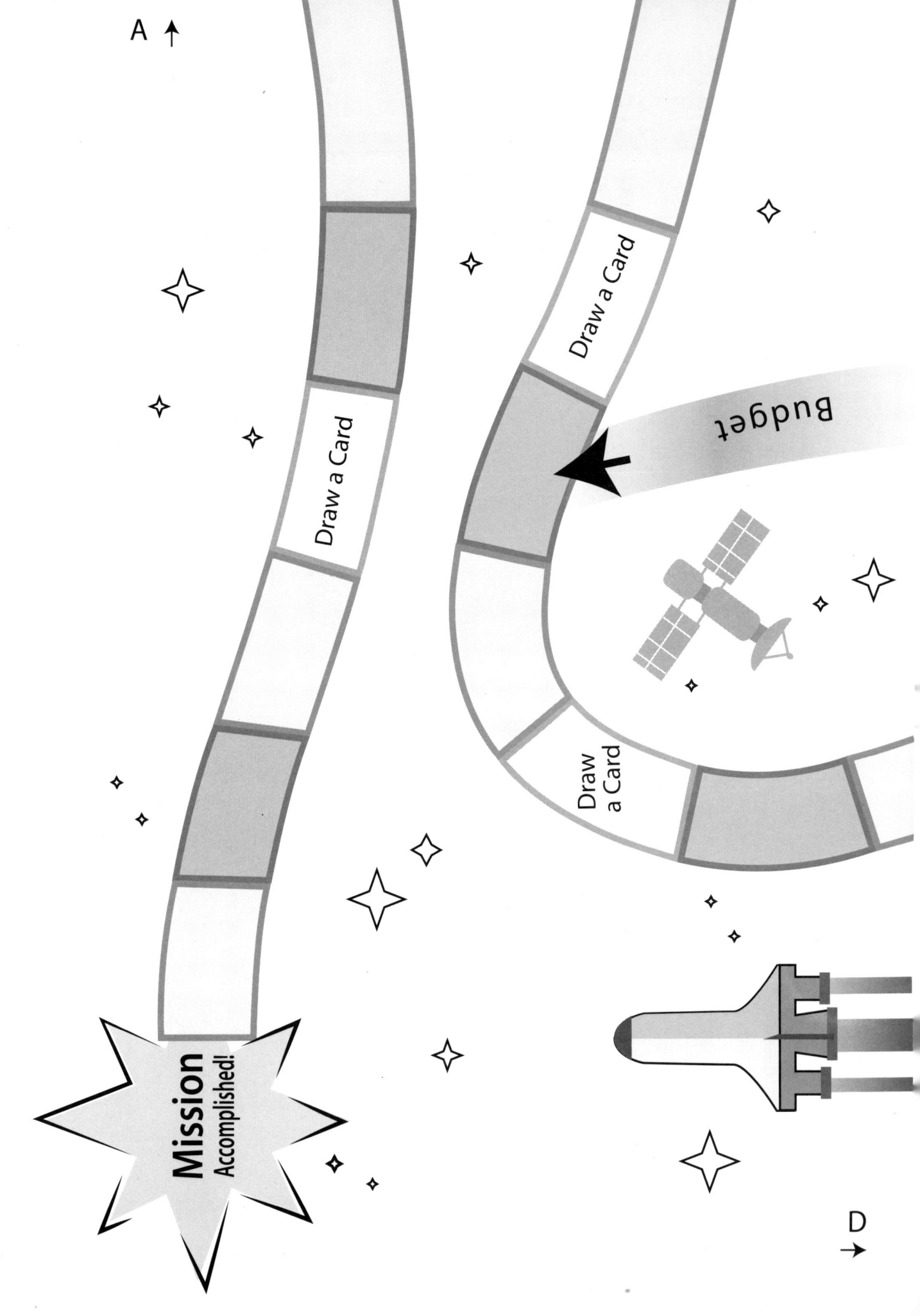

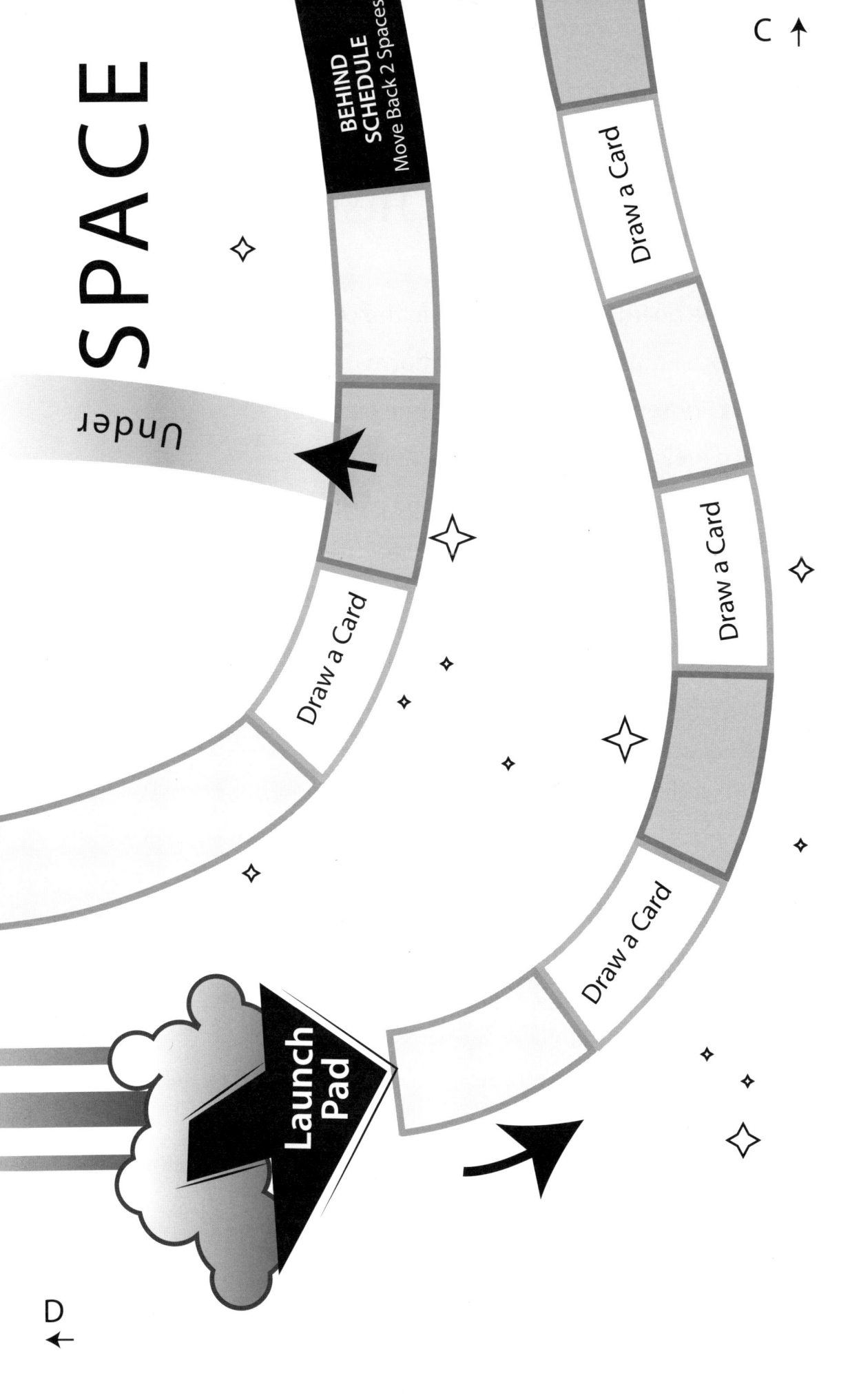

Space Race Game Instructions

HOW TO SET UP

- Place the game board on a flat surface in the center of the players.
- Place the Space Race Game Cards face down next to the board.
- Each player chooses a different color game piece.
- Players place their pieces on the Launch Pad.
- Each player takes turns rolling the die. The player with the highest roll will go first.
- Player order will continue to the left (clockwise).

HOW TO PLAY

- The object of the game is to be the first player to reach "Mission Accomplished!"
- On your turn, roll the die and move forward that number of spaces.
- If you land on a "Draw a Card" space, draw a Space Race Game Card and read it aloud so that all players can hear.
  - If your card says, "CRITERION MET: ROLL AGAIN!," then roll the die and move forward that number of spaces.
  - If your card says, "CRITERION NOT MET: MOVE BACK ONE SPACE!," then move back one space.
- If you land on a black (or purple) space, read aloud what it says and follow the instructions.
- If you land on a shortcut space, read aloud what it says and follow the arrows to take the shortcut.
- Play continues until all players have reached "Mission Accomplished!" or until time is called.

Space Race Game Cards

PROJECT APOLLO

CRITERION: *To land a person on the Moon*

FACT: In 1969, Neil Armstrong became the first man to walk on the Moon.

CRITERION MET: ROLL AGAIN!

PROJECT APOLLO

CRITERION: *To explore the Moon*

FACT: Many experiments were done and hundreds of rock and soil samples were collected during six Moon landings.

CRITERION MET: ROLL AGAIN!

PROJECT APOLLO

CRITERION: *To land on the moon and return safely to Earth*

FACT: Apollo 1 was never launched because of a tragic accident before takeoff.

CRITERION NOT MET: MOVE BACK ONE SPACE!

PROJECT APOLLO

CRITERION: *To land astronauts on the Moon multiple times*

FACT: By the last Apollo mission in 1972, 12 astronauts had walked on the Moon.

CRITERION MET: ROLL AGAIN!

Picture-Perfect STEM Lessons, 3–5

Chapter 18

SPACE SHUTTLE PROGRAM

CRITERION: *To carry large payloads into space*

FACT: Space shuttles have carried the Chandra X-Ray Observatory, the Hubble Space Telescope, and large pieces of the International Space Station into space.

CRITERION MET: ROLL AGAIN!

SPACE SHUTTLE PROGRAM

CRITERION: *To be an orbiting laboratory*

FACT: Spacelab was a reusable laboratory flown on many shuttle missions. Scientists performed many experiments in microgravity while in orbit.

CRITERION MET: ROLL AGAIN!

SPACE SHUTTLE PROGRAM

CRITERION: *To carry people into orbit and return them safely*

FACT: During the 30 years of the shuttle program, tragedies happened on the *Challenger* and *Columbia* shuttles.

CRITERION NOT MET: MOVE BACK ONE SPACE!

SPACE SHUTTLE PROGRAM

CRITERION: *To be a reusable spacecraft*

FACT: 5 different space shuttles flew a total of 135 missions over 30 years.

CRITERION MET: ROLL AGAIN!

Chapter 18

HUBBLE SPACE TELESCOPE

CRITERION: *To be a space-based telescope that can orbit the Earth*

FACT: The Hubble was carried into orbit by a space shuttle in 1990. It is expected to work until 2020.

CRITERION MET: ROLL AGAIN!

HUBBLE SPACE TELESCOPE

CRITERION: *To be a telescope that can be serviced by astronauts so old parts can be replaced*

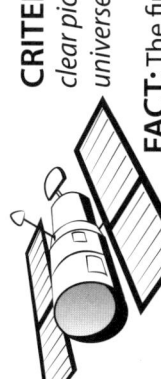

FACT: Astronauts have flown to Hubble five times to fix and replace parts.

CRITERION MET: ROLL AGAIN!

HUBBLE SPACE TELESCOPE

CRITERION: *To take very clear pictures of things in the universe*

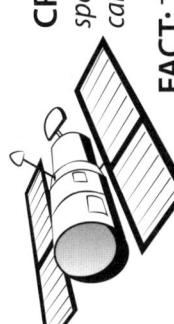

FACT: The first pictures taken by Hubble were blurry. The primary mirror was found to be flawed.

CRITERION NOT MET: MOVE BACK ONE SPACE!

HUBBLE SPACE TELESCOPE

CRITERION: *To take very clear pictures of things in the universe*

FACT: Although the first pictures were blurry, astronauts were able to fix Hubble and get clear pictures of objects humans had never seen.

CRITERION MET: ROLL AGAIN!

Picture-Perfect STEM Lessons, 3–5

Chapter 18

INTERNATIONAL SPACE STATION (ISS)

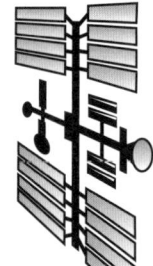

CRITERION: *To carry a six-person crew of men and women from around the world*

FACT: The first international crew arrived in 2000, and the ISS has been continuously occupied since.

CRITERION MET: ROLL AGAIN!

INTERNATIONAL SPACE STATION (ISS)

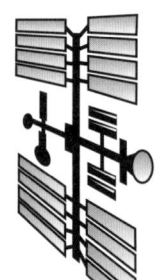

CRITERION: *To be resupplied by U.S. space shuttles and Russian rockets*

FACT: The space shuttle program ended in 2011, and since then some Russian rocket failures have made reaching the ISS more difficult.

CRITERION NOT MET: MOVE BACK ONE SPACE!

INTERNATIONAL SPACE STATION (ISS)

CRITERION: *To be built through an international partnership of space agencies*

FACT: The ISS has been built by NASA, as well as space agencies in the Russian Federation, Japan, Canada, and the European Union.

CRITERION MET: ROLL AGAIN!

INTERNATIONAL SPACE STATION (ISS)

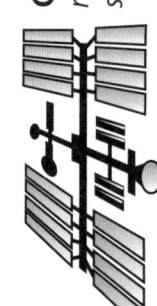

CRITERION: *To be a space-based lab for scientific research*

FACT: ISS crews have done hundreds of experiments in microgravity, including research on the long-term effects of living in space.

CRITERION MET: ROLL AGAIN!

Name : _____

Mission to Space Info Sheet

We learn more about our universe through developing new technologies to explore space. Record the following information about the current or future space mission you are investigating.

1. Name of Mission: _____

2. Primary Destination: _____

3. Launch Date: _____

4. Criteria (desired features or outcomes of the mission):

   - _____
   - _____
   - _____
   - _____
   - _____

5. Constraints (limits on available resources for completing the project or building the technology):

Time Constraints	Budget Constraints	Materials Constraints

Picture-Perfect STEM Lessons, 3–5

Name: _____

6. How will the mission help us understand more about the universe?

_____

_____

_____

7. Other interesting facts:

_____

_____

_____

8. Photographs, drawings, or both:

Name : _____

Mission to Space Presentation Rubric

Create a poster or electronic presentation to share what you learned about the space exploration mission you investigated.

Score	Criteria
____ 4 ____ 3 ____ 2 ____ 1	Name of the mission, primary destination, and launch date
____ 4 ____ 3 ____ 2 ____ 1	Criteria of the mission
____ 4 ____ 3 ____ 2 ____ 1	Constraints of the mission (time, budget, and materials)
____ 4 ____ 3 ____ 2 ____ 1	Description of the types of evidence the mission will collect
____ 4 ____ 3 ____ 2 ____ 1	Interesting facts about the mission, including a photograph or drawing

4—Excellent   3—Above Average   2—Average   1—Below Average

_____ Total/20 Points

Picture-Perfect STEM Lessons, 3–5

Name: _____

STEM at Home

Dear _____,

At school, we have been learning about **space exploration**.

I learned that: _____
_____
_____

My favorite part of the lesson was:
_____
_____
_____

At home, we can use a website called "Trace Space Back To You" to explore the technologies we use every day that are "spinoffs" of space exploration.

Search "Trace Space Back to You" in your web browser to find this NASA website: *www.nasa.gov/pdf/330860main_nasa_city.pdf*.

After we explore the website together, we can make a list of things we use that are spinoffs of technologies invented for space exploration.

- _____
- _____
- _____
- _____
- _____

National Science Teachers Association

Chapter 19

Star Stuff

Description

In this lesson, students read about astrophysicist Carl Sagan and his passion for learning and teaching others about the stars. They complete an investigation to find out how the apparent size and brightness of stars in the night sky depends on the stars' distances from Earth. They also learn why star distance is measured in light-years instead of kilometers. Finally, they share what they have learned by creating their own episode of Sagan's iconic *Cosmos: A Personal Voyage* television show.

Suggested Grade Levels: 3–5

LESSON OBJECTIVES Connecting to the *Framework*		
Science and Engineering Practices	**Disciplinary Core Ideas**	**Crosscutting Concept**
Developing and Using Models Using Mathematics and Computational Thinking Engaging in Argument From Evidence	**ESS1.A:** The Universe and Its Stars	Scale, Proportion, and Quantity

Featured Picture Books

TITLE: ***Star Stuff: Carl Sagan and the Mysteries of the Cosmos***
AUTHOR: **Stephanie Roth Sisson**
ILLUSTRATOR: **Stephanie Roth Sisson**
PUBLISHER: **Roaring Brook Press**
YEAR: **2014**
GENRE: **Narrative Information**
SUMMARY: *This biography of Carl Sagan focuses on his childhood and culminates with the Voyager mission and the Golden Record.*

TITLE: ***Jump Into Science: Stars***
AUTHOR: **Steve Tomecek**
ILLUSTRATOR: **Sachiko Yoshikawa**
PUBLISHER: **National Geographic Children's Books**
YEAR: **2006**
GENRE: **Non-Narrative Information**
SUMMARY: *This book introduces stars and what they are made of, how they shine, their positions relative to Earth, and more.*

Picture-Perfect STEM Lessons, 3–5

Chapter 19

Time Needed

This lesson will take several class periods. Suggested scheduling is as follows:

Day 1: **Engage** with *Star Stuff* Read-Aloud

Day 2: **Explore** with Star Light, Star Bright and **Explain** with Explaining the Model

Day 2: **Explain** with *Jump Into Science: Stars* Read-Aloud

Day 3: **Elaborate** with How Far Are the Stars?

Day 4: **Evaluate** with Your Own *Cosmos*

Materials

For Star Light, Star Bright

- 2 identical, battery-operated lanterns of the same size and brightness (Flashlights can be used instead, but lanterns work better because their light is multidirectional. LE Outdoor LED Lantern Flashlights [2 per pack] are available at Amazon.com.)
- Large, dark room or long, dark hallway

For Explaining the Model

- Index cards (1 per student)

For How Far Are the Stars?

- Calculator (optional)

For Your Own Cosmos

- Art supplies to make models and visual aids
- Recording devices such as tablets or video cameras

SAFETY
- Use caution when working in a dark area
- Make sure there are no slipping, tripping, or falling hazards.

Student Pages

- Stars Extended Anticipation Guide
- *Cosmos* Episode Planning Sheet
- *Cosmos* Episode Scoring Rubric
- STEM at Home

Background for Teachers

Carl Sagan, an astrophysicist and Pulitzer Prize–winning author often described as "the scientist who made the universe clearer to the ordinary person," was arguably the most well-known scientist of the 1970s and 1980s. His PBS television series *Cosmos: A Personal Voyage* taught hundreds of millions of people worldwide about the universe.

Sagan was born in 1934 in Brooklyn, New York. His interest in astronomy developed early in life. He studied at the University of Chicago and worked at the University of California, Berkeley; Harvard University; and the Smithsonian Astrophysical Observatory. He was a professor at Cornell University and worked with NASA on many projects, including the Voyager mission.

The Voyager 1 and Voyager 2 spacecraft were launched in 1977 and have explored the giant planets of the outer solar system: Jupiter, Saturn, Uranus, and Neptune. The spacecraft contain many instruments for collecting information, and each carries a disk known as the *Golden Record*, which contains greetings, sounds, music, pictures, and other information from Earth intended to communicate the story of our planet to extraterrestrials. Sagan chaired the committee that decided what information would be contained on the Golden Record. In 2012, Voyager 1 made history by leaving our solar system and entering interstellar space. It has traveled farther into space than anyone or anything in history. Both Voyagers are still traveling far beyond our solar system and are currently sending scientific information about their surroundings through the Deep Space Network, or DSN. (To view a real-time odometer of their distances from the Sun and Earth, go to the NASA Voyager Interstellar Mission page at *http://voyager.jpl.nasa.gov/where*.)

Sagan co-founded the Planetary Society, an international nonprofit organization focusing on space exploration. Apart from the success of *Cosmos*, he helped to popularize science by writing hundreds of articles and more than two dozen books. He even wrote a science fiction novel called *Contact*, which was made into a movie starring Jodie Foster. Sagan was also well known as a pioneer in the field of *exobiology*, the study of the possibility of extraterrestrial life. Sagan died in 1996, but he left behind a legacy of exploration and is remembered for inspiring people to search for understanding and share science with others. One of Sagan's most recognizable quotes is "Somewhere, something incredible is waiting to be known."

According to Sagan's first episode of *Cosmos*, there are more stars in the sky than there are grains of sand on all of the beaches of the world. This number is greater than anyone could ever count. Stars appear as points of light in the night sky, except for one star that outshines them all—the Sun. In this lesson, students learn that the Sun is considered a small to average-size star but appears larger because it is relatively close to Earth, a mere 93 million miles away. The next nearest star, Proxima Centauri, is nearly 25 trillion miles from Earth. Because these numbers are so large, astronomers use light-years instead of miles or kilometers to communicate astronomical distances. A *light-year* is defined as the distance light can travel in one year. Light travels at an incredible 186,000 miles per second. That makes a light-year equal to almost 6 trillion (6,000,000,000) miles!

A Framework for K–12 Science Education suggests that by grade 5 students learn that "The Sun is a star that appears larger and brighter than other stars because it is closer. Stars range greatly in their size and distance from Earth." (NRC 2012, p. 174) This disciplinary core idea (ESS1.A) is taught in the context of Carl Sagan's childhood quest to learn more about stars. Students incorporate the science and engineering practice of developing and using models by using lanterns as models of stars to compare relative size and brightness at different distances from Earth. Mathematics and the crosscutting concept of scale, proportion, and quantity are also applied as students learn to convert light-years to miles. Finally, students demonstrate their learning by creating their own episode of Sagan's *Cosmos* television show to explain why the Sun appears so much larger and brighter than other stars. That activity incorporates the science and engineering practice of engaging in argument from evidence.

engage

Star Stuff Read-Aloud

Show students the cover of *Star Stuff: Carl Sagan and the Mysteries of the Cosmos,* and *ask*

? Have you ever heard of Carl Sagan? (Answers will vary.)

 Inferring

Ask

? Based on the picture on the cover, what do you think this book might be about? (Answers will vary.)

> **Connecting to the Common Core**
> **Reading: Informational Text**
> KEY IDEAS AND DETAILS: 3.2, 4.2, 5.2

Introduce the author and illustrator, Stephanie Roth Sisson, and explain that she has long been a fan of Carl Sagan, who taught her to look up at the night sky in wonder. Tell students that as you read the book aloud, you would like them to think about the title *Star Stuff* and listen for the reasons she chose this title. Read the book aloud. Then, *ask*

? Why do you think the author decided to title this book *Star Stuff*? (Young Carl was interested in stars; there is a lot of "stuff," or information, in the book about stars; and Carl Sagan taught us that we are made of "star stuff.")

? Did you notice any information in the book that is now out of date? (Young Carl read that "Our sun is a big ball of fiery gas, held together by gravity. Nine planets, including Earth, orbit around it in our solar system." When Carl Sagan read that in his science book, Pluto was considered the ninth planet. Our solar system is no longer described as containing nine planets!)

Explain that in 2006 (10 years after Carl Sagan died), a group of scientists who classify space-related discoveries changed the definition of a planet to be as follows:

- It must orbit a star.
- It must be round.
- It must not be a moon—a solid object that orbits another planet.
- It must have cleared the surrounding area of other large objects through its gravitational pull.

By this definition, poor little Pluto, formerly known as the ninth planet from the Sun, was demoted to a "dwarf planet" because it had not cleared its surrounding area of other large objects.

 Turn and Talk

Star Stuff is a true story and a great example of how a childhood passion can become your life's work. Tell students to discuss the following question with a partner: "What are you passionate about? What steps would you need to take to make this your career, or your 'life's work'?"

 Making Connections: Text to Text

> **Connecting to the Common Core**
> **Reading: Informational Text**
> INTEGRATION OF KNOWLEDGE AND IDEAS: 3.9, 4.9, 5.9

Tell students that you have a video about Carl Sagan to show them, and the video has the same main title as the book—"Star Stuff" (see "Websites" section). Tell students that as they watch, you would like them to compare the video and the book. *Ask*

? What do the video and the book have in common? (They both portray Carl Sagan as a young boy, they both depict when he asks for a book on stars and is given one about Hollywood stars, they both reveal that he was a very curious child, and they both explain his tremendous influence on the study of space.)

? **How are they different?** (The video is told in first-person, with an actor portraying young Carl narrating and with some narration from the television show *Cosmos* by Carl Sagan himself, whereas the book is written in third-person. The video also employs an inspirational soundtrack.)

? **How have the book and the video influenced your ideas or opinions about Carl Sagan and the title *Star Stuff*?** (Answers will vary.)

explore

Star Light, Star Bright

 Making Connections: Text to Self

Ask

? When Carl Sagan was a young boy, he imagined the stars hanging down like lightbulbs on long black wires. Do you think this is true?

? Have you ever looked up at the stars on a clear night?

? What have you observed or wondered about the stars?

? Have you ever wondered why some stars are brighter than others?

Tell students that you can use a model to learn more about one of the reasons some stars appear brighter than others. The following activity must be done in a large, dark room or a long, dark hallway (*Note*: This activity was adapted from the activity on pp. 30–31 of *Jump Into Science: Stars*, which you will read later in the lesson.):

1. Call on two students (volunteer A and volunteer B) to each hold an identical, battery-operated lantern to represent two stars of the same size and brightness. (Flashlights can also be used, but lanterns work better in this model because their light is multi-directional.)

2. Have both volunteers stand side by side, as far away from the rest of the class as possible.

COMPARING THE BRIGHTNESS SIDE BY SIDE

3. Have both volunteers turn on their lanterns and face the class.

4. Turn off the lights to make the room or hallway as dark as possible.

5. Tell the rest of the students to imagine that they are on Earth looking at two stars that are the same size in the night sky.

6. Ask, "How do the size and brightness of the two 'stars' compare when they are side by side?" (They are the same size and brightness.)

7. Next, have volunteer B walk toward the rest of the class. Tell volunteer B to stop walking when you can see a significant difference between the brightness of the two lanterns. Volunteer A should not move. Both volunteers should continue to shine their lanterns toward the class. Then ask, "How do the size and brightness of the two 'stars' compare now?" (Volunteer B's lantern [the closer one] appears larger and brighter.)

Picture-Perfect STEM Lessons, 3–5

COMPARING THE BRIGHTNESS AT DIFFERENT DISTANCES

8. Have volunteer B walk several steps closer to the class. Then ask, "How do the size and brightness compare now?" (Volunteer B's lantern [the closer one] appears even brighter and larger than the other lantern.)

Repeat the activity with different students so that everyone has a chance to observe the differences.

explain

Explaining the Model

After completing the Star Light, Star Bright activity, give each student an index card and ask them to write their answers to the following questions on the card:

1. In this model, which volunteer would represent the Sun? (Volunteer B)
2. Why do you think the Sun appears larger and brighter than all the other stars?
3. What does this model tell you about the relationship between distance and the apparent size and brightness of stars?

Have students share their answers with a partner, and then call on a few students to share with the class. Allow students to revise their answers based on the discussion. You may want to collect the cards to use as a formative assessment. Students should understand the following:

- Volunteer B represented the Sun in this model.
- The Sun appears larger and brighter than other stars because it is so much closer.
- The closer the light source, the brighter the light appears; therefore, the closer the star, the brighter the star appears.

Jump Into Science: Stars Read-Aloud

Connecting to the Common Core
Reading: Informational Text
KEY IDEAS AND DETAILS: 3.1, 4.1, 5.1

 Anticipation Guide

Show students the cover of *Jump Into Science: Stars* and introduce the author, Steve Tomecek, and illustrator, Sachiko Yoshikawa. Have students fill in the "Before Reading" true/false blanks on the Stars Extended Anticipation Guide. Tell them that you will be reading the book to find out if their answers are correct or incorrect. Read the book aloud and have students signal (raise their hands) when they hear evidence from the text for or against any of the six statements. After you read, have students fill in the true/false blanks on the right-hand side of the anticipation guide and complete the "Evidence From the Text" section to support their "After Reading" choices. When students finish, go over each question and ask students to share their answers and evidence from the text. The answers are as follows:

1. True: In the daytime, stars are still shining in the sky, but you cannot see them because the Sun makes the sky so bright (p. 7).

2. False: The Sun is the closest star, but not the largest. Compared with many other stars, the Sun is small or average-size (p. 12).

3. False: Some stars are closer to Earth than others (p. 14).

4. True: Some stars look brighter because they are hotter, some stars look brighter because they are bigger, and some stars, such as the Sun, look brighter because they are closer to Earth (p. 16).

5. False: Stars can be red, yellow, white, or blue, depending on their temperature (p. 18).

6. True: If you watch the night sky for a few hours, the stars seem to move slowly across the sky. The stars all "rise" in the east and "set" in the west because of the direction Earth turns (pp. 24–25).

 Rereading

After reviewing the anticipation guide, reread pages 16–17, which explain, "Some stars look bright because they are very, very, hot. Other stars look bright because they are very, very, big. Some stars, like our sun, look bright because they are close to Earth. This can be confusing. It means that a small, hot star that is far away can look dimmer than a cool, big star that is closer to Earth." Then, *ask*

? Which star appears largest and brightest to us? (the Sun)

? Is the Sun the largest star? (no)

? Is the Sun the hottest star? (no)

? Think back to the activity we did with the lanterns (or flashlights). Why does the Sun appear larger and brighter than all other stars? (It is much closer to Earth than any other star.)

elaborate

How Far Are the Stars?

Connecting to the Common Core
Mathematics
MEASUREMENT AND DATA: 5.MD.1

 Rereading

Reread pages 14–15 of *Stars,* which explain how far away various stars are from Earth. Then, point out the different numbers next to the circle representing various stars. *Ask*

? What do these numbers represent? (light-years)

Explain that a *light year* is the distance light travels in one year. *Ask*

? Why do you think astronomers use light-years to represent astronomical distances such as the distance between the stars and Earth? (Answers will vary.)

? How do we measure very large distances on Earth? (miles and kilometers)

Tell students they can find out why scientists use light years instead of miles or kilometers for the distance of stars by doing some conversions. Reread the last sentence on page 14, which explains that a light-year is equal to almost 6 trillion miles. *Ask*

? How many zeros behind the six do I need to write to represent 6 trillion? (nine zeroes)

Write 6,000,000,000 on the board. Next, have students choose a star from pages 14–15 to convert its distance into miles. For example, the star Betelgeuse is approximately 650 light years from Earth. Demonstrate how to convert that distance to miles by multiplying 650 light years × 6,000,000,000 miles (using a calculator, if desired).

650 light years × 6,000,000,000 miles = 3,900,000,000,000 miles

Repeat this conversion with some of the other stars featured on pages 14–15. Then, *ask*

? Now, why do you think astronomers use light-years when discussing the distances to the stars? (The numbers are too big when using miles or kilometers, making calculations cumbersome to write and confusing to interpret. Light-years are more practical and easier to calculate and communicate.)

Explain that another helpful thing about light-years is that unit of measurement provides an interesting perspective on how long the light from that star takes to reach your eyes. For example, the star Vega (on p. 14) is 25 light-years away from Earth. This means that when you find Vega in the night sky tonight, you are seeing light that left that star 25 years ago!

evaluate

Your Own *Cosmos*

Tell students that Carl Sagan was a brilliant scientist, writer, and thinker, and one of his greatest talents was sharing his knowledge and sense of wonder about the universe with others. Tell them that you would like to show them the introduction to Carl Sagan's *Cosmos: A Personal Voyage* television show, which first aired on PBS in 1980. This video will help them understand Sagan's passion for sharing his knowledge of the universe with the public. The word *cosmos*, as defined by Sagan, means the whole universe—all that is, all that was, and all that will ever be.

Show the 5 min. introduction to *Cosmos* (see "Websites" section). Tell students that each episode of *Cosmos* taught the viewer something about the universe. Explain that the series was rebooted in 2014 to continue Carl Sagan's vision and legacy, adding up-to-date special effects and scientific discoveries to the original concept. The new series, *Cosmos: A Spacetime Odyssey* was hosted by astrophysicist and Sagan admirer Neil deGrasse Tyson and is currently available on Netflix. You may also want to show students the trailer for the new *Cosmos* series (see "Websites" section).

> **Connecting to the Common Core**
> **Writing**
> RESEARCH TO BUILD AND PRESENT KNOWLEDGE: 3.8, 4.8, 5.8

Tell students that you would like them to explain what they have learned about stars in this lesson by making a short video modeled after the *Cosmos* television show. The video will be a 3–5 min. episode titled "Cosmos: The Brightness of Stars." This video should not only teach others why the Sun appears to be larger and brighter than the other stars but also share with viewers some other fascinating facts about stars. Explain that in the *Cosmos* television series, Sagan and Tyson always provided evidence for the scientific ideas they were explaining, so "Cosmos: The Brightness of Stars" should emphasize the scientific evidence for any of the ideas presented. Give each student copies of the *Cosmos* Episode Planning Sheet and the *Cosmos* Episode Scoring Rubric student pages. After students have completed their scripts and planning sheets, they should bring them to you to review and sign at the "Teacher Checkpoint."

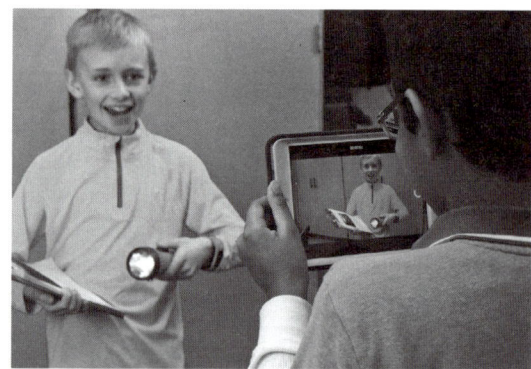

RECORDING A *COSMOS* EPISODE

> **Connecting to the Common Core**
> **Speaking and Listening**
> PRESENTATION OF KNOWLEDGE AND IDEAS: 3.4, 4.4, 5.4

Next, have students use art supplies to make models and visual aids for the show. Provide recording devices such as tablets or video cameras for students to use to film each other performing their *Cosmos* episode (turtlenecks optional!). Use the *Cosmos* Episode Scoring Rubric to evaluate students.

STEM at Home

Have students complete the "I learned that …" and "My favorite part of the lesson was …" portions of the STEM at Home student page as a reflection on their learning. They may choose to do the following at-home activity with an adult helper and share their results with the class. If students do not have access to the internet or these materials at home, you may choose to have them complete this activity at school.

"At home, we can watch a video about Voyager 1 and Voyager 2, which have traveled farther into space than any other spacecraft."

Search *"Voyager, Humanity's Farthest Journey"* in your web browser to find the video at www.jpl.nasa.gov/video/details.php?id=980.

"After we watch the video, we can visit the Voyager website (*http://voyager.jpl.nasa.gov*) to collect the following information:

1. In what year were the Voyager spacecraft launched?
2. Where are Voyager 1 and Voyager 2 now?
3. What things are contained on the Golden Record carried by the Voyagers?

Together, we can discuss what we would put on our own Golden Record!"

For Further Exploration

This section is provided to help you encourage your students to use the science and engineering practices in a more student-directed format. This box lists questions and challenges related to the lesson that students may select to research, investigate, or innovate. Students may also use the questions as examples to help them generate their own questions. After selecting one of the questions in the box or formulating their own questions, students can individually or collaboratively make predictions, design investigations or surveys to test their predictions, collect evidence, devise explanations, design solutions, or examine related resources. They can communicate their findings through a science notebook, at a poster session or gallery walk, or by producing a media project.

Research

Have students brainstorm researchable questions:

? What is the largest known star?

? How many stars are in our galaxy? How many galaxies are in the universe?

? How can astronomers tell if a star has planets orbiting it?

Investigate

Have students brainstorm testable questions to be solved through science or math:

? What stars and planets can you see from your home without a telescope? (Use a star finder app, such as Star Walk.)

? Can you see more stars with binoculars than with the naked eye?

? What is an *astronomical unit*? How do astronomical units compare with light-years?

Picture-Perfect STEM Lessons, 3–5

Innovate

Have students brainstorm problems to be solved through engineering:

- Can you create a model that compares the distance from Earth to the Sun with the distance from Earth to the next nearest star?
- Can you design a time capsule to send into outer space? What would you include to represent our planet?
- Can you design a spacecraft that can land on the surface of a planet without harming the spacecraft's cargo, using an egg to represent the cargo?

Reference

National Research Council (NRC). 2012. *A framework for K–12 science education: Practices, crosscutting concepts, and core ideas*. Washington, DC: National Academies Press.

Websites

Carl Sagan's Introduction to the *Cosmos* Series (video)
https://vimeo.com/26028916

Cosmos: A Spacetime Odyssey Trailer (video)
https://vimeo.com/85772236

"Star Stuff: The Story of Carl Sagan" (video)
https://vimeo.com/136262971

More Books to Read

Becker, H. 2015. *National Geographic Kids: Everything space*. Washington, DC: National Geographic Children's Books.
Summary: This fascinating book will pull readers in like gravity. It is full of fun facts, space maps, infographics, reports from explorers, and more than 100 pictures.

Forest, C. 2012. *The kids' guide to the constellations*. Mankato, MN: Capstone Press.
Summary: This informative book shares the science and the stories behind the stars. It also includes information on how to find each constellation in the night sky.

Tomecek, S. 2016. *Jump Into science: Sun*. Washington, DC: National Geographic.
Summary: This book introduces the Sun as a star and explains the daily and yearly phenomena that result from the position of the Earth relative to the Sun.

Name : _____

Stars Extended Anticipation Guide

Before Reading
True or False

After Reading
True or False

_____   1. You can't see stars in the daytime because the Sun makes the sky too bright.   _____

_____   2. The Sun is the biggest star.   _____

_____   3. Stars are all about the same distance from Earth.   _____

_____   4. The brightness of stars depends on their size, temperature, and distance from Earth.   _____

_____   5. All stars are white.   _____

_____   6. Stars appear to rise in the east and set in the west, just like the Sun.   _____

Evidence from the text:

1. _____
2. _____
3. _____
4. _____
5. _____
6. _____

Cosmos Episode Planning Sheet

Directions: Write a script for a 3–5 min. video called "Cosmos: The Brightness of Stars" that explains why the Sun appears larger and brighter than other stars, even though it is a small to average-size star. Use the prompts below as a guide for your script.

Introduction
Introduce yourself.
_____
_____

State the idea you will be explaining.
_____
_____

Explanation
Explain why the Sun appears larger and brighter than other stars, even though it is a small to average-size star. Use scientific evidence in your explanation.
_____
_____
_____

Visual Aids or Models
What visual aids or models will you include in your video? Will you include music?
_____
_____
_____

Fascinating Facts
What fascinating facts about stars will you include to inspire your audience to learn more?
_____
_____
_____

Teacher Checkpoint ☐

Name : _____

Cosmos Episode Scoring Rubric

Write and record a 3–5 min. video called "Cosmos: The Brightness of Stars" that explains why the Sun appears so much larger and brighter than other stars, even though it is a small to average-size star. Your video will be evaluated according to the criteria in the table below.

Score	Criteria
____ 4 ____ 3 ____ 2 ____ 1	***Introduction:*** The host and topic are clearly introduced.
____ 4 ____ 3 ____ 2 ____ 1	***Explanation:*** A clear and accurate explanation is given for why the Sun appears larger and brighter than other stars. Scientific evidence is included.
____ 4 ____ 3 ____ 2 ____ 1	***Visual Aids or Models:*** Appropriate visual aids or models are used to enhance the explanation. (Music is optional.)
____ 4 ____ 3 ____ 2 ____ 1	***Fascinating Facts:*** Two or more facts about stars are included to inspire the viewer to want to learn more.

4—Excellent   3—Above Average   2—Average   1—Below Average

_____ Total/16 Points

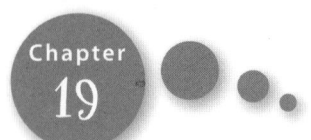

Name : _____

STEM at Home

Dear _____,

At school, we have been learning about **stars**—their sizes, colors, temperatures, brightness, and distances from Earth.

I learned that: _____
_____

My favorite part of the lesson was: _____
_____

At home, we can watch a video about Voyager 1 and Voyager 2, which have traveled farther into space than any other spacecraft.

Search "Voyager, Humanity's Farthest Journey" in your web browser to find the website at *www.jpl.nasa.gov/video/details.php?id=980*.

After we watch the video, we can visit the Voyager website (*http://voyager.jpl.nasa.gov*) to collect the following information:

1. In what year were the Voyager spacecraft launched? _____

2. Where are Voyager 1 and Voyager 2 now?
   _____
   _____

3. What things are contained on the Golden Record carried by the Voyagers?
   _____
   _____

Together, we can discuss what we would put on our own Golden Record!

National Science Teachers Association

From Trash to Treasure

Description

After reading a true story about a woman in The Gambia who started a plastic bag upcycling project that transformed her community, students learn the benefits and risks of using plastic, explore ways that others are solving the plastic pollution problem, and then design their own upcycled plastic product.

Suggested Grade Levels: 3–5

LESSON OBJECTIVES Connecting to the *Framework*		
Science and Engineering Practices	**Disciplinary Core Ideas**	**Crosscutting Concept**
Constructing Explanations and Designing Solutions Obtaining, Evaluating, and Communicating Information	**ESS3.C:** Human Impacts on Earth Systems **ETS1.A:** Defining and Delimiting Engineering Problems **ETS2.B:** Influence of Science, Engineering, and Technology on Society and the Natural World	Cause and Effect

Featured Picture Books

TITLE: **One Plastic Bag: Isatou Ceesay and the Recycling Women of The Gambia**
AUTHOR: **Miranda Paul**
ILLUSTRATOR: **Elizabeth Zunon**
PUBLISHER: **Millbrook Press**
YEAR: **2015**
GENRE: **Narrative Information**
SUMMARY: *The inspirational true story of Isatou Ceesay and the women of The Gambia who turned the problem of plastic bag pollution into a marketable product—purses. As a result, the women changed their community for the better by cleaning up the plastic trash and providing income for themselves.*

TITLE: **Recycling Crafts (Craft Attack!)**
AUTHOR: **Annalees Lim**
PUBLISHER: **Gareth Stevens**
YEAR: **2014**
GENRE: **Non-Narrative Information**
SUMMARY: *This book gives detailed instructions (including photos) for upcycling trash into useful products—from water bottle bracelets to CD case frames.*

Picture-Perfect STEM Lessons, 3–5

347

Chapter 20

Time Needed

This lesson will take several class periods. Suggested scheduling is as follows:

Day 1: Engage with *One Plastic Bag* Read-Aloud and **Explore** with Plastics Everywhere

Day 2: Explain with Interactive Article: Plastics and Exit Slip

Day 3: Elaborate with *Recycling Crafts* Read-Aloud and Trash to Treasure Design Challenge

Day 4 and beyond: Elaborate with Trash to Treasure Design Challenge (continued) and **Evaluate** with Evaluate Your Product

Materials

For One Plastic Bag *Read-Aloud*

- World map or globe

For Plastics Everywhere (per student)

- Piece of scrap paper or an index card
- Clipboard

For Interactive Article: Plastic

- Computers or tablets for each student, or a projector for whole-class viewing
- Index card (1 per student)

For Trash to Treasure Design Challenge

- Clean, used plastic products (bags, bottles, etc.)
- Supplies for upcycling their plastic products

> **SAFETY**
> - Use caution when handling bottles. They may have sharp edges that can cut skin.
> - Have students wash hands with soap and water after completing this activity.

Student Pages

- Interactive Article: Plastics
- Trash to Treasure Letter Home
- Trash to Treasure Instruction Sheet
- Evaluate Your Product
- STEM at Home

Background for Teachers

Plastic is everywhere! We use it every day in so many ways. Single-use plastic products are convenient and more durable than other materials, such as wood, metal, glass, or paper. But plastic is manufactured with petroleum, which is a limited resource, and plastic pollution is becoming a global problem. Many types of plastic are nearly indestructible, so when they end up in a landfill, they can take hundreds of years to degrade.

In this lesson, students learn that there are many possible solutions to the plastics problem. Most of us are familiar with the "3Rs" (reduce, reuse, recycle), but people concerned with plastic pollution are advocating that those three tasks are not enough. Three more *R*s have been added to the recommendations (refuse, reinvent, and replace). *Refuse* means to stop using single-use plastics such as straws, plastic bottles, plastic grocery bags, and plastic zipper bags. *Reinvent*, also known as upcycling, means coming up with ways to use discarded materials to create new products of greater value than the original product. Students learn about a project in The Gambia that upcycles plastic grocery bags into purses. *Replace* means to come up with new technologies to substitute for plastics. Currently, research is being done with nanocarbons, nanoclays, and even mushrooms to replace plastic with biodegradable materials. There are several different versions of the 6Rs in use. Some versions use *rethink* instead of *reinvent* or have added *repair*. But no matter the particular *R*s that are used, the big idea is the same: Cut down on the amount of plastic pollution to make a healthier Earth for all.

This lesson engages students in the idea of reinventing, or upcycling, with the inspirational story of Isatou ("EYE-sa-too") Ceesay. She was a remarkable woman from a small African nation called The Gambia who transformed her community by designing a creative solution to a real-world issue. In the 1980s in villages throughout The Gambia, a switch from using baskets made of natural materials to non-biodegradable plastic bags led to a problem that Isatou noticed: The plastic bags were accumulating in ugly heaps alongside roads. Water pooled in them, bringing mosquitos and disease, and burning or burying them caused problems for humans, animals, and the environment. She led a group of five other women, including Peace Corps volunteer Peggy Sedlak, who together figured out a way to turn plastic bags into beautiful purses that could be sold. Today, Isatou's village is cleaner, and the entire ecosystem is healthier. People from other towns travel to her village to learn the craft of upcycling, the purses continue to be purchased, and the women contribute some of the proceeds to a local community center. In 2012, that center also became the region's first public library. This true story shows how one person's actions really can make a difference in the world!

Isatou's story is a great example of how individuals and communities use science ideas to protect the Earth's resources and the environment. Students continue to unpack this theme with an interactive article in which they explore a variety of *other* ways individuals and communities around the world are working to solve the plastic pollution problem. Then, students are presented with a design challenge during which they invent their own upcycled product, according to specific criteria and constraints. They learn that although no perfect solutions exist, how well a solution meets the criteria and how well it takes the constraints into account determines its success. The lesson also gives students the opportunity to experience the science and engineering practices of constructing explanations and designing solutions and obtaining, evaluating, and communicating information, as well as the crosscutting concept of cause and effect— all in the context of how humans can influence the environment.

Chapter 20

engage

One Plastic Bag Read-Aloud

Making Connections: Text to Self

Before reading, show students a plastic bag. *Ask*

? What do you use plastic bags for?

? Where do you see plastic bags in use?

? What do you do with plastic bags once you are done with them?

? Have you ever seen plastic-bag litter in your neighborhood?

Show students the cover of *One Plastic Bag: Isatou Ceesay and the Recycling Women of The Gambia,* and introduce the author, Miranda Paul, and the illustrator, Elizabeth Zunon. *Ask*

? Where is The Gambia?

Use a classroom globe or world map to show students that The Gambia is in West Africa. It is long and narrow and the smallest country on the African continent. Point out the Gambia River and the village of Njau, where the story takes place. Tell students that English is the official language of The Gambia, but there are several other native languages spoken. One of these languages is Wolof. Several words throughout the book are Wolof words. See the pronunciation guide in the back of the book to familiarize yourself with the pronunciations. You can also share with students an excellent PowerPoint presentation about The Gambia from Miranda Paul's website (see "Websites" section).

> **Connecting to the Common Core**
> **Reading: Informational Text**
> KEY IDEAS AND DETAILS: 3.1, 4.1, 5.1

Read the book aloud, or show students a video from the Library of Congress of the author reading the book (see "Websites" section). After reading, *ask*

? In the beginning of the story, Isatou sees plastic bags as something useful and even beautiful. By the middle of the story, how have her ideas changed? (She sees the plastic bags as ugly.)

? What happened to cause her to change her mind? (The goats were eating the plastic bags and dying. The plastic bags were piling up in trash heaps.)

? How did Isatou solve the problem of the plastic bags? (She found a way to make something useful out of them.)

? How did people treat Isatou when she first started the project? (People laughed, except her friends who helped her.)

? How did Isatou respond? (She worked with her friends in secret.)

? What happened when Isatou introduced the purses at the market? (People laughed at first, but then they started buying them.)

Tell students that people all over the world have been inspired by Isatou's story and have purchased her purses. Read the Author's Note at the end of the book, which explains that Njau is much cleaner, the goats are healthier, and the gardens grow better since Isatou began her upcycling project. This true story shows how one person's actions really can make a difference in the world!

explore

Plastics Everywhere

Tell students that Americans use and throw away an estimated 100 *billion* plastic bags every year. But we also use many other plastic products every day. Give each student a piece of scrap paper or an index card and a clipboard. Tell students that you would like them to walk around the room silently and make a list of all the things they can find in the room that are made of plastic. You may want to play a song while they make their lists and have them return to their seats when the song is over. You could also do this activity in any other room of your school such as the cafeteria, gym, or art room. The goal is

LOOKING FOR PLASTICS IN OUR CLASSROOM

to have a long list of plastic products for students to refer to during the following discussion. After students finish making their lists, *ask*

- ? How do these plastic objects make our lives more convenient? (Answers will vary.)
- ? Which items on your list can be used just once? (Answers will vary. Have students circle these items on their lists.)
- ? Which items on your list can be used over and over? (Answers will vary.)
- ? What happens to plastic when it is thrown away? (Answers will vary.)
- ? Can all plastic products be recycled? (Answers will vary.)
- ? What plastic items can you think of that you used before you came to school today? (Answers will vary, but students may say a toothbrush, milk jug, etc.)

explain

Interactive Article: Plastic

Making Connections: Text to World

Connecting to the Common Core
Reading: Informational Text
KEY IDEAS AND DETAILS: 3.1, 4.1, 5.1

Tell students that although using plastic products can be convenient, plastic trash is creating a worldwide pollution problem. Give students a copy of the Interactive Article: Plastic student page. This article includes several websites and videos for students to explore and seven questions for them to answer, so students will need to have access to a computer or tablet. (If students do not have access to computers, students can read each section on a printed copy of the article and then you can share the links with the whole class on a projector.) Individually or in pairs, students will read each section of the article, and then use the links to open web pages to view a brief video or read a short passage. Pairing websites and media clips with text can help students build background knowledge and access new learning, and it gives them the opportunity to synthesize information. Students will not only read about the topics in the article but also see and hear them as well.

After students have had a chance to read the article and visit the websites, discuss their answers to the questions in the article. The questions and answers are as follows:

1. How does so much plastic end up in the ocean? (It gets into drains and then into waterways that lead to the ocean.)
2. What surprised you about how plastic is made? (Answers will vary, but students may be surprised to learn that plastic is made from oil.)
3. What innovative features do the 4-in-1 shopping bags have? (They have wide handles, one is insulated, and they all fit in a small pouch.)
4. What plastic codes does your local recycling center accept? (Answers will vary.)
5. What changes did Team Marine make in its community? (The team members were able to ban the use of plastic grocery bags in their community.)
6. What products did you see Terracycle upcycling in the video? (Answers may include butter tubs, toothpaste boxes, drink pouches, T-shirts, and plastic bottles.)
7. How are the mushroom packaging materials made? (Biological waste material such

as cotton shells are placed in a mold with the mushrooms, and the mushrooms grow around them filling in all the space.)

Explain that although plastic trash has some harmful effects on communities, this interactive article contains some good examples of people positively influencing their communities by using science and engineering to reduce, reuse, recycle, refuse, reinvent, and replace plastic products.

Exit Slip

Synthesizing

After discussing the article, give students index cards and ask each student to write on a card his or her answer to this question:

? What is the main idea of this article? Keep your answer to 140 characters or less (just like a Tweet!).

Collect the cards and use them as an informal evaluation of students' learning so far.

elaborate

Recycling Crafts Read-Aloud

Making Connections: Text to Text

Now that students have learned about the 6Rs, *ask*

? Which of the *R*s did Isatou Ceesay demonstrate in the book *One Plastic Bag*? (reinvent)

? What is another word for *reinventing*? (upcycling)

? What does Isatou have in common with the individuals and communities featured in the article? (Isatou and the people in the article all came up with creative solutions to help solve the plastic pollution problem.)

> Connecting to the Common Core
> **Reading: Informational Text**
> CRAFT AND STRUCTURE: 3.5, 4.5, 5.5

Features of Nonfiction

Tell students that people all over the world are reinventing with plastics. Show students the cover of *Recycling Crafts*. Read the introduction on page 4. Explain that this is a nonfiction instructional book, so its structure is a little bit different from other nonfiction books. Show students the examples of the upcycled plastic crafts, such as the water bottle bracelets on page 18, the CD case photo frame on page 28, and the plastic bag weaving on page 30. Point out the following features on each spread that are unique to this nonfiction instruction book:

- Name of the craft
- Catchy description underneath the name
- "You Will Need" list
- Numbered, step-by-step instructions for making the craft with photos

Trash to Treasure Design Challenge

Tell students that you have a design challenge for them: Turn trash into treasure—plastic trash to be exact! Challenge students to upcycle something plastic that has already been used, such as a plastic bottle, butter tub, plastic cup, milk jug, or other item. Explain that designers and engineers always have desired features or outcomes in mind when designing solutions to problems or design challenges, whether those solutions are projects, products, systems, or technologies. These desired features or outcomes are known as *criteria*. The criteria for their upcycled projects are as follows:

- Must be of higher value than the used materials it is made of
- Must be fairly long-lasting (not single-use only)
- Must be simple enough for someone else to make

UPCYCLED PLASTIC BOTTLE PIGGY BANK

UPCYCLED PLASTIC BAG JUMP ROPE

Explain that designers and engineers also have to work within constraints when designing solutions, whether they are projects, products, systems, or technologies. *Constraints* are typically limits on time, materials, and money. The constraints for their upcycled projects are as follows:

- Plastic materials must be *used*, not new.
- Plastic materials must be cleaned before creating the product.
- The challenge must be completed by the due date.

Explain that no perfect solutions exists, but how well a solution meets the specified criteria and how well it takes the constraints into account determines

UPCYCLED PLASTIC BAG MAT

UPCYCLED PLASTIC BAG BASKET

Picture-Perfect STEM Lessons, 3–5

its success. Students will need to keep this in mind for their Trash to Treasure project to be a success!

If time permits, show students the video of Isatou explaining in detail how to upcycle plastic bags into purses (see "Websites" section).

Fill out the due date on the Trash to Treasure Letter Home, and sign your name. Then, indicate at the bottom of the page whether the assignment is to be completed at home or at school, and send the letter home. If they need ideas, many online resources have creative upcycling ideas (see "Websites" section for projects for upcycling plastic bottles). Have students make their product at home or bring in the materials to make it at school.

> Connecting to the Common Core
> **Writing**
> PRODUCTION AND DISTRIBUTION OF WRITING: 3.4, 4.4, 5.4

After all students have created their upcycled products, give each student a copy of the Trash to Treasure Instruction Sheet student page. Students will teach others how to make their products by writing down the step-by-step instructions, showing a drawing or photograph of the finished product, and providing a materials list. They will also come up with a slogan for their products. Have them complete their instruction sheets, and then put all of the sheets together to make a class book. Share the book with students, and have the class compare the different ways all of the upcycled products help solve the problem of plastic pollution. (*Optional:* Have students create an instructional video showing how to make their upcycled products.)

evaluate

Evaluate Your Product

> Connecting to the Common Core
> **Writing**
> PRODUCTION AND DISTRIBUTION OF WRITING: 3.4, 4.4, 5.4

After students have completed their projects and instruction sheets, tell them that you would like each student to evaluate his or her product and explain how it helps solve the problem of plastic pollution. Give each student a copy of the Evaluate Your Product student page.

First, students must evaluate whether their products met the criteria and stayed within the constraints. Then, they are asked to explain how their upcycled products help the Earth and the environment. Each student should include the following information in his or her explanation:

- The name of the new, upcycled product and what it is designed to do
- What used material the product is made of (plastic bottle, plastic bag, etc.)
- How upcycling the plastic item helps protect Earth's resources and environment

STEM at Home

Have students complete the "I learned that ..." and "My favorite part of the lesson was ..." portions of the STEM at Home student page as a reflection on their learning. They may choose to do the following at-home activity with an adult helper and share their results with the class. If students do not have access to the internet or these materials at home, you may choose to have them complete this activity at school.

"At home, we can watch a video of a college student who turns e-waste into art."

Search "E-Waste Into Art With Robb Godshaw" to find the video: at www.pbslearningmedia.org/resource/d231cb0a-15c3-43ab-a99d-6e253c5c1e31/e-waste-into-art-with-robb-godshaw.

"After we watch the video, we can answer these questions [answers shown in parentheses]:

1. What is e-waste? (electronic waste)
2. What kinds of things does Robb Godshaw make from e-waste? (rainbow projectors, microscopes, walk/don't walk mats)

3. How is what Robb Godshaw does *similar* to what Isatou Ceesay did in the book *One Plastic Bag*? (They both took discarded items and made something. Robb makes art and Isatou's purses could be considered artistic.)

4. How is what Robb Godshaw does *different* from what Isatou Ceesay did in the book *One Plastic Bag*? (They used different materials and made things for different purposes. Robb's art is intended to be admired and thought about in an art gallery setting, whereas Isatou's purses are intended to be sold and used.)

5. What discarded items do you think you could use to make art? (Answers will vary.)"

For Further Exploration

This section is provided to help you encourage your students to use the science and engineering practices in a more student-directed format. This box lists questions and challenges related to the lesson that students may select to research, investigate, or innovate. Students may also use the questions as examples to help them generate their own questions. After selecting one of the questions in the box or formulating their own questions, students can individually or collaboratively make predictions, design investigations or surveys to test their predictions, collect evidence, devise explanations, design solutions, or examine related resources. They can communicate their findings through a science notebook, at a poster session or gallery walk, or by producing a media project.

Research

Have students brainstorm researchable questions:

? What do plastics engineers do?

? What is the Great Pacific Garbage Patch?

? How does plastic trash affect wildlife?

Investigate

Have students brainstorm testable questions to be solved through science or math:

? Survey your friends and family. How often do they use plastic grocery bags? (Response choices might include "every time I shop," "often," "once in a while," or "never"). Graph the results, then analyze your graph. What can you conclude? You may want to follow up by asking why they don't use reusable bags every time they shop.

? How many items are typically packed into reusable grocery bags by a bagger?

? Take an inventory of your own pantry or food cabinets at home. How many items are packed in plastic, glass, cardboard, paper, and metal? Graph the results, then analyze your graph. What can you conclude?

Picture-Perfect STEM Lessons, 3–5

Innovate

Have students brainstorm problems to be solved through engineering:

? What other products made from upcycling plastic bags can you design?

? What could you design to help people remember to use their reusable shopping bags when they go to the store?

? Go to www.boomerangbags.org to find out about an innovative program for reducing plastic pollution. Can you design your own boomerang bag?

Websites

"39 Amazing Things to Make From Plastic Bottles"
www.tipsbulletin.com/39-amazing-things-to-make-from-plastic-bottles

Miranda Paul Reading *One Plastic Bag* at the Library of Congress (video)
www.loc.gov/today/cyberlc/feature_wdesc.php?rec=7058

One Plastic Bag Teacher Resources Page (includes an author-created PowerPoint presentation about The Gambia)
http://oneplasticbag.com/teacher-resources

Isatou Ceesay Teaching How to Make Purses From Plastic Bags (video)
www.youtube.com/watch?v=1zZw7yrl22M

More Books to Read

Inches, A. 2008. *The adventures of a plastic bottle: A story about recycling*. New York: Little Simon.
Summary: This diary of a plastic bottle follows the bottle's adventures from being made from oil to being used, reused, and recycled.

Knight, G. 2012. *Plastic pollution*. Chicago: Heinemann Library.
Summary: From the *Hot Topic* series, this 64-page book is full of information on the problem of plastic pollution and a variety of solutions for solving the problem.

Newman, P. 2014. *Plastic ahoy! Investigating the great pacific garbage patch*. Minneapolis: Millbrook Press.
Summary: A team of researchers went on a scientific expedition to explore the Great Pacific Garbage Patch, where millions of pieces of plastic have collected. Read about their expedition and the discoveries they made.

Scheunemann, P. 2013. *Trash to treasure: A kids' upcycling guide to crafts*. Minneapolis: Mighty Media Junior Readers.
Summary: With easy step-by-step instructions, this book will help kids get creative and upcycle their trash into handmade treasures.

Ward, D. J. 2012. *What happens to our trash?* New York: Harper Collins
Summary: This *Let's-Read-and-Find-Out Science* book follows the objects we throw away and how they end up in a landfill, incinerator, compost bin, or recycling plant.

Name: _____

Interactive Article: Plastic

Plastic Is Everywhere

Chances are you have used hundreds of plastic items already today. From the toothbrush you used this morning, to the plastic zipper bag that holds your sandwich, to the plastic chair that you sit on—plastic is everywhere!

Fantastic Plastic

Plastic is amazing! It is durable, lightweight, flexible, and fairly inexpensive to make, and it can be molded into any shape or size. Plastic has become an important part of our lives. Plastic products make our lives easier, and some plastic products can even save our lives, such as bicycle helmets, plastic coatings on electrical wires, and plastic insulation in our refrigerators that keeps our food cold and safe to eat.

The Problem With Plastics

Although plastic is incredibly useful, our years of prolific use are creating a global pollution problem. The problem with plastics is that when we throw them away, they never really go away. Americans throw away 30 million tons of plastic per year. Unlike wood, paper, and other natural materials, plastic does not biodegrade, or decompose. Plastic trash is filling up our landfills and waterways, and a huge amount is collecting in our oceans. You can watch this video to find out how so much of the plastic we throw away ends up in the ocean:

Why Is Plastic Marine Debris So Common?
www.pbslearningmedia.org/resource/noaa-interactive-32/why-is-plastic-marine-debris-so-common-ocean-today

1. How does so much plastic end up in the ocean?

   _____
   _____
   _____

Possible Solutions

You've probably heard of the 3Rs: reduce, reuse, recycle. Those are good things to do, but groups who study the plastic pollution problem say those solutions are not enough. They have added three more *R*s to the list of solutions: refuse, reinvent, and replace. Below are descriptions of each of these 6Rs and examples of how individuals and communities are applying each one.

Reduce: One solution to the plastic pollution problem is to reduce the amount of plastics we use. Plastics are made from oil, a fossil fuel. Fossil fuels are limited, and producing plastic creates pollution. Reducing the amount of plastics we use would result in less plastic being made, which means less oil being used and less pollution being generated. Watch this video to learn more about how plastic is made:

How Plastic Is Made
www.youtube.com/watch?v=f3BjWvTT9Ro

2. What surprised you about how plastic is made?

   _____
   _____
   _____

Picture-Perfect STEM Lessons, 3–5

Name : _____

Reuse: Another solution is to reuse plastics. We can do this by washing and reusing disposable plastic containers and bringing reusable shopping bags to the store. Gordon Dancy, the man who invented the cheap plastic shopping bag, did so because he thought replacing paper bags would save trees. But he realized that the plastic bags were becoming an environmental hazard. So when he retired, he started a recycling company. His daughter now has her own company that promotes the reuse of plastics. Check out her company's website to see her innovative 4-in-1 reusable shopping bag:

MyECO Shopping Bags
www.getmyeco.com/shopping

3. What innovative features does the 4-in-1 shopping bag have?

   _____

   _____

   _____

Recycle: When we must use plastics, we can recycle them so that they can be melted and made into something new. Currently, only about 8% of plastic trash is recycled. Not all plastic is recyclable, so it is important to find out what kinds of plastics your local recycling center accepts. You can use the Earth911 website to find out. Visit this website and enter your zipcode in the "Find a Recycling Location" search box:

Earth911 Recycling Center Search **http://earth911.com/recycling-center-search-guides**

4. What plastic codes does your local recycling center accept?

   _____

   _____

   _____

Refuse: Some groups have recommended that one solution to the plastic pollution problem is for people to refuse single-use plastics. These are plastics that are used only once and then thrown away, such as straws, plastic bags, and plastic water bottles. In fact, 33 percent of the plastic that Americans use is single-use plastic. To see how some high school students encouraged their community to refuse the use of plastic bags, check out their video:

Team Marine
www.pbslearningmedia.org/resource/yvcc-sci-teammarine/team-marine

5. What changes did Team Marine make in its community?

   _____

   _____

   _____

Reinvent: We can rethink ways to make something useful out of used plastic objects. *Upcycling* means to reuse discarded objects or materials to create a product of a higher quality or value than the original. In the book *One Plastic Bag*, Isatou Ceesay and other women of The Gambia came up with a way to upcycle plastic bags into purses. People all over the world are coming up with

Name: _____

ways to upcycle plastic trash. Check out this company called Terracycle, which specializes in upcycling:

Terracycle Welcome Video
www.youtube.com/user/TerraCycleVideos

6. What products did you see Terracycle upcycling in the video?

_____
_____
_____

Replace: We need to find ways to replace some of the plastic products we use with other materials. Chemists are working on creating materials that have properties similar to plastic but are made with biodegradable materials. The most promising replacements for plastic are made of *nanocarbons*. Nanocarbons are not made of oil and can do all the things plastics can do. They are currently being used to make bowling balls, golf balls, and sports equipment. The problem with nanocarbons is that they are more expensive to produce than plastics made from oil. Scientists in India are working on *nanoclay*, which is similar to plastic, but biodegrades more quickly. An American designer has come up with a way to use mushrooms to make packaging materials to replace plastics. Check out his TED talk to find out more:

Eben Bayer's TED Talk: Are Mushrooms the New Plastic?
www.ted.com/talks/eben_bayer_are_mushrooms_the_new_plastic?language=en

7. How are the mushroom packaging materials made?

_____
_____
_____

Conclusion

We depend on plastic for so many things. But we need to be aware that once we throw plastic products away, they never really go away. By practicing the 6Rs, we can all pitch in to help solve the problem of plastic pollution. Several groups have different versions of the 6Rs. Some use *rethink* instead of *reinvent* or have added *repair*, making 7Rs. But no matter the particular *Rs* used, the big idea is the same: Cut down on the amount of plastic pollution to make a healthier Earth for all who inhabit it.

Name: _____

Trash to Treasure Letter Home

Dear Parents and Guardians,

We are learning about "upcycling" as one solution to the plastic pollution problem. Your student has been challenged to design an upcycled product out of something plastic that has been used before (e.g., plastic bottles or bottle caps, plastic bags, or disposable plastic containers). More information about the challenge, including criteria and constraints, is listed below:

Challenge: Design an upcycled product out of a used plastic item.

Criteria (desired features or outcomes for the product):

- Must be of higher value than the materials it is made of
- Must be fairly long-lasting (not single use only)
- Must be simple enough for someone else to make

Constraints (limits on time and materials):

- Plastic materials are *used*, not new.
- Plastic materials must be cleaned before creating the product.
- The challenge must be completed by _____.

☐ Assignment is to be completed at home.

☐ Assignment is to be completed at school. Please send in materials by _____.

Thank you,

_____

Name: _____

Trash to Treasure Instruction Sheet

From _____ **to** _____

Designer: _____

Product Name: _____

Description or Catchy Slogan: _____

Materials Needed:

Photo or Drawing:

Step-by-Step Instructions (numbered):

_____
_____
_____
_____
_____
_____

Picture-Perfect STEM Lessons, 3–5

Evaluate Your Product

Name : _____

Challenge: Design an upcycled product out of a used plastic item.

Did your product meet the criteria and stay within the constraints?

Criteria (✓):

☐ Must be of higher value than the materials it is made of

☐ Must be fairly long-lasting (not single use only)

☐ Must be simple enough for someone else to make

Constraints (✓):

☐ Plastic materials are *used*, not new.

☐ Plastic materials were cleaned before creating the product.

☐ Product was finished by the due date.

Explain how your upcycled product helps Earth and the environment. In your explanation, include the following information:

- The name of the new, upcycled product and what it is designed to do
- What used material the product is made of (plastic bottle, plastic bag, etc.)
- How upcycling the plastic item helps protect Earth's resources and the environment

_____
_____
_____
_____
_____
_____
_____
_____

Name: _____

STEM at Home

Dear _____,

At school, we have been learning about how discarded plastic can be **upcycled** into useful things.

I learned that: _____
_____

My favorite part of the lesson was: _____
_____

At home, we can watch a video of a college student who turns e-waste into art.

Search "E-Waste Into Art With Robb Godshaw" in your web browser to find the video at *www.pbslearningmedia.org/resource/d231cb0a-15c3-43ab-a99d-6e253c5c1e31/e-waste-into-art-with-robb-godshaw*.

After we watch the video together, we can answer these questions:

1. What is e-waste? _____

2. What kinds of things does Robb Godshaw make from e-waste?
   _____

3. How is what Robb Godshaw does *similar* to what Isatou Ceesay did in the book *One Plastic Bag*?
   _____

4. How is what Robb Godshaw does *different* from what Isatou Ceesay did in the book *One Plastic Bag*?
   _____

5. What discarded items do you think you could use to make art?
   _____

Appendix

Alignment With the *Next Generation Science Standards*

Chapter 6: The Inventor's Secret

Performance Expectation

3-5-ETS1-3: Plan and carry out fair tests in which variables are controlled and failure points are considered to identify different aspects of a model or prototype that can be improved.

Science and Engineering Practices	Disciplinary Core Ideas	Crosscutting Concept
Planning and Carrying Out Investigations Plan and conduct an investigation collaboratively to produce data to serve as the basis for evidence, using fair tests in which variables are controlled and the number of trials considered. **Constructing Explanations and Designing Solutions** Generate and compare multiple solutions to a problem based on how well they meet the criteria and constraints of the design solution.	**ETS1.C: Optimizing the Design Solution** Different solutions need to be tested in order to determine which of them best solves the problem, given the criteria and constraints. **ETS2.B: Influence of Engineering, Technology, and Science on Society and the Natural World** Engineers improve existing technologies or develop new ones to increase their benefits, decrease known risks, and meet societal demands.	**Structure and Function** Different materials have substructures, which can sometimes be observed. Substructures have shapes and parts that serve functions.

Note: The activities in this lesson will help students move toward the performance expectation listed, which is the goal after multiple activities. However, the activities will not by themselves be sufficient to reach the performance expectations.

Appendix

Chapter 7: Mesmerized

Performance Expectation

3-5-ETS1-3: Plan and carry out fair tests in which variables are controlled and failure points are considered to identify aspects of a model or prototype that can be improved.

Science and Engineering Practices	Disciplinary Core Ideas	Crosscutting Concept
Asking Questions and Defining Problems Ask questions about what would happen if a variable is changed. **Planning and Carrying Out Investigations** Plan and conduct an investigation collaboratively to produce data to serve as the basis for evidence, using fair tests in which variables are controlled and the number of trials considered. **Engaging in Argument From Evidence** Make a claim about the merit of a solution to a problem by citing relevant evidence about how it meets the criteria and constraints of the problem.	**ETS1.C: Optimizing the Design Solution** Different solutions need to be tested in order to determine which of them best solves the problem, given the criteria and constraints. **ETS2.B: Influence of Engineering, Technology, and Science on Society and the Natural World** Engineers improve existing technologies or develop new ones to increase their benefits, decrease known risks, and meet societal demands.	**Cause and Effect** Cause-and-effect relationships are routinely identified, tested, and used to explain change.

Note: The activities in this lesson will help students move toward the performance expectation listed, which is the goal after multiple activities. However, the activities will not by themselves be sufficient to reach the performance expectations.

Chapter 8: Wind It Up

Performance Expectations

3-PS2-2: Make observations and/or measurements of an object's motion to provide evidence that a pattern can be used to predict future motion.

3-5-ETS1-3: Plan and carry out fair tests in which variables are controlled and failure points are considered to identify aspects of a model or prototype that can be improved.

Science and Engineering Practices	Disciplinary Core Ideas	Crosscutting Concept
Asking Questions and Defining Problems Ask questions that can be investigated and predict reasonable outcomes based on patterns such as cause-and-effect relationships. **Analyzing and Interpreting Data** Represent data in tables and/or various graphical displays to reveal patterns that indicate relationships. **Constructing Explanations and Designing Solutions** Construct an explanation of observed relationships.	**PS2.A: Forces and Motion** The patterns of an object's motion in various situations can be observed and measured; when that past motion exhibits a regular pattern, future motion can be predicted from it. **ETS1.C: Optimizing the Design Solution** Different solutions need to be tested in order to determine which of them best solves the problem, given the criteria and constraints.	**Patterns** Patterns of change can be used to make predictions. **Cause and Effect** Cause-and-effect relationships are routinely identified, tested, and used to explain change.

Note: The activities in this lesson will help students move toward the performance expectation listed, which is the goal after multiple activities. However, the activities will not by themselves be sufficient to reach the performance expectations.

Chapter 9: Light It Up!

Performance Expectations

4-PS3-4: Apply scientific ideas to design, test, and refine a device that converts energy from one form to another.

3-5-ETS1-1: Define a simple design problem reflecting a need or a want that includes specified criteria for success and constraints on materials, time, or cost.

Science and Engineering Practices	Disciplinary Core Ideas	Crosscutting Concept
Constructing Explanations and Designing Solutions Apply scientific ideas to solve design problems. **Analyzing and Interpreting Data** Represent data in tables and/or various graphical displays to reveal patterns that indicate relationships.	**PS3.B: Conservation of Energy and Energy Transfer** Energy can also be transferred from place to place by electric currents, which then can be used locally to produce motion, sound, heat, or light. The currents may have been produced to begin with by transforming the energy of motion into electrical energy. **ETS1.A: Defining Engineering Problems** Possible solutions to a problem are limited by available materials and resources (constraints). The success of a designed solution is determined by considering the desired features of a solution (criteria). Different proposals for solutions can be compared on the basis of how well each one meets the specified criteria for success or how well each takes the constraints into account. **ETS2.B: Influence of Engineering, Technology, and Science on Society and the Natural World** Engineers improve existing technologies or develop new ones to increase their benefits, decrease known risks, and meet societal demands.	**Energy and Matter** Energy can be transferred in various ways and between objects.

Note: The activities in this lesson will help students move toward the performance expectation listed, which is the goal after multiple activities. However, the activities will not by themselves be sufficient to reach the performance expectations.

Appendix

Chapter 10: Burn

Performance Expectations

5-PS1-2: Measure and graph quantities to provide evidence that regardless of the type of change that occurs when heating, cooling, or mixing substances, the total weight of matter is conserved.

5-PS1-4: Conduct an investigation to determine whether the mixing of two or more substances results in new substances.

Science and Engineering Practices	Disciplinary Core Ideas	Crosscutting Concept
Obtaining, Evaluating, and Communicating Information Obtain and combine information from books and/or other reliable media to explain phenomena.	**PS1.B: Chemical Reactions** When two or more substances are mixed, a new substance with different properties may be formed. No matter what reaction or change in properties occurs, the total weight of the substances does not change. **ETS2.B: Influence of Engineering, Technology, and Science on Society and the Natural World** Engineers improve existing technologies or develop new ones to increase their benefits, decrease known risks, and meet societal demands.	**Cause and Effect** Cause-and-effect relationships are routinely identified, tested, and used to explain change.

Note: The activities in this lesson will help students move toward the performance expectation listed, which is the goal after multiple activities. However, the activities will not by themselves be sufficient to reach the performance expectations.

Appendix

Chapter 11: From Edison to the iPod

Performance Expectations

4-PS4-3: Generate and compare multiple solutions that use patterns to transfer information.

3-5-ETS1-2: Generate and compare multiple possible solutions to a problem based on how well each is likely to meet the criteria and constraints of the problem.

Science and Engineering Practices	Disciplinary Core Ideas	Crosscutting Concept
Constructing Explanations and Designing Solutions Generate and compare multiple solutions to a problem based on how well they meet the criteria and constraints of the design solution.	**PS4.C: Information Technologies and Instrumentation** Digitized information can be transmitted over long distances without significant degradation. High-tech devices, such as computers or cell phones, can receive and decode information—convert it from digitized form to voice—and vice versa. **ETS1.C: Optimizing the Design Solution** Different solutions need to be tested in order to determine which of them best solves the problem, given the criteria and constraints. **ETS2.B: Influence of Engineering, Technology, and Science on Society and the Natural World** Engineers improve existing technologies or develop new ones to increase their benefits, decrease known risks, and meet societal demands.	**Stability and Change** Change is measured in terms of differences over time and may occur at different rates.

Note: The activities in this lesson will help students move toward the performance expectation listed, which is the goal after multiple activities. However, the activities will not by themselves be sufficient to reach the performance expectations.

Chapter 12: Better Together

Performance Expectation		
3-LS2-1: Construct an argument that some animals form groups that help members survive.		
Science and Engineering Practices	**Disciplinary Core Ideas**	**Crosscutting Concept**
Analyzing and Interpreting Data Represent data in tables and/or various graphical displays to reveal patterns that indicate relationships. **Using Mathematics and Computational Thinking** Organize simple data sets to reveal patterns that suggest relationships **Engaging in Argument From Evidence** Construct and/or support an argument with evidence, data, and/or a model.	**LS2.D: Social Interactions and Group Behavior** Being part of a group helps animals obtain food, defend themselves, and cope with changes. Groups may serve different functions and vary dramatically in size. **ETS2.A: Interdependence of Science, Engineering, and Technology** Tools and instruments are used to answer scientific questions, while scientific discoveries lead to the development of new technologies.	**Cause and Effect** Cause-and-effect relationships are routinely identified, tested, and used to explain change.

Note: The activities in this lesson will help students move toward the performance expectation listed, which is the goal after multiple activities. However, the activities will not by themselves be sufficient to reach the performance expectations.

Appendix

Chapter 13: Spider Science

Performance Expectation

3-LS3-2: Use evidence to support the explanation that traits can be influenced by the environment.

Science and Engineering Practices	Disciplinary Core Ideas	Crosscutting Concept
Constructing Explanations and Designing Solutions Construct an explanation of observed relationships. Use evidence (e.g., observations, patterns) to support an explanation.	**LS3.A: Inheritance of Traits** Many characteristics of organisms are inherited from their parents. Other characteristics result from individuals' interactions with the environment, which can range from diet to learning. Many characteristics involve both inheritance and environment. **LS3.B: Variation of Traits** Different organisms vary in how they look and function because they have different inherited information. The environment also affects the traits that an organism develops. **ETS2.A: Interdependence of Science, Engineering, and Technology** Science and technology support each other.	**Cause and Effect** Cause-and-effect relationships are routinely identified and used to explain change.

Note: The activities in this lesson will help students move toward the performance expectation listed, which is the goal after multiple activities. However, the activities will not by themselves be sufficient to reach the performance expectations.

Chapter 14: Bionic Animals

Performance Expectations

4-LS1-1: Construct an argument that plants and animals have internal and external structures that function to support survival, growth, behavior, and reproduction.

3-5-ETS1-1: Define a simple design problem reflecting a need or a want that includes specified criteria for success and constraints on materials, time, or cost.

Science and Engineering Practices	Disciplinary Core Ideas	Crosscutting Concept
Engaging in Argument From Evidence Make a claim about the merit of a solution to a problem by citing relevant evidence about how it meets the criteria and constraints of the problem. **Obtaining, Evaluating, and Communicating Information** Read and comprehend grade-appropriate texts and/or other reliable media to summarize and obtain scientific and technical ideas and describe how they are supported by evidence.	**LS1.A: Structure and Function** Plants and animals have both internal and external structures that serve various functions in growth, survival, behavior, and reproduction. **ETS1.A: Defining and Delimiting Engineering Problems** Possible solutions to a problem are limited by available materials and resources (constraints). The success of a designed solution is determined by considering the desired features of a solution (criteria). Different proposals for solutions can be compared on the basis of how well each one meets the specified criteria for success or how well each takes the constraints into account. **ETS2.B: Influence of Engineering, Technology, and Science on Society and the Natural World** Engineers improve existing technologies or develop new ones to increase their benefits, decrease known risks, and meet societal demands.	**Structure and Function** Substructures have shapes and parts that serve functions.

Note: The activities in this lesson will help students move toward the performance expectation listed, which is the goal after multiple activities. However, the activities will not by themselves be sufficient to reach the performance expectations.

Chapter 15: From Seed to Tree

Performance Expectation

5-LS1-1: Support an argument that plants get the materials they need for growth chiefly from air and water.

Science and Engineering Practices	Disciplinary Core Ideas	Crosscutting Concept
Engaging in Argument From Evidence Compare and refine arguments based on an evaluation of the evidence presented.	**LS1.C: Organization of Matter and Energy Flow in Organisms** Plants acquire their material for growth chiefly from air and water. **ETS1.A: Interdependence of Science, Engineering, and Technology** Science and technology support each other.	**Energy and Matter** Energy can be transferred in various ways and between objects.

Note: The activities in this lesson will help students move toward the performance expectation listed, which is the goal after multiple activities. However, the activities will not by themselves be sufficient to reach the performance expectations.

Chapter 16: Hurricane!

Performance Expectations

3-ESS3-1: Make a claim about the merit of a design solution that reduces the impacts of a weather-related hazard.

3-5-ETS1.2: Generate and compare multiple possible solutions to a problem based on how well each is likely to meet the criteria and constraints of the problem.

Science and Engineering Practices	Disciplinary Core Ideas	Crosscutting Concept
Constructing Explanations and Designing Solutions Generate and compare multiple solutions to a problem based on how well they meet the criteria and constraints of the design problem. **Engaging in Argument From Evidence** Make a claim about the merit of a solution to a problem by citing relevant evidence about how it meets the criteria and constraints of the problem.	**ESS3.B: Natural Hazards** A variety of natural hazards result from natural processes. Humans cannot eliminate natural hazards but can take steps to reduce their impacts. **ETS1.B: Developing Possible Solutions** Research on a problem should be carried out before beginning to design a solution. Testing a solution involves investigating how well it performs under a range of likely conditions. At whatever stage, communicating with peers about proposed solutions is an important part of the design process, and shared ideas can lead to improved designs. Tests are often designed to identify failure points or difficulties, which suggest the elements of a design that need to be improved. **ETS2.B: Influence of Engineering, Technology, and Science on Society and the Natural World** Engineers improve existing technologies or develop new ones to increase their benefits, decrease known risks, and meet societal demands.	**Cause and Effect** Cause-and-effect relationships are routinely identified, tested, and used to explain change.

Note: The activities in this lesson will help students move toward the performance expectation listed, which is the goal after multiple activities. However, the activities will not by themselves be sufficient to reach the performance expectations.

Appendix

Chapter 17: Solving the Puzzle Under the Sea

Performance Expectation
4-ESS2-2: Analyze and interpret data from maps to describe patterns of Earth's features.

Science and Engineering Practices	Disciplinary Core Ideas	Crosscutting Concept
Analyzing and Interpreting Data Analyze and interpret data to make sense of phenomena using logical reasoning, mathematics, and/or computation. **Obtaining, Evaluating and Communicating Information** Obtain and combine information from books and other reliable media to explain phenomena.	**ESS2.B: Plate Tectonics and Large-Scale System Interactions** The locations of mountain ranges, deep ocean trenches, ocean floor structures, earthquakes, and volcanoes occur in patterns. Most earthquakes and volcanoes occur in bands that are often along the boundaries between continents and oceans. Major mountain chains form inside continents or near their edges. Maps can help locate the different land and water features of Earth. **ETS2.A: Interdependence of Science, Engineering, and Technology** Science and technology support each other.	**Patterns** Patterns can be used as evidence to support an explanation. **Stability and Change** Change is measured in terms of differences over time and may occur at different rates. Some systems appear stable, but over long periods of time will eventually change.

Note: The activities in this lesson will help students move toward the performance expectation listed, which is the goal after multiple activities. However, the activities will not by themselves be sufficient to reach the performance expectations.

Appendix

Chapter 18: Space Exploration

Performance Expectation		
3-5-EST1-1: Define a simple design problem reflecting a need or a want that includes specified criteria for success and constraints on materials, time, or cost.		
Science and Engineering Practices	**Disciplinary Core Ideas**	**Crosscutting Concept**
Engaging in Argument From Evidence Compare and refine arguments based on an evaluation of the evidence presented. **Obtaining, Evaluating, and Communicating Information** Read and comprehend grade-appropriate texts and/or other reliable media to summarize and obtain scientific and technical ideas and describe how they are supported by evidence.	**ESS1.B: Earth and the Solar System** The orbits of Earth around the Sun and of the Moon around Earth, together with the rotation of Earth about an axis between its North and South poles, cause observable patterns. These include day and night; daily changes in the length and direction of shadows; and different positions of the Sun, Moon, and stars at different times of the day, month, and year. **ETS1.A: Defining and Delimiting Engineering Problems** Possible solutions to a problem are limited by available materials and resources (constraints). The success of a designed solution is determined by considering the desired features of a solution (criteria). Different proposals for solutions can be compared on the basis of how well each one meets the specified criteria for success or how well each takes the constraints into account. **ETS2.A: Interdependence of Science, Engineering, and Technology** Tools and instruments are used to answer scientific questions, while scientific discoveries lead to the development of new technologies. **ETS2.B: Influence of Engineering, Technology, and Science on Society and the Natural World** Engineers improve existing technologies or develop new ones to increase their benefits, decrease known risks, and meet societal demands.	**Systems and System Models** A system can be described in terms of its components and their interactions.

Note: The activities in this lesson will help students move toward the performance expectation listed, which is the goal after multiple activities. However, the activities will not by themselves be sufficient to reach the performance expectations.

Chapter 19: Star Stuff

Performance Expectation		
5-ESS1-1: Support an argument that differences in the apparent brightness of the sun compared to other stars is due to their relative distances from Earth.		
Science and Engineering Practices	**Disciplinary Core Ideas**	**Crosscutting Concept**
Developing and Using Models Develop and/or use models to describe and/or predict phenomena. **Using Mathematics and Computational Thinking** Describe, measure, estimate and/or graph quantities such as area, volume, weight, and time to address scientific and engineering questions and problems. **Engaging in Argument From Evidence** Support an argument with evidence, data, or a model.	**ESS1.A: The Universe and Its Stars** The Sun is a star that appears larger and brighter than other stars because it is closer. Stars range greatly in their distance from Earth.	**Scale, Proportion, and Quantity** Natural objects and/or observable phenomena exist from the very small to the immensely large or from very short to very long time periods.

Note: The activities in this lesson will help students move toward the performance expectation listed, which is the goal after multiple activities. However, the activities will not by themselves be sufficient to reach the performance expectations.

Chapter 20: From Trash to Treasure

Performance Expectations

3-5-ETS1-1: Define a simple problem reflecting a need or a want that includes specified criteria for success and constraints on materials, time, or cost.

5-ESS3-1: Obtain and combine information about ways individuals and communities use science ideas to protect the Earth's resources and environment.

Science and Engineering Practices	Disciplinary Core Ideas	Crosscutting Concept
Constructing Explanations and Designing Solutions Generate and compare multiple solutions to a problem based on how well they meet the criteria and constraints of the design solution. **Obtaining, Evaluating, and Communicating Information** Obtain and combine information from books and/or other reliable media to explain phenomena or solutions to a design problem.	**ESS3.C: Human Impacts on Earth Systems** Human activities in agriculture, industry, and everyday life have had major effects on the land, vegetation, streams, ocean, air, and even outer space. But individuals and communities are doing things to help protect Earth's resources and environments. **ETS1.A: Defining and Delimiting Engineering Problems** Possible solutions to a problem are limited by available materials and resources (constraints). The success of a designed solution is determined by considering the desired features of a solution (criteria). Different proposals for solutions can be compared on the basis of how well each one meets the specified criteria for success or how well each takes the constraints into account. **ETS2.B: Influence of Engineering, Technology, and Science on Society and the Natural World** Engineers improve existing technologies or develop new ones to increase their benefits, decrease known risks, and meet societal demands.	**Cause and Effect** Cause-and-effect relationships are routinely identified, tested, and used to explain change.

Note: The activities in this lesson will help students move toward the performance expectation listed, which is the goal after multiple activities. However, the activities will not by themselves be sufficient to reach the performance expectations.

Index

Page numbers printed in **boldface type** indicate tables, figures, or illustrations.

A
acquired traits, 205, 208, 209–210, **210**, 218
advertising. *See* "Mesmerized" lesson
A Framework for K–12 Science Education
    Common Core State Standards for English Language Arts (CCSS ELA), 25, 27, 29, **30–36**
    Common Core State Standards for Mathematics (CCSS Mathematics), 25, 27, **37–45**
    crosscutting concepts, 25–27, **26**
    disciplinary core ideas, 27, **28**
    and *Next Generation Science Standards (NGSS)*, 25–27, **26**
    See also science and engineering practices
"A Green Way to Fight Fires" video, 147
Alva Award, 168
amplification, 165
analog technology, 163, 167
An Ambush of Tigers, 179, 180, 182–183, 189
Animals That Live in Groups, 179, 180, 186, 188
anticipation guide
    about, 13
    How Mountains Are Made read-aloud, 296–297
    Jump Into Science: Stars, 338–339
arguments, 312
ash, 140

atoms, 139

B
Barretta, Gene, 161
Beaty, Andrea, 61, 64
Beginning Reading and Writing, 2
Berann, Heinrich, 292
"Better Together" lesson, 51, 179–202, 371
    background for, 180–182
    BSCS 5E Model for, 182–189
        elaborate, 187–188
        engage, 182–183
        evaluate, 189, **189**
        explain, 184–187, **184**
        explore, 183–184, **184**
    Common Core connections, 164–170
    description of, 179
    Framework connections, **179**
    further exploration suggestions, 170–171
    materials for, 180
    objectives of, 179
    performance expectations, 371
    picture books for, 179
    student pages for, 180, 193–202
    time needed for, 180
    websites and suggested reading resources, 191–192
Biological Sciences Curriculum Study (BSCS) 5E Model, 5, 17–23
    as constructivist cycle of learning, 17, **17**, 19
    elaborate, 18, **18**, **20**, **21**
    engage, 17, **18**, **20**, **21**
    evaluate, 19, **19**, **20**, **21**

    explain, 18, **18**, **20**, **21**
    explore, 18, **18**, **20**, **21**
    and *Next Generation Science Standards (NGSS)*, 22
    phases of, 17–19
    student roles, 19, **21**
    teacher roles, 19, **20**
    and using picture books for STEM education, 22–23
    See also specific lessons
biomedical engineering, 225, 226
Biomedical Engineering and Human Body Systems, 223, 224, 232–233, 234
"Bionic Animals" lesson, 52, 223–244, 373
    background for, 225–226
    BSCS 5E Model for, 226–235
        elaborate, 232–234
        engage, 226–227, **228**
        evaluate, 234–235, **235**
        explain, 230–232, **231**
        explore, 227–230, **230**
    Common Core connections, 226–235
    description of, 223
    Framework connections, **223**
    further exploration suggestions, 236–237
    materials for, 224–225, **224**, **225**
    objectives of, 223
    performance expectations, 373
    picture books for, 223
    safety precautions for, 224
    student pages for, 225, 238–244
    time needed for, 224
    websites and suggested reading resources, 237

Index

blind taste test, 83–84, **83**, **84**
blind tests, 83, 87
boiling of water, 139
Boy, Were We Wrong About the Solar System, 307, 308, 310–311, 312
brainstorming, Balloon Car Design Challenge, 65
breached levees, 265–266, **266**, 269–270
Bruno, Iacopo, 79, 81
Building Dikes and Levees, 263, 264, 269–270
bulls, 180, 185, 197
Burleigh, Robert, 289, 293
Burn: Michael Faraday's Candle, 137, 138, 140, 142, 144
"Burn" lesson, 26, 137–160, 369
    background for, 139–141
    BSCS 5E Model for, 142–148
        elaborate, 145–147
        engage, 142, **142**
        evaluate, 147–148
        explain, 144–145, **144**
        explore, 142–143, **143**
    Common Core connections, 142–148
    description of, 137
    Framework connections, **137**
    further exploration suggestions, 148–149
    materials for, 138
    objectives of, 137
    performance expectations, 369
    picture books for, 137
    safety precautions for, 138, 142–143, 150
    student pages for, 138
    time needed for, 138
    websites and suggested reading resources, 149
Bybee, Rodger, 22

C

calls, 183
calves, 180, 197
capillary action, 140, 144, 153
card sequencing
    about, 13

Music Player Info Cards, 166–167
card sort
    about, 13
    Boy, Were We Wrong About the Solar System read-aloud, 311–312
Carroll, Kevin, 225, 226, 228, 232, 233, 234
Cassels, Jean, 263, 267
Ceesay, Isatou, 349, 350
chemical changes to matter, 139–140, 145, 153
chlorophyll, 246, 251
chloroplasts, 246, 251
chunking
    about, 14
    Biomedical Engineering and Human Body Systems read-aloud, 232
    Let's Think About the Power of Advertising read-aloud, 85
civil engineer, 270, 271, 272, 284
claims, 81, 92
clans, 180, 197
Clink, ix, 4, 99, 100, 102, 107
clock springs, 101
cloze strategy
    about, 14
    How Things Work: Lightbulbs read-aloud, 120–121
coil springs, 101
Colon, Raúl, 289, 293
combustion, 140, 145, 147
    See also "Burn" lesson
Common Core State Standards for English Language Arts (CCSS ELA), 9–10, 25, 27, 29, **30–36**
Common Core State Standards for Mathematics (CCSS Mathematics), 25, 27, **37–45**
compact fluorescent lamp (CFL), 117–118, **117**, 121–124
compounds, 139
condensation, 139
conservation of matter, 140–141, **141**, 145
constraints, 65, 230, 231, 270, 353

consumers, 81, 92
Contact, 335
controlled variables, 81, 92–93
conversion of energy, 119, 125
Cosmos television show, 333, 335, 337, 340
criteria, 65, 97, 230, 231, 270, 352

D

Deep Space Network (DSN), 335
defensive circles, 181, 197
degradation, 167
designing
    Balloon Car Design Challenge, 65–66
    See also "Bionic Animals" lesson; "From Edison to the iPod" lesson
Design Process, 64, **64**, 67–68, **68**, 74
    See also "The Inventor's Secret" lesson
Design Squad Global video, 64, 65, 67
determining importance
    about, 13
    "Elephant Behavior" video, 183
    If You Hold a Seed read-aloud, 248
    "Kratt's Creatures" video, 186, 187
    Next Time You See a Maple Seed read-aloud, 249–250
    structures and functions T-chart, 227, **228**
    The Inventor's Secret read-aloud, 66
    Timeless Thomas read-aloud, 164
    Wildlife Tracking Technology, 188
digital technology, 163–164, 167
dikes, 269–270
DiPucchio, Kelly, 99, 102
DNA, 204–205
Don't Be Fooled article, 86, 92–94
dual-purpose information books, 6

Dugan, Christine, 307, 312

E
Edison, Thomas, 63, 66–67, 68–69, 117, 161, 163, 166, 168
electricity. *See* "Light It Up!" lesson
electrodes, 117
elephants. *See* "Better Together" lesson
energy
    and changes in states of matter, 139
    conversion/transformation of, 119, 125
energy-efficient lightbulbs, 117–119, **118**, **119**
engineer, 270, 271
ethograms, 179, 181, 184
ethologists, 181
evaporation of water, 139
evidence, 81, 92, 312
exobiology, 335
experimental variables, 81, 92–93

F
Fadell, Tony, 161, 168, 169
fair tests, 81, 86–87, 92–93
family groups, 197
Faraday, Michael, 137, 140, 142, 144
Federal Trade Commission, 81, 93
Federal Trade Commission (FTC), 118–119
filaments, 117
fire lines, 141
fire triangles, 141, 146, 156
5E Instructional Model. *See* Biological Sciences Curriculum Study (BSCS) 5E Model
"Food Ad Tricks" video, 85–86, 89
food stylists, 85
Ford, Henry, 63–64, 66–67
Franklin, Ben, 79, 81, 82, 83
freezing of water, 139
friendship gate, 63
"From Edison to the iPod" lesson, 51, 161–178, 370
    background for, 163–164
    BSCS 5E Model for, 164–170
        elaborate, 168–169
        engage, 164, **164**
        evaluate, 169–170, **169**
        explain, 165–167, **166**, **167**
        explore, 164–165, **165**, 166, **166**
    Common Core connections, 164–170
    description of, 161
    Framework connections, **161**
    further exploration suggestions, 170–171
    materials for, 162
    objectives of, 161
    performance expectations, 370
    picture books for, 161
    student pages for, 162, 172–178
    time needed for, 162
    websites and suggested reading resources, 171
"From Seed to Tree" lesson, 22, 245–261, 374
    background for, 246–247
    BSCS 5E Model for, 248–252
        elaborate, 251–252, **252**
        engage, 248, **248**
        evaluate, 252
        explain, 249–251, **251**
        explore, 249, **249**
    Common Core connections, 248–252
    description of, 245
    Framework connections, **245**
    further exploration suggestions, 253
    materials for, 246
    objectives of, 245
    performance expectations, 374
    picture books for, 245
    student pages for, 246, 255–261
    time needed for, 246
    websites and suggested reading resources, 254
"From Trash to Treasure" lesson, 6, 53, 347–363, 379
    background for, 349
    BSCS 5E Model for, 350–355
        elaborate, 352–354, **353**
        engage, 350
        evaluate, 354
        explain, 351–352
        explore, 350–351, **351**
    Common Core connections, 350–355
    description of, 347
    Framework connections, **347**
    further exploration suggestions, 355–356
    materials for, 348
    objectives of, 347
    performance expectations, 379
    picture books for, 347
    safety precautions for, 348
    student pages for, 348, 357–363
    time needed for, 348
    websites and suggested reading resources, 356
fuel, 140, 153
Furgang, Kathy, 137, 145

G
gallery walks, 54, **54**
gas states of matter, 139
genes, 204–205, 207, 208
genetics, 205, 208
geologist, 270, 271, 284
gestures, 183
glucose, 246
Goodall, Jane, 183
Google Earth, 297–298, **298**
Gorongosa National Park, 183, 184

H
Hale, James Graham, 289, 296
Hatkoff, Craig, 223
Hatkoff, Isabella, 223
Hatkoff, Juliana, 223
heatsinks, 118
Heezen, Bruce, 289, 291–292, 293–294

Index

herds, 180, 197
Hooke, Robert, 101
horns, **163**, 165, 166
How Mountains Are Made, 289, 290, 296–297
How Things Work: Lightbulbs, 115, 116, 120, 121, 129
"Hurricane!" lesson, 263–287, 375
  background for, 265–267, **265**, **266**
  BSCS 5E Model for, 267–272
    elaborate, 270–272, **271**
    engage, 267
    evaluate, 272
    explain, 268–270
    explore, 267–268, **268**
  Common Core connections, 267–272
  description of, 263
  Framework connections, **263**
  further exploration suggestions, 273
  materials for, 264–265
  objectives of, 263
  performance expectations, 375
  picture books for, 263
  safety precautions for, 270
  student pages for, 265, 275–287
  time needed for, 264
  websites and suggested reading resources, 273–274
hydrocarbons, 140
hydrologist, 270, 271, 284
hypotheses and the scientific method, 82, 83

I
identifying problems, Balloon Car Design Challenge, 65
If You Hold a Seed, 22, 245, 246, 248
incandescent lightbulb, 117, **117**, 121–124
inferences, 143, 184
inferring
  about, 13
  An Ambush of Tigers read-aloud, 182–183
  Animals That Live in Groups read-aloud, 186
  Clink read-aloud, 102
  Nefertiti, the Spidernaut read-aloud, 211
  Star Stuff read-aloud, 336
inherited traits, 205, 207, 208, 209–210, **210**, 218
innovate (further exploration)
  "Better Together" lesson, 191
  biomedical engineering, 237
  "Burn" lesson, 149
  design process, 71
  "From Edison to the iPod" lesson, 171
  hurricanes and levees, 273
  lightbulbs, 127
  mountains, 300
  plant growth, 253
  product testing, 89
  recycling, 356
  "Space Exploration" lesson, 317
  "Spider Science" lesson, 214
  wind-up toys, 108
instinctive behavior, 205
International Space Station (ISS), 205, 206–207, 245, 247
interval sampling, 181, 185
investigate (further exploration)
  "Better Together" lesson, 190
  biomedical engineering, 236
  "Burn" lesson, 148–149
  design process, 71
  "From Edison to the iPod" lesson, 171
  hurricanes and levees, 273
  lightbulbs, 127
  mountains, 300
  plant growth, 253
  product testing, 89
  recycling, 355
  "Space Exploration" lesson, 317
  "Spider Science" lesson, 214
  wind-up toys, 108
iPod and Electronics Visionary Tony Fadell, 161, 162, 168, 169

J
Jago (illustrator), 179, 182
Jump Into Science: Stars, 23, 333, 334, 337, 338

K
Kudlinski, Kathleen V., 307, 310–311

L
Larson, Kirby, 263, 267
law of conservation of matter, 141, 154
L'Engle, Madeleine, 11
Let's Think About the Power of Advertising, 79, 80, 84–85, 86
levees, 265–266, **266**, 269–270, 277, **277**
light-emitting diode (LED), 118, **118**, 119, 121–124
"Light It Up!" lesson, 19, 54, 115–136, 368
  background for, 117–119
  BSCS 5E Model for, 119–126
    elaborate, 124–125, **125**, **126**
    engage, 119
    evaluate, 126
    explain, 120–121, 122–124, **122**, **123**
    explore, 121–122, **121**
  Common Core connections, 119–126
  description of, 115
  Framework connections, **115**
  further exploration suggestions, 127
  materials for, 116–117
  objectives of, 115
  performance expectations, 368
  picture books for, 115
  safety precautions for, 116, 125
  student pages for, 117, 129–136
  time needed for, 116
  websites and suggested reading resources, 128
light-year, 335

Index

liquid states of matter, 139
locomotion, 226, 229–230
looking for patterns, *How Mountains Are Made* read-aloud, 297
lumens, 122

M
MacKay, Elly, 245, 248
making connections
    about, 12
    blind taste test, 83–84
    Interactive Article: Plastic, 351–352
    looking at ads, 84, **84**
    Mesmerized read-aloud, 82
    Next Time You See a Spiderweb read-aloud, 206
    One Plastic Bag read-aloud, 350
    Recycling Crafts read-aloud, 352
    Rosie Revere, Engineer read-aloud, 68, 69
    Solving the Puzzle Under the Sea read-aloud, 293
    Space Exploration read-aloud, 312
    Star Stuff read-aloud, 336–338
    structures and functions T-chart, 227, **228**
    The Inventor's Secret read-aloud, 69
    Timeless Thomas read-aloud, 164–165
    "Wild About Animals" video, 228, 232
    Winter's Tail read-aloud, 226–227
Making Machines With Springs, 4, 99, 100, 104–105
maple seeds. See "From Seed to Tree" lesson
matriarchs, 180–181, 185, 197
matter
    burning of, 140
    changes to, 139–140, 153
    conservation of, 140–141, **141**, 154
    defined, 139
Mattern, Joanne, 115, 119
Mesmer, Franz, 79, 81, 82
Mesmerized, 79, 80, 81–82, 83, 87
"Mesmerized" lesson, 52, 79–98, 366
    background for, 81
    BSCS 5E Model for, 81–88
        elaborate, 86–87
        engage, 81–82
        evaluate, 87–88, **88**
        explain, 84–86, **84**
        explore, 83–84, **83**, **84**
    Common Core connections, 81–88
    description of, 79
    Framework connections, **79**
    further exploration suggestions, 89
    materials for, 80
    objectives of, 79
    performance expectations, 366
    picture books for, 79
    safety precautions for, 80
    student pages for, 80, 90–98
    time needed for, 80
    websites and suggested reading resources, 89
microgravity, 210
Milano, Mariel, 47
Mohammed, Amr, 205, 210–212
molecules, 139, 153
Morgan, Emily, 203, 206, 211, 245, 249
Mozambique, 183
Myers, Matthew, 99, 102
mystery objects, 119

N
narrative information books, 6
NASA (National Aeronautics and Space Administration), 205, 251, 260, 309–310, 312, 313, 332, 335
National Geographic Kids: Wildfires, 137, 138, 145–147, 156–157
National Oceanic and Atmospheric Administration (NOAA), 275, 276
natural hazards. See "Hurricane!" lesson
Nefertiti, the Spidernaut, 203, 204, 210–212
Nethery, Mary, 263, 267
new vocabulary list
    about, 14
    Building Dikes and Levees read-aloud, 269–270, 281
Next Generation Science Standards (NGSS)
    and *A Framework for K–12 Science Education*, 25–27, **26**
    and BSCS 5E Model, 22
    lesson performance expectations, 365–379
Next Time You See a Maple Seed, 245, 246, 249–250, 256
Next Time You See a Spiderweb, 203, 204, 206, 208–209, 211
nonfiction features
    about, 15
    Building Dikes and Levees read-aloud, 269
    Recycling Crafts read-aloud, 352
non-narrative information books, 6

O
observations, 143
observations and the scientific method, 82, 83
ocean floor. See "Solving the Puzzle Under the Sea" lesson
One Plastic Bag, 6, 347, 348, 350, 352, 355, 358, 363
orb web video, 207
Orion and the Dark, 115, 116, 124
Ortiz Catalan, Max, 233, 234
overtopped levees, 265–266, **266**, 269–270

P
pairs read
    Don't Be Fooled article, 86

Plants in Space, 251–252
The Chemistry of a Candle article, 145
The Social Lives of Elephants, 187
Pattison, Darcy, 137, 142, 203, 210, 211
Paul, Miranda, 347, 350
Pets Evacuation and Transportation Standards (PETS) Act, 266
phosphor, 118
photosynthesis, 246, 250–252
physical changes to matter, 139–140, 145, 152–153
picture books and STEM instruction, 1–2
picture walk
   about, 14
   Biomedical Engineering and Human Body Systems read-aloud, 233
Planetary Society, 335
plant growth. See "From Seed to Tree" lesson
plant pillows, 251, 260
plastics. See "From Trash to Treasure" lesson
Poole, Bob, 183
Poole, Joyce, 181, 183, 184
predators, 197
prescribed burns, 141, 145
product claim tests, 81, 92
product comparison tests, 81, 92
products, 141
product testing. See "Mesmerized" lesson
profiles, 295–296
prosthetic limbs, 225, 233–234
   See also "Bionic Animals" lesson
prototype, 70

Q
questioning
   about, 12
   animal locomotion videos, 229–230
   Animal Prosthesis Design Challenge, 227–228
   Biomedical Engineering and Human Body Systems read-aloud, 232
   Birthday Candles, Part 1, 144–145
   Elephant Behavior Graph, 185–186
   Let's Think About the Power of Advertising read-aloud, 84–85
   Mesmerized read-aloud, 81–82
   Nefertiti, the Spidernaut read-aloud, 212
   Next Time You See a Spiderweb read-aloud, 208–209
   Orion and the Dark read-aloud, 124
   Rosie Revere, Engineer read-aloud, 64–65
   Solar System app, 310
   Solving the Puzzle Under the Sea read-aloud, 293–294
   Space Exploration read-aloud, 312–313
   The Chemistry of a Candle article, 145
   The Inventor's Secret read-aloud, 66–67
   "The Known Universe" video, 309
   The Social Lives of Elephants, 187
   Two Bobbies read-aloud, 267
   "Wild About Animals" video, 228, 232
   Wildfires read-aloud, 146–147
   Winter's Tail read-aloud, 230–231

R
range, species range, 182
reactants, 141
reactions, chemical, 139, 153
reading aloud, 9–15
   comprehension enhancement tools, 13–15
   picture books and comprehension enhancement, 15
   reading comprehension strategies, 11–13
   reasons for, 9–10
   tips for, 10–11
Recycling Crafts, 347, 348, 352
recycling. See "From Trash to Treasure" lesson
Reinhardt, Jennifer Black, 61
reproducibility, 81, 86–87, 92
rereading
   about, 14
   How Things Work: Lightbulbs, 122–124
   Jump Into Science: Stars, 339–340
   Next Time You See a Maple Seed read-aloud, 250
research (further exploration)
   "Better Together" lesson, 190
   biomedical engineering, 236
   "Burn" lesson, 148
   design process, 71
   "From Edison to the iPod" lesson, 170
   hurricanes and levees, 273
   lightbulbs, 127
   mountains, 300
   plant growth, 253
   product testing, 89
   recycling, 355
   "Space Exploration" lesson, 316
   "Spider Science" lesson, 213–214
   wind-up toys, 107
Roberts, David, 61, 64
Rocco, John, 307, 310–311
Rockliff, Mara, 79, 81
Rosenthal, Betsy R., 179, 182
Rosie Revere, Engineer, 5, 61, 62, 64, 68, 69, 72
Rs of recycling, 349
rubber bands, 101, 104, 105–106

S
safety precautions
   "Bionic Animals" lesson, 224
   "Burn" lesson, 138, 142–143, 150

Index

"From Trash to Treasure" lesson, 348
"Hurricane!" lesson, 270
"Light It Up!" lesson, 116, 125
"Mesmerized" lesson, 80
safety practices overview, xvii
"Solving the Puzzle Under the Sea" lesson, 290
"Spider Science" lesson, 204
"Star Stuff" lesson, 334
"The Inventor's Secret" lesson, 62
"Wind It Up" lesson, 100, 103
Saffir-Simpson Hurricane Wind Scale, 265, **265**
Sagan, Carl, 333, 335, 336, 337, 340
samara (maple seed), 248, 249
satellite transmitters, 182
science and engineering practices in the 3–5 classroom, 48, **49–50**, 51
and *A Framework for K–12 Science Education*, 25, 47–48
differences between, 51, 52, **52**
engineering design process, 52–53, **53**
and *NGSS*, 48, 52–53, **53**
student-directed Further Exploration, 53–55, **56–58**
scientific method, 82, 83
seafloor spreading, 292
Sedlak, Peggy, 349
seismicity, 298
Sisson, Stephanie Roth, 333, 336
Sjonger, Rebecca, 223, 232
sketch to stretch, Derby the Bionic Dog, 234–235
sky Jell-O, 146
Slade, Suzanne, 61
social behavior in animals, 180–181
See also "Better Together" lesson
Solar System app, 308, 309–310, **310**

solid states of matter, 139
Solving the Puzzle Under the Sea, 22, 289, 290, 292, 293, 295–296
"Solving the Puzzle Under the Sea" lesson, 22, 51, 289–306, 376
background for, 291–292
BSCS 5E Model for, 292–299
elaborate, 297–298, **298**
engage, 292–294
evaluate, 298–299
explain, 296–297, **297**
explore, 294–296, **295**, **296**
Common Core connections, 292–299
description of, 289
Framework connections, **289**
further exploration suggestions, 299–300
materials for, 290–291, **291**
objectives of, 289
performance expectations, 376
picture books for, 289
safety precautions for, 290
student pages for, 291, 301–306
time needed for, 290
websites and suggested reading resources, 300
sonar devices, 291
soot, 140
sounding box, **291**, 295
soundings, 291, 295
Space Exploration, 6, 307, 308, 312–313, 315
"Space Exploration" lesson, 6, 307–332, 377
background for, 308–309
BSCS 5E Model for, 309–316
elaborate, 312–315, **313**
engage, 309
evaluate, 315–316
explain, 310–312, **311**
explore, 309–310, **310**
Common Core connections, 309–316

description of, 307
Framework connections, **307**
further exploration suggestions, 316–317
materials for, 308
objectives of, 307
performance expectations, 377
picture books for, 307
student pages for, 308, 318–332
time needed for, 308
websites and suggested reading resources, 317
species range, 182
spiderlings video, 207–208
"Spider Science" lesson, 26, 52, 203–221, 372
background for, 204–206
BSCS 5E Model for, 206–213
elaborate, 210–212
engage, 206
evaluate, 212–213, **212**
explain, 207–210, **210**
explore, 206–207, **207**, 209, **209**
Common Core connections, 206–213
description of, 203
Framework connections, **203**
further exploration suggestions, 213–214
materials for, 204
objectives of, 203
performance expectations, 372
picture books for, 203
safety precautions for, 204
student pages for, 204, 216–221
time needed for, 204
websites and suggested reading resources, 214–215
spring steel, 101
Star Stuff, 333, 334, 336–337
"Star Stuff" lesson, 22–23, 333–346, 378
background for, 335
BSCS 5E Model for, 336–341

Index

elaborate, 339–340
engage, 336–337
evaluate, 340, **340**
explain, 338–339
explore, 337–338, **337, 338**
Common Core connections, 336–341
description of, 333
Framework connections, **333**
further exploration suggestions, 341–342
materials for, 334
objectives of, 333
performance expectations, 378
picture books for, 333
safety precautions for, 334
student pages for, 334, 343–346
time needed for, 334
websites and suggested reading resources, 342
STEM at Home
"Better Together" lesson, 189–190, 202
"Bionic Animals" lesson, 235, **236**, 244
"Burn" lesson, 148, 160
"From Edison to the iPod" lesson, 170, 178
"From Seed to Tree" lesson, 252, 261
"From Trash to Treasure" lesson, 354–355, 363
"Hurricane!" lesson, 272, 287
"Light It Up!" lesson, 126, 136
"Mesmerized" lesson, 88, 98
"Solving the Puzzle Under the Sea" lesson, 299, 306
"Space Exploration" lesson, 316, 332
"Spider Science" lesson, 213, 221
"Star Stuff" lesson, 340–341, 346
"The Inventor's Secret" lesson, 70, 78

"Wind It Up" lesson, 107, 113
STEM instruction
context for concepts, 3–4
and correction of science misconceptions, 4
depth of coverage in picture books, 4
described, 2–3
disciplines, **3**
and picture book genres, 5–7
selection of picture books for, 4–5
using picture books for, 1–7
stop and try it
about, 15
Candle Observations, 142–143
capillary action demonstration, 144–145
Making Machines With Springs read-aloud, 104–105
storybooks, 5–6
Strategies That Work, 11
support for test results and the scientific method, 82, 83
synthesizing
about, 13
Comparing the Design Process, 232
Comparing Maps, 298–299
Fadell's TED Talk, 169
"Giving the World a Helping Hand" video, 234
Hurricane Katrina Investigation Journal, 268–269
Interactive Article: Plastic, 352
Making Machines With Springs read-aloud, 105
Next Time You See a Spiderweb read-aloud, 209
Spool Car Racer Instruction Manual, 11–13, 106–107
Wildfires read-aloud, 147–148

T
talkie tapes, 163, 165

tectonic processes, 292
TED Talks, 169
tentative nature of science, 312
tests and the scientific method, 82, 83
Tharp, Marie, 289, 291–292, 293–294, 296, 299
The Gambia, 349, 350
The Inventor's Secret, 61, 62, 63–64, 66–67, 69
"The Inventor's Secret" lesson, 5–6, 48, 51, 61–78, 163, 365
background for, 63–64
BSCS 5E Model for, 64–70
elaborate, 68–69
engage, 64–65
evaluate, 69–70, **70**
explain, 66–68, **68**
explore, 65–66, **66**
Common Core connections, 64–70
description of, 61
Framework connections, **61**
further exploration suggestions, 71
materials for, 62–63
objectives of, 61
performance expectations, 365
picture books for, 61
safety precautions for, 62
student pages for, 63, 73–78
time needed for, 62
websites and suggested reading resources, 72
"The Known Universe" video, 309
three-dimensional (3-D) printing, 223, 225, 226, 234
Timeless Thomas, 161, 162, 164, 165
Tisnés, Valeria, 203, 210
Tjernagel, Kelsi Turner, 179, 186
Tomecek, Steve, 333, 338
"Toyologist" video, 107, 113
traits, 204
transformation of energy, 119, 125
Translating the NGSS for Classroom Instruction, 22

Index

trash. *See* "From Trash to Treasure" lesson
trunk (elephant), 198
turn and talk
   about, 15
   An Ambush of Tigers read-aloud, 183
   Animals That Live in Groups read-aloud, 186
   Biomedical Engineering and Human Body Systems read-aloud, 232–233
   "Food Ad Tricks" video, 85–86
   "How to Map the Ocean Floor" video, 294
   "Kratt's Creatures" video, 187
   Let's Think About the Power of Advertising read-aloud, 85
   lightbulbs, 119
   Making Machines With Springs read-aloud, 104, 105
   photosynthesis video, 250–251
   Star Stuff read-aloud, 336
   Wildfires read-aloud, 145–146
   Wildlife Tracking Technology, 188
Two Bobbies, 263, 264, 267

U

undersea mountains. *See* "Solving the Puzzle Under the Sea" lesson
upcycling, 349, 352, 353

V

vaporization, 140, 153
variables, 81, 92–93
visionary person, 168
visualizing
   about, 12
   Burn read-aloud, 142
   iPod and Electronics Visionary Tony Fadell read-aloud, 168
Voyager 1 and Voyager 2 spacecraft, 335

W

water vapor, 139
watts, 122
wildfires
   described, 141
   National Geographic Kids: Wildfires, 137, 138, 145–147, 156–157
Willis, Peter, 137, 142
"Wind It Up" lesson, ix, 4, 26, 99–113, 367
   background for, 101
   BSCS 5E Model for, 102–107
      elaborate, 105–106, **106**
      engage, 102
      evaluate, 106–107
      explain, 104–105, **105**
      explore, 102–104, **102**, **103**, **104**
   Common Core connections, 102–107
   description of, 99
   Framework connections, **99**
   further exploration suggestions, 107–108
   materials for, 100
   objectives of, 99
   performance expectations, 367
   picture books for, 99
   safety precautions for, 100, 103
   student pages for, 101, 109–113
   time needed for, 100
   websites and suggested reading resources, 108
"Wind Up Racer" video, 105–106, 108
wind-up tub toys, 224–225, **224**, **225**
Winter's Tail, 223, 224, 226–227, 229, 230–231, 232
writing
   Animals in Space Research Proposal, 212–213
   Better Together Booklet and Scoring Rubric, 189
   Evaluate Your Design, 272
   Nightlight Instruction Manual, 126

Y

Yarlett, Emma, 115, 124
Yates, David, 223
Yellow Brick Roads, 9
Yoshikawa, Sachiko, 333, 338

Z

Zoehfeld, Kathleen Weidner, 289, 296
Zunon, Elizabeth, 347, 350

FOR REFERENCE

Do Not Take From This Room